WITHDRAWAL

Embracing Mathematics

"This book is likely to have a substantial impact on the landscape of pedagogical thinking. The pedagogy is a model for how to be engaged in life, in this sense presenting a personal vision that informs more than how to teach."

David Kirshner, Louisiana State University

"*Embracing Mathematics* offers mathematics teachers other possibilities for teaching, in practical and very tangible ways. Simultaneously, [it] challenges the current field and provides possibilities for how these 'different' ways of considering teaching can work effectively in our current situations."

Sarah Smitherman Pratt, University of North Carolina at Greensboro

"*Embracing Mathematics* presents a wealth of educational ideas. It provides lenses for seeing classroom practice as open to fascinating changes. Emerging as a powerful collective achievement, it provides a dialogical opening of post-modern trends in critical mathematics education, seen as practice, as research, and as learning for everybody."

Ole Skovsmose, Aalborg University, Denmark

"This is a book that helps the reader think about his/her practice. It guides teachers in a quest for teaching methods that will enable students to develop a critical mind, which should be the goal of all mathematics education. But it is also an extraordinarily practical book that along with stimulating ideas provides numerous applications, which are immediately useful in the mathematics classroom."

Corinne Hahn, Advancia-Negocia and European School of Management; President, International Commission for Study and Improvement of Mathematics Teaching

This alternative textbook on teaching mathematics asks teachers and prospective teachers to reflect on their relationships with mathematics and how these relationships influence their teaching and the experiences of their students. Applicable to all levels of schooling, the book covers basic topics such as planning and assessment, classroom management, and organization of classroom experiences. It also introduces novel approaches, such as psychoanalytic perspectives and post-modern conceptions of curriculum. Traditional methods-of-teaching issues are recast, provoking new ideas for making mathematics education meaningful to both teachers and their students. Coauthored by a university professor and several practicing elementary, middle, and high school mathematics teachers, this text is a collaboration across all educational levels, making it ideal for teacher discussion groups K-16.

Embracing Mathematics is intended as a methods text for undergraduate- and master's level mathematics education courses and more specialized graduate courses on mathematics education, and as a resource for teacher discussion groups.

Peter Appelbaum is Associate Professor, Arcadia University, Pennsylvania, where he teaches Mathematics Education, Curriculum Theory, and Cultural Studies, and is Coordinator of Mathematics Education and Curriculum Studies Programs, Director-at-Large of General Education, and Director of the sTRANGELY fAMILIAR mUSIC gROUP.

David Scott Allen teaches 7th, 8th, and sometimes 9th grade math at Pennbrook Middle School in North Wales, Pennsylvania. He and his wife Marie live in Souderton, Pennsylvania. Three of their four children know *pi* to 35 decimal places or more because they memorized a song their father composed.

Embracing Mathematics

On Becoming a Teacher and Changing with Mathematics

PETER APPELBAUM
with DAVID SCOTT ALLEN

Routledge
Taylor & Francis Group

NEW YORK AND LONDON

First published 2008
by Routledge
270 Madison Ave, New York, NY 10016

Simultaneously published in the UK
by Routledge
2 Park Square, Milton Park, Abingdon, Oxon OX14 4RN

Routledge is an imprint of the Taylor & Francis Group, an informa business

© 2008 Taylor & Francis

Cover photo and concept by Sophia Appelbaum with Noah Appelbaum

Typeset in Sabon and Neue Helvetica by Book Now Ltd, London
Printed and bound in the United States of America on acid-free paper by Edwards Brothers, Inc.

Library of Congress Cataloging in Publication Data
Appelbaum, Peter Michael.
Embracing mathematics: on becoming a teacher and changing with mathematics / by Peter Appelbaum with David Scott Allen. — 1st ed.
    p. cm.
1. Mathematics—Study and teaching—Evaluation. 2. Mathematics teachers—Psychology. I. Allen, David Scott. II. Title.

QA11.2.A625 2008
510.71—dc22   2007044909

ISBN-10: 0-415-96384-2 (hbk)
ISBN-10: 0-415-96385-0 (pbk)
ISBN-10: 0-203-93024-X (ebk)

ISBN-13: 978-0-415-96384-8 (hbk)
ISBN-13: 978-0-415-96385-5 (pbk)
ISBN-13: 978-0-203-93024-3 (ebk)

For Belinda, on our 25th anniversary.

Peter Appelbaum

For my father
Coulter Boileau Allen
who struggled with math
but convinced me that I could do *anything*
and for my mother
Margaret Jane Allen
who played card games with me
and who has lived to see this book published.

David Scott Allen

Brief Contents

Contents

While teachers struggle to teach their students how to be critical thinkers, this chapter offers numerous ideas for how a teacher may organize activities in their classroom by assuming that students would think critically if the classroom did not prevent such behaviors. Peter writes about critical points in changing his teaching practice that have pushed his way of thinking forward.

Response to Chapter 4: It is Critical to Think

David reacts to all of the "way-out" ideas of this chapter with his thoughts about a triangle of tension among the standardized test criteria for success, the problem-solving-as-goal approach to mathematics education, and the students' own personal journeys. The only thing we can be sure of is change, and that is exacerbated by the shifting sands of pedagogy, curriculum, and social attitudes and politics within states and districts. He offers his own set of critical points on how he has changed his teaching in recent years to promote problem solving and problem posing within an accountability era.

Action Research 4 — Ada Rocchi, Lesson: World Population and Wealth

Ada describes a "replacement unit" she has developed for her class, which has the interesting challenge of being comprised of students studying several different years of mathematics in the same room, for unspecified lengths of time.

MathWorlds 4: Pitching Questions at Various Levels

Our fourth strategy of inventing and exploring MathWorlds is to compose your own mathematical questions for others, and to design these questions so that they are meant for mathematicians with varying types of previous experiences with mathematical inquiry.

AFTERWORD: What will You Write in *Your* Chapter? 296

David suggests that teaching is the closest profession to parenting, and that much of this discussion comes down to values. It is important for a district to state their values. There are modern values and post-modern ones, and they are in conflict.

APPENDICES 298

This is where we put stand-alone documents which will help the reader to implement some of our ideas.

Preface: How Can I (Better) Embrace Mathematics?

David Scott Allen

In this book, we seek to make whole the apparently opposing ideas which have been in the forefront of the school-fought "math wars." Each chapter has its own set of opposite forces, ideals, goals, pedagogies, and so on. We seek to give you a starting point for change by exploring alternatives to your current teaching practices. You will not be comfortable with everything that is suggested, but perhaps this is the best reason to try something different.

Dr. Peter Appelbaum is a teacher who is very much a student. I am a student of his and also a teacher. He thinks very deeply about the teaching of mathematics, and then passes it on to his students for them to grapple with and gain from. I hope to pass along to you my journey through this material, much of which was forged in graduate classes I have taken.

My goal is to give you, after each chapter, the reflections of one educator's trajectory from total ignorance to first practice of new ideas. I will share with you my victories, defeats, and differences of opinion. For me, these chapters represent a summary of my recent work in public school as well as starting points for new investigations with my students. For now, I give you my thumbnail sketch of the chapters to come.

Each chapter cluster addresses a central question which serves as an anchor – a central point of reference for all that is discussed.

PROLOGUE: What (Exactly) do *You* Think it Means to be a Teacher of Mathematics?

This chapter sets the tone by presenting activities for teachers to do in the classroom *with* their students. We submit that learning with your students can be at least as much of an adventure as helping your students to learn. You are encouraged to do some or all of the activities by yourself and/or with your students. You may use them as the basis for further investigation, modify them to suit your students' abilities, and use the websites and your own originality to find other challenging questions.

CHAPTER 1: Planning and Assessment
How Can I Engage my Students in Meaningful Mathematics?

Here, we seek to describe the issue of irony in the teacher/student relationship. Our students have a completely different world view than we have. Ironically, this can be a springboard to richer educational experiences. A five-part inquiry process is described, along with discussion on rubrics and how all of the above affects the classroom experience and is affected by the goals of the teacher.

CHAPTER 2: A Psychoanalytic Perspective
What Does Mathematics Education Ask of Us?

This perspective is perhaps the most foundational to all of this writing. Teaching is not psychology, but has powerful lessons to learn from the interplay between psychologist and patient. Education requires us to change our fundamental ideas of how the world is structured. This is especially true in mathematics. We must help our students come through their awkwardness with the material to maturity in knowledge of new structures and relationships. In the process, we learn much about ourselves, our students, and teaching. Interviewing is introduced as a technique for learning about students and their ideas about mathematics. At the core of all of this are the relationships of the students and the teacher to the objects of mathematics.

CHAPTER 3: You Are a Mathematician
How Does a Mathematician Work?

Beginning with the premise that everyone is a mathematician, we conclude that what a mathematician *is* is largely determined by what a mathematician *does*. Mathematicians largely quantify and qualify abstractions, which may or may not collide with ordinary objects or experience. Using descriptors, mathematicians probe the familiar unknown, and ask questions worth answering for the discovery of the answer. Methods of connecting all this to our students in the classroom are discussed.

CHAPTER 4: Critical Thinkers Thinking Critically
How am I to Think of "Critical Thinking" in my Classroom?

Peter takes us through eight critical junctions of the student/teacher relationship. That is to say, he reframes the things that a teacher should be doing in the classroom. Neither the old approach nor the new seem completely satisfactory, but rather the next step in the process is deemed the most important. Communication is viewed as the key to making this happen.

CHAPTER 5: Consuming Culture: Commodities and Cultural Resources
How Should I Relate to Mathematics?

At some point all districts bought textbooks. These omnipresent resources require us to think within them at the same time that we think without them. The Published Curriculum has a huge impact on how the Proscribed Curriculum is delivered. Yet in order to deliver it, these packages are useful. The assumptions of the ideologies behind a curriculum are important factors in deciding whether or not to use it.

CHAPTER 6: Metaphors for the Classroom Space

What Does it Feel Like to be in a Mathematics Classroom?

Time is the critical component here. If we are to teach mathematics, we must take time to do so. What is the purpose of that time? How do we structure that time? How will the students view what has just occurred? What do we do next? All these are legitimate questions with perhaps a wider variety of answers than we have been comfortable accepting.

CHAPTER 7: Places where People Learn Mathematics

How am I to Interact with Mathematics?

Students need to interact with mathematics in order to learn it, but they also need to learn mathematics in order to interact with it. The teacher's primary job is not to tell students what to do or how to do it, but to get them to want to do it and value doing it. College does not prepare teachers for this. Perhaps this chapter will help fill the gap.

CHAPTER 8: When Students Don't Learn

How am I to Think about my Students?

All of us have stories to tell about our own students. Each of us has our own story to tell about being a student. Learning is a choice. Sometimes it is a struggle, and sometimes it comes with ease, but it is always an act of the will. Each of us, as well as our students, must choose with care the direction of the race we wish to run. We must also make sure that we are prepared to run it or we will have to deal with the consequences.

EPILOGUE: Becoming a Teacher and Changing with Mathematics

How am I to Move Forward as a Teacher?

All of this may seem confusing. We cannot have as much control as we are told we should. We may feel we need to take sides in the "math wars." Peter's final word suggests that such dichotomies are artificial, and that a mechanistic model of education is a dream which we need to confront as not compatible with reality. We are struggling to design real mathematical work with our students. We must take our own highly personal position and move forward with our own creative vision in this age of constant transformation.

I encourage you to mine this book for the diamonds you can use. In my sections, I will show you the diamonds I have found, describe if and how each was used, and describe how you may borrow them. You might interpret the writing in an entirely different way, or choose to pursue another aspect of what is presented. Read the text several times, each time looking at it a different way, or looking for something different. Don't read to justify a previously held notion, but to find a new one, or to note contrasts between old thinking and new thinking, and so on. In other words, have a purpose as you read, and you will get more out of it. The book does what it expects you to do: explore. So it may seem confusing as we break new ground, or redundant as we revisit territory in a new light. All the principles here stated or implied also apply to your reading situation. For now, the text is your classroom. Self reference is a powerful tool, so be prepared to be critical of your own thinking. Embrace!

Acknowledgments

Peter Appelbaum

This book has been many years in the making, and owes its existence to numerous supporters who believed in the need for an "alternative methods text," a guide to teaching and learning mathematics for all grades and ages, not pitched at elementary, middle, secondary, or post-secondary alone, but striving to foster dialogue across the levels, a book that can be read by experienced and novice teachers together. The first version owed its birth to Joe Kincheloe, who cornered me at an American Educational Studies Association conference, and told me I had to write this book. Years, versions, and numerous publishers later, I have Naomi Silverman to thank for believing such an "alternative" could sell well enough to warrant its publication; and Mary Hillemeier at Routledge/Taylor Francis for steering the final copy through numerous last-minute layout and chapter-ordering confusions through to the actual publication. Along the way, generous friends have read versions of chapters, encouraged me to continue, and otherwise helped me to think of further topics that "belonged" in this book, including Astrid Begehr, Alan Block, Paolo Boero, Julie Burke, Stella Clark, Suzanne Damarin, Erica Davila, Belinda Davis, William Doll, Diana Erchick, Jayne Fleener, Uwe Gellert, Rachel Hall, Stephen Herschkorn, Eva Jablonka, Rochelle Kaplan, Christine Keitel, David Kirshner, Bob Klein, Christophe Kotanyí, Michael Meagher, Elijah Mirochnik, Josh Mitteldorf, Nicholas Ng-a-Fook, Pedro Palhares, Julia Plummer, William Pinar, Sarah Smitherman Pratt, David Pushkin, Bernard Robinson, Bill Rosenthal, and Judy Sowder.

Most important have been my students at Arcadia University, who have taken courses with me for the last six years, suffering through drafts of chapters in rough or later stages. These include the over 420 elementary education majors who experienced the "mega-course" entitled "Refining and Integrating Curricular Practices," and who used some or all of the chapters included here as a guide for leading groups of eight to ten first–eighth grade students through mathematics inquiries for a semester. I am indebted as well to the responses and insights of the approximately 60 graduate students at Arcadia who have worked with me in each of the past six iterations of "Mathematics and the Curriculum," and "Clinical Mathematics Education," some of whom summarized their action research projects for inclusion in this book. The following Arcadia students have read drafts of this work and made important editorial and other changes in the text: Yela Awunyo-Akaba, Bernadette Bacino, Kia Karaam, Sandy Morrash, Petal Sumner, and Shuang Zheng. Thanks to Jinell Smithmyer, for introducing me to The Books.

David Allen started out as one of these students. But even in the first course we experienced together, he stood out as a teacher-researcher – indeed, as someone who takes risks in his teaching in ways that never put his students at risk, in ways that make it possible for his students to take the risks necessary for learning mathematics and becoming mathematicians. Our collaboration here has been critical to my developing ideas about teaching and learning mathematics. I look forward to further collaborations, and to more of David's insights that challenge my assumptions.

I am grateful for my colleagues in Education and Mathematics at Arcadia. They made it possible for me to try out every idea in this book over and over again in my own teaching. In particular, Louis Friedler, Carlos Ortiz, and Ned Wolff are the most remarkable colleagues in mathematics one could ask for; they consistently encourage me to teach mathematics in the ways described in this book, and challenge me to articulate my ideas more clearly and with useful examples. My department chair, Steve Gulkus, and my co-teaching-partner, Leif Gustavson, have supported me in figuring out how postmodern, psychoanalytic, and post-colonial thinking can be translated into taking small groups of children out of their regular classrooms, and placing them in odd spaces around the school in order to develop their own inquiries, and to demand that our student-teachers document how they are meeting the district core curriculum in mathematics and literacy in the process. Leif and I have been co-teaching for so long that the best aspects of my understanding of inquiry and of criteria for understanding are his. I extend my gracious thanks as well to the staff and students of F.S. Edmonds Elementary School (and especially Ms. Sharen Finzimer, Principal), C.W. Henry Elementary School (and especially Ms. Caren Trantas, Principal), and J.F. McCloskey Elementary School (and especially Mr. John Underwood, Principal) in Philadelphia, who have welcomed 70 elementary education students per year into their classrooms, hallways, libraries, auditoria, cafeterias, and other nooks and crannies, for the past six years; and to Marilyn Bentov, for telling me to go to graduate school and study curriculum.

Important theoretical grounding for this book was developed through my work with Robert Kravis and David Rackow, my mentors during a two-year fellowship at the Psychoanalytic Center of Philadelphia; Bob's bi-weekly meetings, in particular, and the monthly meetings of the City Schools Forum, organized by the Center, enabled me to research, explore, and translate psychoanalytic concepts and practices for mathematics education contexts. Further theoretical development of my ideas was made possible by invitations from the International Commission for the Study and Improvement of Mathematics Education, Encontro Matemática, and the Ohio State Mathematics, Science, and Technology conferences to share my current work; by the supportive and cutting edge discussions of the Gender and Mathematics Working Group of the North American Chapter of the International Group for the Psychology of Mathematics Education; by the questions and discussions with generous and inventive colleagues in curriculum studies not intimidated by my insistence that we examine disciplinary contexts at the *Journal of Curriculum Theorizing* Conferences on Curriculum Theory and Classroom Practice and the American Association for the Advancement of Curriculum Studies; by my visits to the University of Cape Town Language Development Group and Centre for Higher Education Development, supported by the Spencer Foundation; and with the Mathematics Education Group, Arbeitsbereich Grundschulpädagogik of the Freie Universität Berlin, directed by Professor Christine Keitel-Kreidt.

I thank my family, Sophia Appelbaum, Noah Appelbaum, and Belinda Davis, for all of the discussions of school and real-world mathematics that we have had for as long as we have been together. I can't count the times when Noah or Sophia have made sure that my ideas about teaching and learning mathematics made sense in terms of their own school experiences. Nor can I recall each of the innumerable times when Belinda has helped me to think about mathematics and mathematical ways of thinking in new and important ways. These experiences and the conversations that ensued have made indelible impacts on my interpretation of current school mathematics practices, and enabled me to imagine different possibilities for teachers and their students.

Acknowledgments

David Scott Allen

This book has indeed been many years in the making. Peter was the one who invited his students to contribute to this book. After taking several classes with him, I can honestly say that he practices what he preaches. He has opened the door for me to teach at Arcadia, and I look forward to passing on some of my experience.

I owe much to my two student teachers over these years. Joe Adams took on the challenge of three grade levels and five different classes, while allowing me to interview students about their projects and other mathematical thoughts. I have enjoyed his insights, sense of humor, and cut-to-the-chase approach to teaching. Sharon Miller was always willing to try new things in the classroom. She has impressed me with her caring ways and her knack for squeezing every last ounce of teachable moment out of any situation. They are both now colleagues in a sister middle school.

The math and art projects would not have been possible without administrative support. Many thanks go to the art department for allowing the use of materials and giving us space to paint, and to building services for help with hanging the finished products.

My students are the real stars of this show. They continue to amaze me with their analysis of the reality that is school. They remind me daily that we need to honor their intelligence by educating them well. They challenge me, the teacher, to lead the change which is a necessary means to that end. The Math Teams have especially encouraged me with their dedication to learning and doing mathematics.

Finally, I thank my family. My mother-in-law Bonnie McDonald watched our kids as I wrote much of this. As a Special Education teacher, wife, and mother Marie is a model of what it means to serve children and their families. My children Mikayla, Noah, Nason, and Nevin remind me of what it is like to discover something every day. I hope I never lose that childlike wonder. I thank God for their presence in my life.

Other Acknowledgments

Lyrics for "Smells Like Content," from the album *Lost and Safe*, by The Books, courtesy of the artists.

Prologue

Stop.

Before you go on, what (exactly) do you think it means to be a teacher of mathematics? No book can tell you—this, of course, is something only you can determine *for yourself*. What it means for *me* to be a teacher of mathematics is a personal project: it is *me*, as a teacher of mathematics. My "job" is to figure out what that could mean. Who/what/ where am I? Who/what/where/when is a teacher? Who/what/where/when is mathematics, if there even is such a thing as mathematics? Indeed, I have, in my own time, come to find solace in a way of defining things suggested by Ludwig Wittgenstein—that "meaning" is often found in *use*. Who/what/where am I? I am only what I make of myself, what I do, where I choose to be, and how I choose to reflect on the range of options for my actions. As a teacher, the meaning of my profession is determined in the ways that I "teach." For example, I choose to find no sense in teaching distinct from student-ing: teaching is learning, and learning is teaching. Mathematics, for me, is found in its use; mathematics is the act of *mathematizing*. To mathematize is to embrace mathematics, and in so doing, to *be* mathematical. Pattern. Quantity. Shape. Possibility. Relation. Information. To seize opportunities for interpreting experience, listening to others, articulating and representing for others, in mathematical ways, is to mathematize. When I do this, I change. I do not remain static. I learn from others, I see, feel, think, breath, taste, differently. I am always becoming a new person.

As soon as I seem to know who I am, and as soon as I am comfortable and secure in my understanding of mathematics, I cease to be a teacher of mathematics. It is at this point that I *must* find *new ways* to rekindle my questioning, my searching, my enthusiasm. If I can say what it means to be a teacher of mathematics, I am no longer growing, no longer changing, no longer *becoming* a teacher of mathematics.

I do have some strong opinions about how you can begin to embrace mathematics, and in the process, become, with your students, a teacher. These are based on my own experiences, and I hope you find them useful. One idea is to explore *with your students*—

and with others who aspire to teaching—some mathematical situations that place you in unfamiliar terrain. By embracing the mathematics that these situations present, and by creating inquiries that grow out of these situations, you will be able to recall some of the joys, fears, excitements, disappointments, and complexities of mathematics.

In what follows, I would like you to work on these "problems": you will be able to figure out what they are asking; to "specialize" (or try particular instances); to "generalize" (or to look for patterns, relationships, and connections to other situations); and to reflect on the phases of the inquiry process. We will go into these aspects of mathematical experience in more depth in later chapters, and also question how useful they can be for a teacher of mathematics to think about, given their limitations. For now, I invite you to act mathematically in order to be able to later examine how someone can encourage such mathematizing when they are with others.

1. Calculator Patterns

This is an exploration that is good to use when someone asks you for your position on the use of calculators in early mathematics. My general response is, "I like to use calculators to make it possible to do things that are not so easily done without them, rather than to use them to replace important mathematics learning and teaching." Here we can explore stuff about the ways numbers work that would be really hard without a calculator. The idea for this inquiry arose in a second grade classroom. "Kaseem"[1] had been playing around with a calculator and noticed a strange thing: it seemed that pushing the "=" sign would make the calculator repeat the last operation it had performed. Kaseem had entered a 3, then + 4 =, and had gotten 7; he then pushed "=" again and got 11. Absent-mindedly pushing buttons, he continued to press the equals sign, getting 15, 19, 23, 27, 31, 35, 39, 41, . . . Try this yourself on a calculator and you are likely to see what Kaseem saw: look at the 1s digit as you press "=" over and over; it seems like a pattern is repeating!

In our class, this became the subject for the whole group, and the beginning of a class inquiry. On Thursday, Kaseem asked the class if they thought they would always get a pattern with this, if they started with a "starting number," and then entered an "adding number," and used the equals button to continue to add the starting number over and over. Molly thought there would be a pattern, because she had noticed a pattern last year when her class had been "skip counting" (counting by any number—2s, 3s, 4s etc., for example, 3, 6, 9, 12, 15, 18, 21, 24, . . . is counting by threes; 10, 20, 30, 40, 50, . . . is counting by tens). Molly repeated that this seemed a little bit like skip counting. Most of the class really thought they needed to try some numbers themselves before they could even begin to understand what Kaseem meant. We agreed to do this over the next couple of days whenever we had a chance, for example if we completed other work early or chose to work on this during a free-choice time, so that we would be ready to talk about it on Monday.

---

1 Pseudonyms have been used throughout this book.

Hi again,

What can you say you have learned? What do you know about the calculator patterns? What questions do you have now that you have explored the calculator for a while? This is what we talked about in my class. On the blackboard, we recorded everything we were able to notice about patterns in the 1s digits. On a large piece of paper, we recorded the questions we *now* wanted to find out about. Members of the class formed groups based on the question they were most interested in, and began investigating. You may want to choose a question of your own yourself, or one from our class list. As before, it would be far more helpful if you could talk to someone as you are working on your investigation. If you can, give yourself a couple of days and take a break at some point. Our class worked on this for about a half hour in class on Monday and on Tuesday, with each group assigning itself homework each day to prepare for the next day. On Wednesday, the class talked again about what we now knew, and about what we now wanted to know. We formed new groups based on our new questions, and on Thursday and Friday, held a short group session where anybody could share good insights or ask for advice or help on their particular question. In the end, we had some clear generalizations and some unanswered questions that students chose to work on as they felt like it over the rest of the school year. These more challenging open questions made for some satisfying discussions at various points in the year as one or two students would suddenly announce that they had come up with something new and important to say about our old inquiry.

This is a little unfair to you. You are likely to find this investigation more challenging than a typical second grader who has been working steadily with skip counting, grouping numbers in myriad ways, trading base-ten blocks, and so forth. But my point with these experiences is to get you to be able to look back on a few fresh examples of times when you did not really know what you were supposed to be thinking about, and to consider the types of ways that you learned by yourself and with others. By reflecting on these experiences, you can start to construct for yourself a set of propositions about what constitutes a successful classroom environment. In our class we came up with several ways to explain what was going on with these patterns. Skip counting fans, most comfortable with counting by tens (they were the easiest to remember), noticed that the adding number is always "falling short of ten more" by 10 minus the amount being added. For example, if I start with 4 and use 7 as my adding number, when I add 7, I fall

Know	Want to Know
Can get different patterns.	Which is more important, the starting number or the adding number, in determining the length of the pattern?
Not all same length	
■ add 5 get patterns of only 2 numbers (length = 2)	Why does adding even numbers give those results?
■ add 0 get patterns of only 1 number length	
■ add 1, get pattern of every digit in order (length = 10)	If length is not 1 or 2, do you always use all ten digits in the pattern before it repeats? Why? (or why not?)
Start with even number, add even number, always get even numbers	Could we predict the length if we knew the starting and adding numbers?
Start with odd number, add even number, always get odd numbers	Could we predict a pattern if we knew the starting number? If we knew the adding number?
There are patterns in the 10's digits too	
If you subtract instead of add, you also get patterns	

short of 10 by 3 and get 11, whereas I would have gotten 14 if had added ten. So each time I add 7, I drop in my 1s digit by 3! That is, this makes sense if I can picture a number line of 1s digits going on forever and repeating over and over (which is what the number line we had made on our classroom wall looked like . . .).

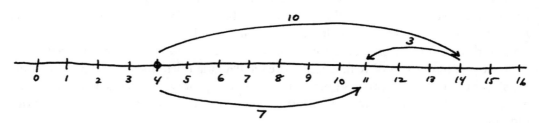

In a way, adding 7 is like subtracting 3! (As long as you keep track of the other digits, like the tens digits in your sum . . . but for these patterns we are not all that interested in the actual sum). And, adding any "adding number" is related to its relationship to 10, because, as Isy pointed out, "we do, after all, use a base-ten system."

Isy's point was that we could see what was going on a little differently. She liked using a 100-chart a lot. On this chart, which she had used in her investigation, she could see that adding an "adding number" is like adding the part of it to the end of the row (to the end of that set of tens), and then adding the rest of it onto the next row.

Starting at 4, to add 7 we first could add 6 to get to 10, and then add 1 more to get to 11; the next 7 takes us to 18; for the next 7, we first add 2 to get to 20, and then 5 more to get to 15; for the subsequent 7, we first add 5 to get to 30 and then two more to get to 32. On Isy's chart, she could put her finger on any number that is in the pattern, and find another one by moving to the right one square and down two. Filling out the whole 100-chart, there is a circle in every column from 1 to 10 (zero in the 1s digit) by the time you get to 67. And then the pattern starts repeating with a 4 in the 1s digit for 74.

Marty and Agnest understood Isy at once because they could see how much her method had in common with what they had done with the base-ten blocks. Adding 7 cubes for them was very much like adding the seven cubes, then trading ten cubes for a "long" or 10-rod and seeing how many individual unit cubes were left over after the trade. For example, starting with 4 and adding 7, they could trade ten of their eleven unit cubes for a long rod, and see one unit cube left over. The 1s digit on their calculator corresponded to the cubes they had left over after trading. They quickly realized that adding 5s meant a long 10-rod every other time, going back and forth between two possible sets of 1s digits, or numbers of individual unit cubes. Adding by 2s meant going to 10 every five times one added, so there would be five possibilities of cubes left over before coming back to the same number of cubes "left over" that one started with. Adding by 4s, 6s and 8s also meant only five possibilities. Odd numbers seemed hard to follow, but they appeared to always leave left-over cubes for 1s until they went through every possible number of cubes left over. The order of how many individual unit cubes was hard to follow, they said; they wanted to know how we could predict the order of the 1s digits for odd adding numbers.

Josh had started out by recording the 1s digits in order (see figure P.4). He then shortened his list to the circle of numbers in the pattern, and, inspired by the circular picture, looked at the patterns around a "clock" of numbers from 0 to 9.

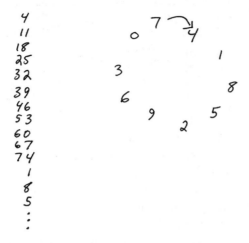

Trying this same diagram for other combinations of starting and adding numbers, he figured out patterns that did not involve every number, later switching to circles that used the numbers from 0 to 9 in order around his clocks.

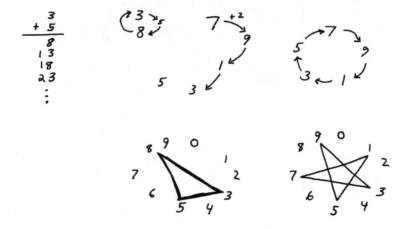

When Josh and Rosa had shared their work, Rosa noticed that you could put the numbers at any point on Josh's circle, so "any of his pictures would work with any of his starting numbers." What she meant was, we could start at any number on the circle and get the same picture, just rotated, given the particular adding numbers. That is when she and Josh decided that the *adding number* was more important than the starting number, and that it alone would determine the length of the pattern. You either got all ten digits, the star or pentagon (with five digits), or the line, with two digits, in your pattern. They wanted to know if it is possible to get any other possible pattern lengths, but offered the "conjecture" that it is not possible to get any other results.

2. Envelope

Here is another situation for you to explore. See if you can create a mathematical language for what you are doing. Imagine you have no envelopes at home but you need to mail your letter. Can you make one of your own? What shape paper might you use? How should you fold your letter? How will you make the envelope? Do you need a separate piece of paper?

When my sixth-graders worked on this one, they first tried a range of shapes for the letter: square, rectangle, triangles, even circular paper folded into a semi-circle. And they explored different paper shapes to fold the envelope from: rectangles, squares, circles, triangles, trapezoids . . . They found inventive options for most of these. Then the fun began: What was the minimum size and shape paper for the envelope given a certain letter? What designs would require the least amount of tape or glue to make them work? Would each be "legal" and conform to postal regulations?

A good way to test out a new design was to telephone or email another student with directions for how to make your envelope. No faxing or pictures was allowed. Soon we began sending sets of new directions in our old hand-made envelopes!

Box P.2

Do this investigation now, preferably with at least one other person. Do not stop until you have come up with two new questions based on your work.

Notice that I have asked you to stop your work only when you have come up with your own new questions. This is an important phase of mathematical inquiry in a dynamic classroom. Old-fashioned classes stopped with answers. Now we stop with new questions. We always leave ourselves more to think about. You may want to reflect on how this makes you feel before you go on. Some people experience discomfort with this idea of inventing new questions rather than tying things up with a neat final answer. This book invites you to challenge many of the assumptions about good teaching and learning experiences that we have come to expect over the years, in order to develop more successful educational environments for learning and teaching mathematics. The next exercise pushes this idea even further, making even the original question itself unclear and poorly framed, because it is more like a real question in life than a carefully fine-tuned textbook problem.

3. Paper Knot

Now that you are vexed with paper folding, think about this one: take a long thin strip of paper and slowly fold it into a knot. Squash it flat. It makes a pentagon. Can you see the pentagon? *Why does it make this shape?* What other shape(s) can you make by tying knots in paper strips?

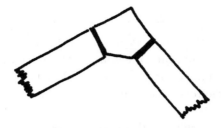

Box P.3

Do this investigation now, preferably with at least one other person. Do not stop until you have come up with two new questions based on your work.

4. Magic Trick

Astound friends and family. Have someone choose 21 cards from a deck, shuffle the 21 cards, pick one card to remember, and then return that card to the deck of 21. Shuffle the 21 cards yourself a lot, and offer your friend the chance to shuffle them some more. Now deal the cards face up into three piles, asking your friend to check that they can remember their card by noticing which pile it is in. After you have dealt all 21 cards, test them by having them point out which pile their card is in without saying what card it is. Say, "If you can remember the card's pile two more times, I'll be able to concentrate on your brain waves and figure out your card." Each time (three times in all), pick up the cards, put the pile with their card in the middle, and remember not to shuffle before dealing them face-up into three piles. Now: have you figured out how to tell which card is your friend's? If you can figure out how this trick works, then you can vary it to amaze your audience forever. Change the number of piles, where you place the pile with their card (in the middle? on top? the bottom?) . . . Can every number of cards work, or is there something special about 21? What other numbers of cards or piles might work for such a trick?

My elementary age children love to investigate magic tricks because they have a fascination with knowing "the secret." This exploration taps right into their sense of awe: stumping someone with a new variation and then sharing the secret behind it is the greatest thrill of all, especially the sharing of the secret behind the trick.

Box P.4

Try to figure out how the trick works. It will be helpful to come up with a way to represent what is going on, with pictures, charts, or manipulative materials. What will stand for the chosen, unknown card? What will represent cards that *might* be the chosen card? What will stand for the cards that you know cannot be the chosen the card? Remember, talking with someone else is an important part of the experience.

Playing this trick a few times will help you reassure yourself that the chosen card ends up being the eleventh one in the pile. So you can pretend to be focusing on brain waves, consider several cards briefly, but finally zero in dramatically with a sudden psychic flourish on the eleventh card. But why and how does this trick work? Only through this understanding will you be able to change the number of cards or piles and make up your own version of the trick.

Lester suggested drawing a picture of the three piles.

He numbered the cards in the order they were dealt. Now, once the friend points out a pile, Lester knows the card is one of the seven in that pile, which he then puts in the middle of the stack. So these cards would now be numbered 8 through 14 in the dealing order.

Suppose your friend points to the first pile during this second round. Then you know the card will be either number 10 or 13, which is about to be placed in the deck to be dealt as number 11 or 12 in the third round. Can you see how Lester's diagrams help him to determine this? Since we have represented those cards in the first round that are the contenders for the chosen card, and because we can see them as a subset of this new identified set of cards, we have narrowed down which cards could possibly be the one our friend picked at first. But there are other possibilities. Suppose your friend points to the second pile; in this case the card is either 8, 11, or 14 in the second round, which become 10, 11, or 12 in the third round when this pile is placed in the middle of the stack and after all the cards have been dealt once more. But, if your friend points to the third pile, then the chosen card is going to be either 10 or 11 when the pile is placed in the middle of the deck. O.K.: so the card that was chosen is going to be either 10, 11, or 12 in the third round of dealing:

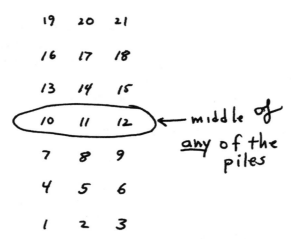

So now, no matter what pile was selected during the second round, the chosen card is *in the middle*. When Lester placed the pile in the middle for the third time in his mind, he was able to predict that the middle card would have to be the eleventh card in the whole stack!

Box P.5

Your job now: Try changing the number of cards and/or the number of piles. What can you say? Find a different picture that you can use to explain which combinations work and why.

Before you go on: Invent your own investigation that is based on something you thought of when working on this exploration. Share your questions with someone else, and work on each other's questions together.

5. String Around the World

Fred Goodman, my beloved advisor at the University of Michigan, used to use this question to get people thinking about the relationship between common sense and abstract knowledge. Imagine somebody has managed to tie a string tightly around the equator of the Earth. Now: you come along, cut the string, and add precisely one foot of string to it, so that it is not taut anymore. The question is, how far from the Earth can the string be pulled? Could you fit your hand under it?

Could you do the limbo under it? Drive a car under it?

Make sure you have talked about this with someone before you read the rest of this paragraph! But even if you are "cheating" and want to work alone, you may find it helpful to model the situation. Try a ping-pong ball, a basketball, and/or a soccer ball, and some other Earth-shaped

Box P.6

Write or email a friend or enemy about this question. What is your explanation for your conclusion? Are you using mathematics to determine your conclusion?

objects. Tie a string around the ball and then add one foot to the string. See how much space is available. How does the size of the ball ("Earth") affect this space?

Did you use mathematics to predict the results? Or, *could* you have predicted this mathematically? If you did not yet use mathematics, try to. The real question is, why do so few people even think about this mathematically and what does this say about the state of school mathematics? What does it say about mathematics and common sense?

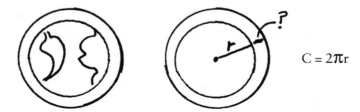

The new diameter is always a mere $1/\pi$, which is less than but very close to one-third of a foot more than the old diameter—*no matter how big or small the ball is*! You might be an amazing, very skinny, limbo dancer, but most likely one-third of a foot will not add much space for driving a car under the string!

Why doesn't the size of the ball matter in this question, and could there have been any way that we might have realized this without working with algebra? There's a way in which some people do not believe the result because they do not trust the algebra more than their common sense. Philip Davis has thought about math and common sense (Davis 1995). He asks, if mathematics makes sense, then wouldn't simple mathematization be used to control traffic lights according to the traffic that actually exists? If you walk when the pedestrian "walk/don't walk" sign is green without paying any attention to the traffic you can easily get killed; if you pay attention to the ambient traffic you can often walk in safety when the sign tells you not to. Davis also asks why we can often find a round-trip, two-way air-fare that is less expensive than a one-way fare to the same destination. Bill Rosenthal (2004) asks why we can argue that 0.9999 . . . is and is *not* equal to 1. Shouldn't we be able to use our common sense to decide this question one way or the other? But it turns out it is not so easy to decide, even though mathematicians are mostly convinced of one way to answer it.

Box P.7

Think of three examples of your own where mathematics does not make sense. For each example, write or tell a story about how this confusion between mathematics and common sense could lead to a surprising or strange situation in the world.

6. String Figures ("Cat's Cradle")

While we are on the subject of string, find a nice piece about three to five feet long, tie it into a loop with a tiny knot, and learn how to make some string figures. Play cat's cradle with a friend until you can do it without ever messing up. Can you explain how to get from any figure to any other one in cat's cradle? Why is this so hard for us, even when cat's cradle is not? Again the point is for you to imagine what it is like for a student of mathematics to be asked to explain or

represent something mathematically. In the same way that explaining directions for moving from one string position to another in cat's cradle is a challenge for us older folks, a child in second grade explaining why he or she is multiplying numbers in order to make a decision, or a youth's explanation about exponentiation in high school is just as weird and new for them. It is important that we challenge ourselves in these new and peculiar situations, that we demand of ourselves that we invent a way of representing our ideas and a language for what we identify as important in the situation at hand.

One important feature of cat's cradle string figures is that they can be passed back and forth on the hands of several players, who add new moves in the building of complex patterns. Thus cat's cradle invites a sense of collective work, of one person not being able to make all the patterns alone. . . . It is not always possible to repeat interesting patterns, and figuring out what happened to result in intriguing patterns is an embodied analytical skill (Haraway 1989). In other words, cats cradle is a serious game about complex, collaborative processes for making and passing on culturally interesting patterns.

We will continue to play with mathematics throughout this adventure. For now, I just want to highlight a few themes. To begin with, I have been trying to communicate several important properties of a vibrant mathematics classroom. In such a classroom, the mathematics is flexible and can appear in forms that are unpredictable and sometimes even unrecognizable. For example, we could mathematize anything we encounter, simply by creating a representation for what we encounter, and then by searching for relationships that are evident in that representation. When we do this, we can usually bring those relations back to the original experience and make conclusions. We can do this because we have created a *model* of something. The model is something we can analyze.

Box P.9

In using cat's cradle as a metaphor for what we observe in a classroom, we seek to understand the processes of constitution and unraveling of diverse, fragmented cultures. Your challenge: (a) to further apply the metaphor of cat's cradle to a classroom you are working in or observing; or (b) to find another analogy for a classroom, explain why you think it is apt, and then invent a mathematical investigation about the relationships that are in the analogous situation.

And our analysis is something that can be applied to what we were thinking about when we came up with the model in the first place. Note though that we may come up with a model that does not quite fit a real life situation and in such an event our conclusions could be bizarre or ridiculous. Also, our model may share important properties with something but nevertheless also emphasize properties that make no sense for what we are trying to model. In any of these cases, the mathematics is still useful and fun, and may one day be applicable to something else, so it is never a waste of our time— especially if we are able to understand the limitations of our model! This too is part of mathematics.

Another important element in a vibrant mathematics classroom is the propensity to explore. Students (and teachers) should always be looking for ways to ask a new question about what might be knowable. This entails formulating a conjecture and then going ahead and testing this conjecture. A conjecture is something like a hypothesis in science. So in our mathematics classrooms we want to provide an environment that supports rather than inhibits experimentation, including the collection of data that can help us to test our conjectures, and develop conclusions or new experiments based on our data.

Important for all of this to occur is a general atmosphere where folks are able to share their findings and solicit advice from others, much in the manner of a professional research community. A classroom will come alive if there are periodic "conventions" or "professional meetings" during which students report to each other on their mathematical research, and get peer feedback, publish their results in peer-reviewed journals, and serve as reviewers of each other's efforts. Before students are ready for such meetings, they need ample time for collecting data in investigations. They need to organize this data as information that they can analyze and interpret, in order to be able to draw conclusions from their efforts. They will likely reach moments when they are not sure what to do, but for which there is no clear-cut best method for proceeding.

How can this take place when we are simultaneously trying to "cover" the curriculum and guarantee the attainment of certain skills? The better way to phrase this last question is as follows: The *only* way to cover the curriculum and foster the attainment of certain skills is to build an infrastructure that supports meaningful learning. Those standards and district objectives are not our enemies but our friends: They help us to determine whether or not our infrastructure is successfully accomplishing our goals, indeed they help us to articulate for ourselves as teachers what the students need to be doing. This is different from what the *students* need to be thinking about. Students may say they are inventing new magic tricks, whereas I as their teacher would describe our work as building concepts of multiplication and division through an application and investigation that captures my students' interest. Others may say that we are preparing a report on a traffic intersection near our school building; I would describe this same project as developing sophisticated skills in advanced statistics and data analysis while also using geometric models. The following chapters describe examples of ways that teachers have found entries into such a mathematics classroom, and in the process, become teachers of mathematics.

Response to Prologue:
Be a Student of Mathematics Learners

David Scott Allen

We must look at these problems from at least two perspectives: the teacher's and the students'. If we are not students, then we cannot be teachers. If we do not minimally "monitor and adjust" by observing our students' strengths and weaknesses, then we cannot teach them so that they will learn. I conclude that unless we are learners, we cannot teach, and if we are to be learners, then let us attempt to learn the way our students learn, as it helps us to get inside their mathematical worlds. Most people who become teachers do so because they found success in school in at least one subject area. But most students in school are not necessarily in love with any particular subject until something or someone sparks that love. Until then, they look at school as a series of irrelevant tasks imposed upon them by old people who just don't get the big picture. We need to convince them that some of us understand an even bigger picture within which their big picture fits. They will then move beyond executing our tasks with minimal effort. For some, what they value is a good grade, but for others the value is in seeing the teacher uncomfortable with the student's noncompliance.

I never felt as (mathematically) lost as the (recent!) day that I realized I had no idea what area really was. I tried to define it myself. I tried looking up definitions in regular dictionaries and mathematics dictionaries, and only got more puzzled in the process. Of course, I know how to calculate areas of figures by using formulas or combinations of formulas. I can draw an area by shading a closed figure. I can explain these to anyone who asks, and I teach the topic regularly in just about all my courses. But I still really don't know what it *is*. This was scary at first, but it helped me to get a glimpse of what my students go through each day. I present them with ideas and combinations of symbols which they must be able to manipulate like the professionals, and they do so with varying degrees of competence and consistency, but many still have no clue what it is they are doing or why they are doing it. Mathematics is found in its use, though not necessarily its usefulness. Students know this intuitively. Hence their perennial question: "When am I ever gonna have to *use* this?" Well, we use area when we buy paint or carpet, or even plan birthday parties (is this room big enough for 24 five-year-olds?). We even use it on standardized tests! This may be circular reasoning, but the fact is that mathematics is a gateway subject for almost any career, and many careers have some sort of test to take. Alas, mathematics can be the original Society to Perpetuate Itself.

And so this book begins with a chapter of activities which put you in the role of the student: at times you are wondering why you are doing this silly activity or thinking about that easy question, and at other times you are so absorbed in the task at hand that you forget where you are. Even if you have seen some of these activities before, they are never part of you until you do them yourself. You can never master them, but you can own them. You can say, "I've been there and this is what I've learned." We place this section first in order to help you make this claim.

These types of pursuits are exactly my dilemma as I am writing. My mind turns to baseball every October, both because I love baseball and because I love manipulating data. I am intrigued by the fact that one franchise wins championships very regularly, while others win occasionally, and still others never win. Which teams are which? Why does this happen? Should I root for a perennial "loser" or "winner"? In 2004, the Boston Red Sox broke the "Curse of the Bambino" and won the World Series after an 86-year draught. I turned to my record books to see which other teams have had long draughts without a championship. Noting that both Chicago teams had gone the longest without a championship, I pondered why and decided to root for both of them. In 2005, the Chicago White Sox won their first championship since before the Black Sox scandal in 1919. So I am now a Cubs fan. It's been a long time for them! It's true that some of the newer teams have never won a championship, but they also haven't waited as long as Chicago. Then in 2006, St. Louis won their tenth championship, the most by a National League team. Every fall, I enjoy inventing a new statistic to add to the classic litany of ERA and RBI. My research goes on, and I write this chapter while watching the games. I mathematize baseball because I love it.

A common argument against using the kinds of activities in this chapter is that these types of problems don't show up on State Standardized Tests. I have decided that these tests present an easy target for most, if not all, of my students to hit, and so if I find a balance of these exploratory activities and those omnipresent multiple choice items, both my students and I can feel more sane and balanced as we study math. For now, I am compartmentalizing these new activities into a day or maybe a week at a stretch and then returning to the tested curriculum. I plan to take the next step, which for me would be to teach both a little technique and a little thinking each day in order to break up the daily grind of one more notational rule to drill. The problems I use to prepare my math team for competition are actually rich with the required standards. I use them in the classroom whenever I can.

Recently, I actually took time to immerse my students in a situation using the board, pieces, and moves of chess. They were to discover the relationships between the moves of the pieces by asking questions about how they relate to each other, as well as to the board. The first 20 minutes of the first day were pedagogical chaos, with some students not knowing the ways the pieces move, some students having no clue where to begin, and others pursuing truly abstract problems. Just then my building principal arrived to do a walk-through of my classroom. In her report, she commented to all of the above activities, but noted how all the students were interested in doing the activity even though they were at various stages of being able to do so. The second day I had them work on some of the questions which we had discussed on the first day. The third day I had them post questions of their own on the board for the entire class to see, and encouraged them to choose which question(s) to pursue and how much time to spend on each. Some of the questions mimicked earlier ones, but some broke new ground. Most of the questions concerned configurations of six chess pieces, one of each type.

One girl thought so outside the class's set of questions that she inadvertently stumbled upon a set of questions which I had researched myself. She selected one of the pieces (the knight), and decided that multiple copies of the same piece would be used to create a chain or loop of moves which come back to the starting point (so each move by the piece is a side of a polygon). I am very interested in the knight's move and the patterns it forms on the board. It was difficult to not give her answers, but she was discovering my work all over again by the end of the third day.

When I asked the class how they wanted to culminate the week on the last day, they said that they wanted to play chess. Wanting to reward their forbearing efforts at such an unusual task cluster, I allowed them to do an informal tournament, and they did so with a spirit of teamwork that had not been as evident before the week began. Finally, I asked them to write down the week's events and experiences from their point of view.

The overall tone of the write-ups was that the students felt that the problems we did in class were problems worth doing. They were more interested in the game of chess and felt that it related more to math than they thought. They expressed a range of emotions about the week, but most of them felt they had accomplished something. For my part, I was amazed at how persistent they were to meet the challenges set up by other members of the class. They either tried to be the first to solve a puzzle or they kept looking for ways to improve upon a current best solution. These are all mathematical activities, the work of mathematicians. Perhaps next year I will challenge them to describe how they go about solving the questions, and what patterns they find among them. They should write about what they find and what they would like to find and then publish a mathematics newsletter for the school.

I have not tried the calculator patterns in class in a strictly numerical, ones digit way, but I have tried a variation on Isy's visualization of the project, which uses the hundred-board. I challenged my students to use different sized boards and to see if the patterns were the same for each "adding number." I asked them to classify the patterns and then compare the results from various board sizes. This is definitely worth taking class time to pursue.

I have not tried the envelope activity, but I have emphasized to my students the importance of coming up with a mathematical language for what they are doing at any given point in such an activity. More recently, I have decided that the primary goal of school mathematics seems to be to train students to communicate mathematical operations using the correct symbol in the correct location at the correct time. We are tacitly communicating that if they want to be mathematical they have to say it a certain way. You show that you really know what you are doing (proper doing and proper communicating are tightly intertwined here) if you write it this way. I remind myself that if all of algebra can be reduced to a set of conventions, then it is doomed to be forgotten. Skill at symbolic manipulation is not an attractive goal at any level, especially middle school.

I asked the students on our Math Team to write questions similar to the ones which are used in competition. What they produced were questions that were more like the ones found in older (and not so much older) textbooks. They are constructed artificially so they can be solved mechanically, and one joins the club of good math students when certain problem types are mastered. Clearly, these mathematically precocious children need instruction outside this box. I stumbled upon this just weeks later as we were doing challenging problems after school. We had finally attempted some problems which nobody could solve during the course of our two-hour meeting that day. I put them aside as Unsolved Mysteries. By the end of the book, we had thirty-six Unsolved Mysteries. During one of our meetings, I gave them copies of the thirty-six, and they set to work solving sixteen that day, and more before our next meeting. Good problems can be an inspiration and a motivation.

I have tried the paper knot question, but I have gotten zero results. Perhaps my seventh- and eighth-grade students are too self-conscious to come up with off-the-wall explanations, and too young to come up with intelligent ones. But my main conclusion to this failure is that my students don't have the vocabulary to describe what they are seeing in or how they are thinking about the problem. My goal is to help them to build such a vocabulary. As for variations, I think I would need a Boy Scout to really execute them. My students are just too young, and so this may not work in the middle school as well as it might in high school.

My students have lost their youthful desire to know the secret of the magic trick and explain it to others. They think it is interesting but a bit trivial. This is the kind of thing I might do during study hall with my students to see their reactions to it, and then if I get enough students there interested in the "answer" to the problem, try it in class. This is one advantage I have in my situation: 40 minutes of study hall with a few dozen students at various levels of ability who also are in my math class. Another set of students is right next door. I can lay out puzzles and games and tricks in a very non-threatening, non-demanding, invitational environment, and then later pursue the ones which catch their interest. I believe students do have at least one interesting question on their minds (if they would admit it), it's just that they are never deeply satisfied by the answers they get from the mathematical establishment (which includes me). I myself am intrigued by the "legal" question of envelopes conforming to postal regulations. Perhaps I will get a chance to do a follow-up with my students someday.

What I am getting at here is that some activities work better in some situations and some work better in others. Your students may love problems which mine just don't get, and vice versa. Much of this can be the age of the students. I expect that we will agree on some issues and not on others. Mathematics is vast. It can withstand such scrutiny.

For example, as I am typing this on my laptop, my long fingers accidentally strike the keys to bring up Insert Diagram. I have never tried this before, and I am intrigued by the fact that each one of the six types of diagrams I can insert has a little sentence suggesting how it can be used. My hunch is that my students have not seen any of these types of diagrams in this context. I ask myself what would happen if I gave them the diagrams and asked them to list potential uses for each. Would they think of other ideas besides the suggestions? Are there other classes of diagrams that they would think of which are not included in this apparently time-tested set of six? How are these diagrams related and unrelated?

Perhaps most importantly, can they devise a mathematical vocabulary to communicate these ideas? Mathematics cannot be bounded. You do not need books like this to challenge your assumptions about "math class." You can challenge them for yourself in ways that only you can do. Mathematics is not just about the course and the material and the expectations. It's about values, environments and relationships.

As for string around the world, this would put my students in a position where they would have to choose to overcome their temptation to say that it is boring in favor of saying that it is an intriguing problem. I would have to bait the hook enough times to get them to the point where they would not be able to resist taking up the problem. I could wave the formula for circumference of a circle in front of their noses just enough to get them to attack it on the right day. Maybe they'd bite on Pi Day.

Where does mathematics not make sense? From the point of view of the students, the answer is: everywhere! In their world, mathematics and arithmetic are synonymous. Why know how to calculate a tip when you can just leave what you want or nothing at all? Why bother doing your taxes when you can pay someone to do them? Even I pay a friend—it's just not worth my time. So we must remember that the symbolic manipulation we teach is a skill, a means to an end. Don't lose sight of why you are teaching what you are teaching. It is partly to allow your students to solve real problems which can be abstracted using variables or diagrams.

Remember that each chapter after this one represents a lens through which to view yourself, your students, and the relationships and interactions among everyone and everything in the classroom. If you are a student of mathematics learners you will begin to embrace the fact that you and your students *will* view class differently, because you are *supposed* to view class differently.

one
Planning and Assessment

Two ironic insights provide the buoyancy and verve I need for my work with students. I say *ironic* because they may sound negative or distancing when my goals include a satisfying and proximal relationship with my students. Yet it is in the very recognition of differences between myself and my students that a possibility for fruitful encounters emerges. First, there is, in general, a fundamental difference in the ways I encounter the world and derive meaning within my experiences and the ways that my students encounter the world and construct meanings within experiences. Second, there is a difference in what I and my students attend to in a pedagogical encounter, mostly due, of course, to our different roles in the encounter. The irony here is that these sound like "problems" to overcome when in fact an understanding of these differences genuinely helps in thinking through the kinds of encounters that we can and should (or might) enjoy together. I will discuss each of these ideas briefly and then offer a model for how I use a basic "unit" of instruction to guide my planning and assessment.

Irony and the "Problem of Culture"

Elijah Mirochnick (2003), an artist and teacher of educational research at George Mason University in Virginia, uses the *Pulp Fiction* "test" to understand generational changes in our comprehension of events and images. Can you remember your reaction to this film? It was nominated for six academy awards back in 1994. If you have not seen it or do not remember it anymore, please rent or download it and view it before going on to read this chapter.

Most people fall into one of two categories when they discuss *Pulp Fiction*, according to Mirochnick. Some of us primarily focus on the violence in the film. We may mention blood splattering on a car window, and go on to describe other depictions of explicit violence. Others of us may recognize that violence is part of the film, but we may focus mostly on how *funny* the movie was. Elijah suggests: If you talk about the violence in the film, you probably interpret the film within a "modernist" paradigm. Modernism tries to reveal a raw truth about the world, and to tell the true story about an event. The scientific revolution in one respect led to modernism as the dominant way of interpreting experience in the nineteenth century, as the scientific method is designed to lead to a clear and vivid picture of the way things work. Very accurate representations in works of art are also modernist, as are novels that try to represent in words a picture of what is happening in the story—as if the novel were translatable into a photograph of the "real" characters, dressed exactly as they would be dressed, holding their bodies precisely as they would be posed, in a setting that is true to the time and place of the scene in the novel.

If you thought *Pulp Fiction* was *funny*, though, then you "got it" about the movie: you understood the irony and, Elijah suspects, you are interpreting this film and many of your life experiences in a *post-modern* way. Some of us can not even begin to watch this film long enough to "get" the irony. This may be because, as modernists, we take in the film literally, at face value, as a depiction of something really happening (or, that might really happen). Youth in our mathematics classes are the first generation to be so fully immersed in a "culture of irony" where nothing is necessarily what it seems to be. A film is just a film; the violence in a videogame or in a TV wrestling match is all pretend. Madonna and Britney Spears dress and present themselves as sluts, Marilyn Monroes, Marlene Dietrichs, and so on, but in doing so, they are understood to be communicating that they are none of these; they instead communicate a very different message, about being intelligent artists and business people. Bart is the anti-hero in *The Simpsons*, the "bad boy"; yet he is, to us, the real hero of the series. If you are shaking your head in bewilderment at these last few examples, you are definitely someone who views the world as a modernist. If you are nodding in agreement, or laughing, then you are a post-modernist like almost all of the youth now in school. The "problem" is, school practices are designed for modernist people, not for the post-modern youth in our classrooms. So we need to reinvent the experiences that happen in school if we are to work with youth who perceive most things they see and are asked to think about in irony.

We also need to enter these encounters with the understanding that we ourselves, as teachers, are understood and experienced by our students with the same irony. We may think we are projecting a certain image or taking on a clear and legitimate role, but our students might see our image and role as just that: an image or role, an image that may be hiding another image, one that may change from one moment to the next, or one that may be standing in for another one. Furthermore, mathematics, too, is a potentially ironic collection of images and representations with no fixed truths or messages. It is pointless to try to explain our modernist "truth" to a post-modern child who will merely parrot what s/he knows you want to hear only to be saying something ironic in the parody. I raise all of this, remember, not to lead us to despair, but to help us to recognize something important about our work: we need to find ways to tap into and engage with youth who are highly skilled with and hypersensitive to irony, parody, satire, and collections of knowledge that may have no necessary grounding in a foundation of first facts or starting points.

My second recognition of difference was once nicely described by Stieg Mellin-Olsen, a mathematics educator who taught in Bergen, Norway, until his sad death in the nineteen-eighties. Mellin-Olsen wrote of "the problem of culture" in education. When he used this phrase he was thinking about the ways that teachers focus on the conceptual goals of an activity while the students focus on the activity itself. Here I am, facilitating an event that I am expecting to help some of my students think about, use, and apply mathematical ideas and techniques, perhaps to help them examine their own use of mathematical strategies and so on. I have long-term goals that have to do with a sort of "growing of mathematicians." At the same time, my students are—simply—focused on the task at hand: how to do the task, what they are making or creating with the task. The traditional roles of teacher and student have meant for most classrooms that the teacher and the students are pretty much never talking about the same thing—unless the teacher abandons the content and makes the rhetorical move to speak *with* the students about the task or about mathematics *through* the task. Because all of the meaning and interest, all of the purpose of the activity is for the teacher in the long-term, content-related goals on the teacher's mind before the shift to the task, it is no wonder that everyone perceives mathematics class as mindless, rote tedium devoid of intrinsic interest. Somehow the activity has to have a purpose in both the teacher/long-term-goal and student/what-I-am-doing-now realms, so that the teacher and the student can share purposes and thus generate a meaningful interest in the events of the classroom. Otherwise, the only way for students to engage with the task through meaning is to invent their own. Some successful mathematics students do this. Unfortunately, what they invent is only arbitrarily likely to have something to do with the teacher's long-term goals. What I am searching for in my own work with youth are processes of interaction and participation that recognize this pedagogical situation and either use the lack of communication in useful ways or change the roles and responsibilities so that the dysfunctional tradition is reconstituted in productive ways.

My desire is to embrace post-modern irony with my students, so that we can together work with narrative mathematical discourses and practices while also avoiding the teacher–student split so endemic in contemporary schooling. The purposes of these mathematical narratives should not be represented in the usual, common sense way that assumes mathematics is describing reality. This means a loss of security and certainty that was a gift of Platonic and applied mathematics. In Plato's mathematics, which is pretty much the mathematics we have inherited in school today, mathematical ideas and methods are "out there" for us to discover; mathematical relationships would be true or false, and verifiable through clear and logical techniques. We lose the one truth in a post-modern way of life, but we gain a multiplicity of mathematical stories that can be told and retold. In applied mathematics, mathematical concepts and methods become the "queen of the sciences," a tool of a modernist scientific story about a true and verifiable, "real world" that we can accurately and completely represent to ourselves. Richard Rorty (1981) called this dream of representing nature in an accurate and "true" way the "mirror of nature." We lose the "mirror of nature" that mathematics generously provided when we live and act without a presumption of modernism, but we gain the potential for mathematics to be much more than a queen. Instead of being led on a wondrous tour of the land of mathematics, and instead of being trained as technicians or engineers, our students must be freed to experience mathematics as ironic representation, as satiric parody, and as offering by itself conflicting and parallel stories (about the world and about mathematics itself). I further hope for mathematics to become a tool and medium for other tales and images—say, of poetry, self-reflection, indignation, aesthetic expression, comedy, tragedy, and so on. I believe I have found one possible structure for this in the "unit" of instruction that follows.

The Five-part Inquiry Unit

Intro/ Opening	Project /Investigation			Archaeology
	Develop a project	Do the Project	Put work back out into the world	

- ongoing assessment

- clinics and mini-lessons

My favorite version of the mathematics thematic unit is adapted from Baker, Stemple and Stead's (1990) *How Big is the Moon?*. This unit has five parts across three main sections: the opening introduction, the period of working on projects, and the important archaeology at the end. Each part might last for about one week. The idea, then, is to think of the first week of the unit as introducing the theme. The second, third and fourth weeks constitute the middle components of the unit devoted to projects: first, helping students to identify and begin to work on a project; second, to delve into their projects in depth; and third, to put together a way of sharing important aspects of their work with people beyond their immediate classroom. In the last week of the unit, the archaeology, the teacher helps the students to identify for themselves the skills and concepts that they have been working with, and to recognize that they can easily use them in new contexts beyond those of their projects and the specific theme at hand.

The middle part of the unit is where serious work unfolds. In this phase of the inquiry, I help students to articulate their projects and support their carrying out of their work. But I also implement important assessment strategies that help me to identify mini-lessons and clinics on specific skills and concepts that I can see the students need in order for their work to be successful. In this latter part of my teaching I often use traditional direct-instruction pedagogies on specific topics. But in general, the feel of the classroom is one of serious investigations and sharing of information and resources. My assessment work in the middle of the unit is very helpful as well in identifying the key foci of the final period of archaeology, where we look back at what we have done and pull out those specific skills and concepts. The last week is spent on applications of that knowledge to smaller projects and activities, but also to puzzles, games and other consolidation experiences that can help the students become aware of what they have accomplished during the unit. Students are left with a clear picture of the skills and concepts with which they have worked, and how they can be used in new contexts. They do not get the feeling that they have merely done a fun project; instead, they also can describe the powerful skills and concepts that they are bringing into new phases of their lives.

Week one: Opening. I need to plan for the first day: how will we introduce the unit theme? I need to find an open-ended activity that can raise issues and questions and at the same time serve as an assessment tool for me as the teacher. After the activity I ask my students: What do we know about this theme? What do we want to know? What do we need to learn in order to find out what we want to know? (Yes, this is a K-W-L activity.) The class then breaks into groups to work on finding out those things that they have identified as need-to-know; they find out and teach the rest of the class in an entertaining way for the last couple of days of the week.

What else do I need to plan? I think about equipment that the students might need for their initial investigation this week. I plan books for the classroom library, I arrange permissions for and access to whatever is necessary (for example, the library, a telephone, a computer connection to the internet . . .); I arrange for visitors who will be coming this week.

Weeks Two, Three, Four: Individual and Group Projects. I cannot plan too much for what the students will actually do as an inquiry, because it is up to them to think of a project. I will be conferencing with them as they come up with a topic or goal, do their research, and put together some sort of presentation, event, or exhibit. What I *can* do is brainstorm potential projects students *might* choose for this unit. Based on their interest, what do I imagine they might think of? I may suggest alternatives or gently steer them in a direction that would be more fruitful since I have done this initial spadework. As I brainstorm projects that students might come up with on their own, I think about how the unit will work for the curriculum. For each project I think of, I list conceptual, procedural and factual knowledge goals that the project can serve. I think of connections across the curriculum for each potential project. I do this within the subject area or across subject areas as appropriate. I look through my district curriculum guide or textbook to find particular objectives that I can have in mind and later use to ensure that I actually am accomplishing what I am supposed to do through this inquiry. Once I have these lists, I can begin to organize in my mind potential clinics and mini-lessons that I may need to teach if I see that the students need them.

During these three project weeks, I schedule clinics of direct instruction as needed. One idea is to tape large sheets of newsprint on the wall. When I see a need for people in the class to learn or practice a skill or concept, I announce that I will offer a mini-lesson if at least five people sign up. Having a minimum number lets folks be good sports and sign up for help even if they do not want to admit they really need it. They do it as a favor to others who do. As people sign up, it is *then* that I can plan the actual lesson based on what I know about the unit, what I have observed and documented about the students' work so far, and my and the district's goals for mathematics. But before the unit actually is happening, what I need to do is think about what lessons I *might* plan, just so I can be sure that I will be ready with various resources, materials, and so on. During the unit I may announce that everyone should attend a clinic session because I want all students to practice something; the flavor of choice still permeates the atmosphere though, because in general the students are planning and monitoring their own work on their projects on their own, with my guidance and feedback.

One more thing to be planning and organizing during the middle of the inquiry unit is needed materials. Here, too, I want to prepare for what I believe students *may* need, such as equipment, books, supplies, and so on. Once the students are involved in particular investigations, I have a better sense of what they really need. I also arrange ahead of time for professionals (e.g., architects, bakers, masons, ad execs, beauticians) who may be called during the month, and I plan with the students for permissions that are needed for trips or for visitors. It is very exciting if the students can act in roles of particular types of professionals! For example, if students are using mathematics to study the ocean, some of them may be marine biologists, others might be economists or ecologists, still others might be taking on the characteristics of statisticians, and so on. Or, if my students are studying circles in geometry, some may be working with Geometer's Sketchpad® to explore sectors; others may be testing conjectures with straightedge and compass; one group may be exploring circles in architecture only to come back to those using Sketchpad® with useful information about sectors; still others may be working with the use of arcs in sports medicine. If students work in these sorts of ways, it is especially intriguing if men and women who do such work can communicate with the students and advise them on how to go about their own investigations, or to comment on what they have been able to learn about. Students should take on the styles of work that the professionals use—for example, carry around a portfolio of sketches, document their data in a field notebook, organize their reflections into a magazine article, and so on. If they are highly focused in a mathematical investigation, then they should work like mathematicians, with a portfolio of starts and stops, ideas and brainstorms. They should occasionally stop to organize their work into a presentable form. They should also get a chance to share their work with "real" mathematicians and other students and adults who can appreciate how they were able to move from the confusion of their initial ideas to a final, well-organized statement of what they now know and can do.

During the first week of the middle of the inquiry (week two), students spend most of their time finding a project to pursue, starting it and rethinking it, until they are pretty sure that they have found a doable and reasonable project to explore for the next week. I conference with individuals and groups as they plan, helping them to find others with common interests, and organizing how the groups will form for the inquiry to proceed. I may offer a couple of early mini-lessons if I see a need for the projects to be more sophisticated or to push the students to think in new directions.

During the second week of the middle section (week three), students pursue their projects. I spend much of my time observing students and carrying out various strategies of assessment that help me monitor what the students know and can do. I offer clinics and mini-lessons and plan further assessments based on my own goals for the unit.

The third week of this middle section (week four) involves students putting together a way of sharing what they have been working on beyond the immediate classroom. This could be a polished performance, but most likely there really is not enough time to plan and present a show. The important goal here is for the students to identify an interesting and important aspect of their work, and to experience articulating what they have been doing for others. Speaking with an architect or statistician and getting advice is one form of this kind of "performance." Setting up an informal "museum" and giving tours to other students in the school is another. Interviewing a beautician about how she determines the proper dyes for coloring hair might be another. The quick preparation for this experience forces my students to think about the material in new ways that help them to understand the key ideas and the essential elements of their work.

Archaeology (week five): Sometimes teachers love it when students are learning and they do not even know it. I think it is actually fundamentally important for students to reflect on their learning after the process has occurred in order to realize that they did indeed do more than just the project that they worked on. It is my responsibility to make sure that my students can describe those skills and concepts that they have developed through this unit, and see for themselves that they can apply these skills and concepts in new ways and in seemingly unrelated situations. Most of the planning for this week happens as the unit unfolds, but there are things that I can start to do ahead of time. By imagining possible activities that I hope my students could be able to do, I can get a better sense of my objectives for the unit, and I can use these goals as benchmarks during the assessments that I carry out throughout the unit. The basic premise is that I want students to be able to recognize for themselves that they have grown through the experience of the unit. So they should be able to do things that they could not do before; they should be able to appreciate subtleties that make them feel like they have special and sophisticated knowledge, and they should enjoy challenging others to figure out puzzles that are only understood by people who have gone through the special experience of studying the material of this unit. Students might be asked to use their skills in a new mini-project that is very different from what they have just been doing, but which I know they can do because they have been developing the skills and using the concepts that are required to be able to do this project. Perhaps a mathematical game or puzzle could become the focus of an investigation. Students might design their own challenges for each other. An artist's work may be analyzed using the concepts of this unit. A community issue might lead to a report for a town council meeting. To prepare for the unit ahead of time, what I like to do is collect games, puzzles, and activity ideas from teacher resource books or the Internet, and to look for high-quality performance assessment activities that might serve me in planning for the next, upcoming unit, or for future topics.

Assessment

One central feature of the unit approach to teaching and learning mathematics is assessment; that is, collecting information about how my students learn, what they seem to know and are able to do, and what they are interested in, so that I can make informed decisions about what they should or could be doing next. Assessment in this sense is very different from "evaluation," in which a teacher judges the quality of student work, scores it, or ranks it compared to other students. While there may be a system of scoring or grading involved in assessment, the primary goal is not to evaluate, but to make professional choices from among options. In my own classroom experiences, I choose from seven basic strategies of assessment in order to collect useful information about my students: testing or quizzing students, interviewing students, observing students while they are working, comparing their work to set criteria in a performance task, reading what they write, or asking students to reflect on their own experience with mathematics. Based on what you think it means to be doing mathematics in your classroom, you need to choose from such a list as well. Your choices directly influence the atmosphere in the room, the perceived purposes of the activity in the room, and the ways in which students make sense of what they are learning. The information I collect also helps me to decide whether I am happy with how things are going; based on the information I collect, I may decide that I am not pleased with the impact of a particular approach, or that things need to change in my classroom.

Many teachers view their role in the classroom as a trainer or coach who drills their students on certain skills. The classroom is mostly a place where students practice certain

Box 1.1

Strategies of Assessment

Test/obstacle course
Interview students
Observe students
Performance tasks
Writing
Student self-assessment and
    peer assessment
Portfolios

techniques. In these classrooms, teachers set up an obstacle course or test to challenge their students to meet the goals of high performance on specific learning outcomes. Correct answers on a quiz would be an indication that students can get these answers, that is, that they have succeeded in mastering the skills that they have been trained to perform. Current thinking in mathematics education suggests that even if this is indeed our goal, it is very helpful if the students understand what the mathematics means in order to be able to get the correct answers. Displaying an ability to use algorithms that have been demonstrated to them on prior class days does not mean that they have any sense of when to use these algorithms, why they work, or how they are related to questions they might want to answer in their everyday life. Because of this, mathematics teachers are increasingly drawn to methods of facilitating conceptual understanding. The belief here is that students who understand mathematical ideas often do not even need to be taught standard procedures for obtaining answers; if they understand the mathematics, they will be able to figure out the algorithms themselves. Teachers like this approach because it can help them figure out how to support students' independence in learning mathematics as well: if I can figure out what a student is thinking about a mathematical idea, then I can ask a question or suggest that they investigate a relationship or situation that will lead them to a different or richer understanding. This way of teaching is much *easier* because it gives me as a teacher a way to enter into a meaningful exchange of ideas with my students. It gives me and my students something to talk about besides rules of behavior in the classroom.

If I need to know more about my students' thinking processes and the ways they are making sense of things, then tests will not help me much. A wrong answer might be a sign of clever thinking rather than a lack of understanding. A correct answer might be obtained using a nonsensical method. I need methods of understanding what my students really are thinking. The rest of the options of assessment are attempts to support such work. In order to know more about what is going on in students' minds, I need to have my students talk—either to me or to each other, or otherwise communicate with or about the mathematics. I suggest you try these ideas if you have not yet done so. Consider interviewing students about a mathematics question or situation. Here you would gain valuable insight into techniques by watching a lot of television talk-shows or listening to talk radio. How do these hosts generate detailed responses from the person being interviewed? Do you like Barbara Walters, Ellen, Rikki Lake, or John Stewart? Why or why not? The point of an interview is to find out as much about the other person as possible— to learn about how the students are constructing conceptual knowledge for themselves, and how they are connecting procedures for obtaining answers, mathematical facts, and other mathematical ideas, through these concepts. This is a time when I suspend teaching if at all possible. What I mean here is that an interview is not a time to be correcting the student or helping him/her to understand something; instead, I really want to know what he or she is thinking, so that I can respond at another time with an experience that will be especially appropriate. Videotaping interviews can be very helpful as you refine your skills; tapes of interviews are often interesting to the students as well, so you should offer students a chance to view them during a free-time or choice period soon afterwards. Some teachers interview one student for about five to ten minutes every day; in this way, they get through the whole class a few times every year. Others interview in groups of three and observe the way group members interact as they discuss the mathematics. I like to experiment with new ways to do this, but no matter what, I record some notes later in a file so that I can remember what I learned about my students. I can use this information in my planning for the following week or month, and have specific examples to share with families during conferences. Good questions to start a conversation include: How might you explain how you got your answer to someone if they had been absent from school when we were learning about this? What if they didn't understand you, could you explain it a different way? How might you teach this to someone in a grade before ours? What questions does this raise for you? Can you use this material or draw a picture to help you describe what that means? In a group interview, I like to ask if they can recognize what strategies they have in common when they are working, how they approached things differently, and what they think about the similarities and differences.

When I am using a lot of collaborative or group work, or when my students are working with manipulative materials, I can gather a great deal of anecdotal information about my students by simply observing them as I walk around the room. I like to carry a clipboard; friends of mine use an index card for each student, and jot down a few comments on five cards per day or week, collecting information on the whole class over time. I might choose to focus on the strategies I identify students using on their own, the sorts of language choices they are making in their mathematical conversations with their peers, how well they are working together as a group, which peer explanations seem more successful than the teacher's, or simply whether or not the students appear to be enjoying mathematics in school. A strong advantage of such systems of documentation is that they translate the more complex assessment tasks I am doing in the classroom into data that can be analyzed. I can develop a picture of a student's work that is descriptive rather than "normed" (that is, based on a standardized set of assumptions or expectations, as in traditional tests). I can then turn these descriptions into a way to report student growth or change over time. This also allows me to use more open-ended activities in my classroom, because I can gather information from them through such records, rather than viewing the open-ended experiences as preparation for later assessment. Moon and Shulman (1995) recommend a "problem solving record sheet" for each student observed. The teacher records the activity, date, and descriptive information on the following: 1) Demonstration of confidence in the process of solving problems (volunteering, being enthusiastic with purpose, and communicating with others—talking to another group member about the mathematics, or talking or listening to the teacher); 2) Illustration, description, and modeling of a variety of problems (using objects, illustrating with graphs, drawing pictures, and repeating the problem in his/her own words); 3) Verifying, interpreting, and justifying solution strategies (explaining the problem solving process, seeing relationships between problem types and solution strategies, and checking the reasonableness of solutions in appropriate mathematical units); and 4) Constructing problems from everyday life with a variety of mathematical concepts (forming mathematics problems and thinking about them, telling it in her/his own words, forming a mathematics problem in a story, writing to others through illustrating with pictures, graphs, and charts, and understanding how a problem can be solved or understanding the solution strategy).

Interviews and observational records expand our repertoire for reflecting on how our teaching/learning strategies are meeting our goals while documenting students' progress over time. Yet they are often criticized for drowning assessment in description to the detriment of careful evaluation of our pedagogical work. Sometimes it feels good to be able to register progress on a scale. In fact, many people will demand some sort of measurement of success, so it is good to have such a component as a regular part of school experiences. One compromise teachers have comfortably adopted, and in many cases found very useful, is the use of rubric scoring of open-ended performance tasks. In Chapter 2 we will consider some critical concerns with such an approach to assessment, but for now I want to stress the kinds of information that a teacher can potentially collect through careful analysis of what his or her students are saying and doing as they think through mathematical ideas. A rubric scoring system can help you to maintain a holistic assessment approach as you assign a rating to indicate the level at which a student product meets predetermined performance standards. This translates the more complex assessment tasks in your classroom into numerical data that can be used to report student growth over time, and to manage information gathered from the open-ended tasks. One way to start is to have students write about mathematics. Sort the papers into three piles: the papers that show the student understands and can communicate the mathematical reasoning involved, the papers that appear inadequate in important ways, and those that are in-between. Then you could assign papers a 3, 1, or 2, respectively. If a large clump of papers ends up in the middle pile, you could sort that pile into two separate piles and assign ratings of 1 to 4. Or, you could extend the sorting of each of the three piles and assign numbers from 1 to 6. Once you do this, you should look over the papers in each category and try to identify what distinguishes each pile from the ones above and below, so that you could state a description of the difference in performance standards or in terms of the conceptual, procedural, or factual goals of the mathematical activity. After a few times through this process, try planning a scoring rubric *before* the papers are sorted: my favorite way to do this at this point in my work is to make a chart for six categories, so that I will be scoring from 1 to 6. I assign a 0 to an empty paper; a 1 should indicate that the student has attempted to be involved in the activity in some way. I think of a 3 as satisfactorily completing my expectations for the assigned task. Then I imagine what a 4 and a 2 will be: a 4 should be completing the expectations *very well*, while a 2 should mean somehow falling short of the mark of satisfying the assignment in some key way or ways. I work on this seriously, thinking of examples of what I would take as an indicator of *very good* work for the 4, and what kinds of things that students might do would suggest that I am just not satisfied that they should get a 3. Finally, I think of what goes into a student performance that is beyond a 4, which for me means going beyond expectations. Such a student might be especially innovative, coming up with a new insight or creating an original approach to the task; she or he might ask new questions that place the original task in a new context, or perhaps break the rules of the task for a good reason. When I use this rubric, I am hoping that the majority of my students are obtaining 3 and above; if too many are in the 1 to 2 range, then I need to consider that the task was far too challenging; if too many are in the 4 to 5 range, then perhaps it was not challenging enough. Sometimes I can use such tasks prior to a unit, to judge at what level my students are entering. In this case, a task that students are performing at the 1 to 2 range might be a good motivating activity for learning about the subject; we can return to it later in the unit to see that we now can approach it at a new level. Once you get comfortable with scoring student papers, you should try moving on to scoring students' performance in group work as you observe them in class; videotape students for a performance-based assessment portfolio. An important use of rubric scoring is to work in groups of teachers. I like to get together with others and plan a common rubric. We talk about student papers together or view a video of a group of students working, and try to figure out what the rubric should be. We bring multiple perspectives to the discussion; clashing differences in cultural, racial, and ethnic backgrounds challenge our ideas and force us to clarify our own thinking about mathematics.

Suppose, however, that you are mainly interested in assessing your students' development in mathematics as an evolving literacy (reading, writing, listening, and speaking mathematically), and not so much in the process as in an interview or in group classwork. I sometimes realize that I mostly want to understand how they can put their ideas into words after they have had time to digest and reflect. At other times, I want to give them a chance to construct ideas through their own reflection, rather than through interaction with others. For these purposes, I like to use writing and speaking as an assessment opportunity. You may wish to adopt techniques of writers' workshop approaches from language arts. Journal writing, for example, can be a superb vehicle for students to develop and articulate mathematical concepts, and then, via various structures of peer feedback and teacher conferencing, to rewrite their journal entries in more public formats. Students enjoy publishing mathematical writing despite the complications of reproducing equations and diagrams on school computers. In the early nineteen-eighties, Brown and Walter (1983, new edition 2005) suggested having groups of 3 to 5 students form editorial boards for mathematics magazines published by the class. The board decides on a type of submission to advertise for (e.g., interesting descriptions of a solution to a problem, personal reflections on problem solving experiences, fiction, entertaining math puzzles or tricks, discussions of math in everyday life); students review the ads and create submissions for the other groups. Editorial boards provide feedback to authors on how to improve their writing for publication. Many teachers find open-ended structures for personal reflections to be useful as well.

Student analyses of classroom activity are interesting in the form of "media reports." Before a mathematical experience, we select several students to be "the media." These students observe the class as roving reporters. Later in the day the media perform a television news program or publish a newspaper about the mathematics: news items (any major discoveries about prime numbers today? etc.); fashion reports (what was fashionable as a problem solving strategy today, or in what students chose as the means for reporting their progress?); critiques of the class as if it were a film or videogame, gossip columns, or satires are thanked for their insightful or provocative commentary on the day's events.

The last two strategies for assessment allow even greater student input into the assessment process. For a portfolio, I prefer when students select items from over a period of time that provide evidence of their growth in mathematics. Journal entries, mathematical autobiographies, letters, or emails to the teacher, older students, or to mathematicians at universities, and reflective pieces, are examples of written materials that are easily placed in a portfolio. But you will also want to stretch the students to include items such as non-routine problems, revisions, and peer assessments. A non-routine problem is a question made up by the students that demonstrates an understanding of the mathematics needed to solve the problem, or a significant effort toward solving the problem. Revisions include drafts of the mathematics writing, work-in-progress, and final versions of work on a complex question; diagrams, graphs and charts, and so on should be included as partial progress on a problem, as well as reflective comments ("here is where I got stuck"; "here is where I got the idea to look at threes, because . . ."). One way to organize the portfolio is chronologically; another is according to which materials demonstrate work that the student is proud of, which demonstrate areas that the students feels she or he needs to work more on, and which materials show areas that the student has an interest in pursuing further. I encourage students officially with special forms, or informally, to obtain feedback on items chosen for their portfolios, before they finalize the collection of sample work. I also require an orientation guide to the portfolio, so that readers/viewers will be able to understand how to interpret the content. One more useful section is a concluding statement regarding what they think is most important about the mathematics in the portfolio, their current questions, and tentative plans for how they expect to pursue these questions on their own as working mathematicians.

The type of portfolio I suggest is a type of personal self-assessment. I believe the best teachers guide students into processes of reflection on portfolios. On my peer-feedback forms, and in conferences with students, as students are assembling their portfolio over time, I ask students why they selected a particular item, what they learned in doing a specific mathematical task that they did not know before, how their ideas changed as they revised a piece of writing, what parts of an activity they think they need practice on, as well as skills they would like to feel more secure about. Once students begin to internalize these prompts, I often decide to stop using them or to introduce others.

In my classroom, projects extend over a period of days or weeks, and tend to be most effective when they result in a formal presentation or exhibition of the material in group or individual formats; feedback on the quality of the presentation as entertainment and education for the classroom community is as crucial as feedback on the mathematics content. Peer assessment involves impressions from class members on another student's work, in relation to performance standards that have been established, or in terms of the students' contributions to others' learning. As students are revising mathematical writing, or developing questions to pursue as an investigation, I encourage them to work with peers and obtain peer feedback. Can you understand why I believe this is an important mathematical question? Can you understand how I solved this problem? Can you help me to figure out an interesting question? To help students get better at this kind of conversation, I like to discuss a sample of student work from another class or school. I ask why they think the student chose this item, can they offer any advice on making the writing clearer, could the student have used a chart or diagram, and so on. When I begin to facilitate such student reflection, however, I must keep in mind that my students may have never done this before. Indeed, my students are likely to have been "taught" not to trust their own opinions about their work or about others' work, or to view a desire to improve as a symbol of failure. Many students have been schooled into the notion that the teacher is the only legitimate judge of quality. It may take time before students adapt. In the meantime, collections of student work are readily admired by family members.

Note that strategies of assessment are simultaneously forms of classroom activity, or, in other words, lesson plans. Interviews, portfolio construction, observation or group problem solving, media presentations, and so on, require that the day-today allocation of time be appropriate. In most cases, this means that I am constructing a way of being in my classroom as well as a way of thinking about mathematics. Susan Ohanian observed outstanding mathematics classrooms, and identified ten foundational questions routinely asked by teachers in these classrooms as part of their ongoing assessment (see Box 1.2 overleaf). Teaching in these classrooms was at the same time also a form of assessment. This is of course the most efficient use of class time. Key here is how lesson planning involves making sure that the lessons allow the teacher to accomplish the kinds of assessment deemed important, both regarding specific skills and concepts, and about larger, long term goals related to student attitudes and interests.

Teaching Journeys, Infrastructure, and the Stories We Tell

David Whitin, a professor of mathematics education at Wayne State University in Michigan, believes that teaching is a "journey" (Whitin and Cox 2003). Key to this idea is that teaching "is never easy, never well marked, and never over." "In fact," he writes, "teaching is inquiry, and the closer we look at what we do as teachers, the more questions we have about what to do next. Feeling a bit unsettled comes with the territory" (Whitin and Cox 2003: 1). I bring this up in this first chapter because it is already time that we face some serious issues in the nature of teaching and learning. Much of this chapter might give the impression that you can figure out the best thing to do: Just understand where your students are coming from, plan well based on assessment, and everything will fall into place—the perfect classroom is just around the corner. Unfortunately, even very experienced teachers who have been acclaimed for their successes—like David Whitin and Robin Cox, a teacher that Whitin works with and with whom he authored a book recently from which that last quote comes—still see their work as a journey, as a work in progress, and still feel like they never really know what to do next. I am sorry to spoil the party, but that's just the way it is. Teaching is confrontation with the limits of what we can know—about ourselves, about our students, about the content of the curriculum, and even about how we should try to figure out what we should do. This does not mean that we should "wing it." Instead I believe I have to challenge myself to always explore new ideas about teaching and learning, and to embrace the very process of interacting with the limits of what I can know about knowing, in order to be the responsible and ethical teacher that I hope to become.

The next chapter introduces ideas from psychoanalysis in order to grapple with these very issues. At its heart, psychoanalysis can be said to be about this very conundrum of not being able to know some things that might be the most important things to know about. One way we will describe this is about always feeling that we have lost what we knew, always feeling like we can almost see what we want to see, yet always recognizing that our efforts to find knowledge about teaching, about our students, and about teaching–learning processes may be the very actions that are preventing us from obtaining the knowledge that we believe we are seeking! We do not need to feel like we are doomed to fail, though. Once we understand our circumstances we can enjoy the challenges of teaching and learning, and appreciate that our work is that much more interesting and rewarding. This may seem daunting at first, but if we plan infrastructure instead of planning precise activities, I believe most of what appears initially traumatic about teaching becomes more exciting and approachable.

If teaching is inquiry, as David Whitin and Robin Cox suggest, then we should take seriously the five conditions for inquiry that they cite in the beginning of their book; they take these conditions from another book that they read as part of a monthly discussion group, written by Dorothy Watson, Carolyn Burke, and Jerome Harste:

1. Teacher and student accept vulnerability and see it as a necessary part of real learning.
2. Teacher and student experience a sense of real community in their learning.
3. Teacher and student insist that their learning be generative; that is, it must lead to further insights, connections, and action.
4. Teacher and student demand democracy, insisting that all voices be heard.
5. Teacher and student recognize that inquiry is reflexive; they see themselves and each other as instruments for their own learning.

(Watson *et al.* 1989: 12; quoted in Whitin and Cox 2003: 2)

An important strategy for teaching that we can take from Whitin and Cox is to reframe goals as questions that guide our practice. For each of these five conditions of inquiry, they try to identify specific things they can do to make them possible. In the next chapter, we will refer to this approach to teaching as "crafting the conditions of libidinality." What this will mean is that we do not plan what we will do or what the students will do, but instead we plan the structures for how work will get done in the classroom. Throughout this current chapter we thought of this as infrastructure. Let's take a peek at how Whitin and Cox design the infrastructure that will make it possible for teaching as inquiry to come to life. This may seem odd to you unless you free yourself to consider that the way work gets done in most schools is never reflected upon or challenged in any way. But there are unlimited possibilities for how we can work together in groups, if we would only imagine the possibilities.

Whitin and Cox take each condition of inquiry or goal for relationships and experiences and turn it into a question. For the first condition, they ask, "How can we develop an environment that encourages learners to share their vulnerability?" Here are their answers to their question:

a. Encourage multiple interpretations of an event.
b. Value tentativeness, uncertainty, being unsure.
c. Support children in sharing unanticipated results.
d. Foster a spirit of risk taking.
e. Respect and nurture difference in all that children do.

(Whitin and Cox 2003: 2)

Those five answers sound really nice, but they still do not tell me what to do in the classroom, or what the students should be doing! This is the interesting part of being a teacher—you get to try out how to make things possible. The key thing is to establish ways of working so that it is more likely that these things will occur than not. Be explicit with the students, and explain that you are working together in the particular ways that you are working precisely to make such things happen. Do they feel like multiple interpretations of events are encouraged in this classroom? Can they point to specific instances where tentativeness, uncertainty, and being unsure have been valued in this classroom? Can they remember a time in the last week when they shared an unanticipated result? Students may not see what you see; if they can't provide these examples, you can share your own perspective by noting such instances when the infrastructure goals have been met, or times when it seems like the class has come closer to these goals than they have in the past. I find that students usually see these things more than I do myself, because I am so caught up in the moment of the educational encounters that I personally can't always see these things happening. A class discussion helps me to check in with the students and get their perspective. We talk about ways to work harder as a group to make these things happen. And if they do not seem to be happening, I talk with my students about why. In a whole-class discussion, we decide to try something new and test to see if it helps to achieve the infrastructure goals.

How can teachers develop an environment that encourages learners to be reflexive? By encouraging their students to represent and retell their problem solving strategies; by supporting their students in revising their initial ideas; and by asking students to evaluate their own efforts. We will explore this more fully in the next chapter. It is an interesting irony, Whitin and Cox suggest, that "as students share their vulnerability (such as their uncertainty), they can become more confident in their mathematical thinking" (5). Their own experiences as teachers helped them to understand that they could only realize such goals in a community of teacher/learners that supported each person's thinking, that supported the active participation of all students, and that allowed students to be reflective about their thinking. Here is a crucial point for assessment.

When I plan, one of the most important things I build into my plans is a routine for looking back on what has happened in the last week or two. I demand of myself that I can tell several stories about what I and my students did together that indicate we have supported each person's thinking; I do the same for active participation of all students and for allowing students to be reflective about their thinking. If I can't think of more than two stories for each of these infrastructure goals, then I really have to do one of two things: I have to either bring this back to the class and talk about it with them, as I suggested above, or I have to re-think my plans for what we are doing together on my own. Most of the time I prefer the first option, but I always consider the second as well. After all, I am the teacher, and it is at rock bottom my responsibility to make sure the class is going well; so I need to consider both what my students can contribute to the classroom as well as the effects that my own authority and power have on the way that the infrastructure comes to life for the group. This duality of knowledge—that this may be something outside myself, better handled by the whole class, but it also might be something that I myself am doing that prevents my goals for the group and for myself from becoming a reality—is another opening for psychoanalysis and thus another entry point for the next chapter.

Planning Infrastructure

(a) For each of the five answers that Whitin and Cox give to their question about how to create an environment that fosters the sharing of vulnerability, think of at least three specific ways that teachers and students could work together in a mathematics classroom. One way to do this is to describe what you would see people doing if this were evident. Be as wacky and open as possible, so that you are not limited by your assumptions regarding what must take place in school. Share your suggestions with others to get their reactions. What aspects of typical schools will you need to be prepared to challenge in order to make your ideas a reality?

(b) Here are Whitin and Cox ideas for how to develop a sense of community in the classroom, and for how to foster the generation of knowledge. For each of these, again, list at least three ways of working together or as individuals that would support the infrastructure goals.

    a. Recognize students as teachers, and teachers as students.
    b. Give students a voice in curriculum investigations.
    c. Develop experiences that recognize individual talent and group potential.
    d. Establish a sharing time for students to discuss their work.
    e. Provide students with multiple ways to express their ideas.
    f. Demonstrate that questions often lead to further questions.
    g. Recognize how our peers help us to generate new ideas.

(Whitin & Cox, 2–3)

Response to Chapter 1:
Engage Yourself in Meaningful Observation

David Scott Allen

Explaining this chapter is a bit like watching the movie mentioned at its beginning. You will interpret what we say differently depending on whether you have a modern or postmodern point of view. Suffice it to say that we should not confuse classroom practices with the material we are attempting to present. We as teachers are focused on the goals of the activity, while the students are focused on the activity. All they really know is what they are doing. The thing that is in front of them is a means to an end for us, but the end itself for them. The carefully prepared elements we present in order to lead them in one direction, may in fact be interpreted by them in such a way as to lead them in another direction.

This chapter explores the irony of the teacher–student relationship. The irony is that neither the goals nor the motives of these two are in sync, yet they are supposedly moving in the same direction. The teacher's goal is meaning, but the students' goal is the experience. Modern teachers ask "Don't you see the Truth?" and the post-modern students answer "Don't you see I'm bored? I want to experience your truth."

Perhaps the reason we have so much philosophical conflict is that we have a cross-wiring of methods and points of view. To use a Punnett square:

Modernist view of Modern method	Modernist view of Postmodern method
Postmodernist view of Modern method	Postmodernist view of Postmodern method

People representing the upper left and lower right can ultimately limit themselves by ignoring the other's ideas, while the upper right and lower left can seem to have nothing in common. And therein lay the tension of today's math landscape, and the foundation of the math wars. But if I as an educator can draw from both approaches, and think through both (all four?!) sides of the issue, I will be richer as a teacher and a more flexible example for my students.

School practices are definitely designed for modernist people (i.e. most teachers and administrators), but part of the reason this is perpetuated is because of the testing structures instituted by the modernist adult authors of NCLB. I could do much more important thinking in my room if I didn't have benchmarks to teach to within a certain timeframe. My building is very supportive of me as a teacher, and my administration gives me the luxury of controlling my students' academic destiny, but the bottom line is that they still have to perform well on both the State Standardized Tests and our own "standardized" quarterly tests (written by the teachers) for each course I teach. And the tests are virtually all multiple-choice. This doesn't exactly shadow real life, it's just convenient to grade. Not that I blame the test creators for wanting convenience. The teachers are doing the best they can to cope with the fact that the test is a priority, but the quality of the test is not. The reliability and validity of the test or the items is not even discussed, except by those who would dare to rock the boat. I blame the system for mandating a test which practically requires that it be multiple-choice. I would grade the open-ended items myself if I could get an insight into students' thinking. One good item explained is worth many multiple-choice items. Even textbooks are written in a modern way, subtly implying the mechanistic philosophy that doing 50 exercises will increase reasoning ability. Students see so many combinations of symbols, they forget what they mean. This paradigm is shifting, but only slowly.

I have my seventh and eighth grade students do projects which allow them to demonstrate knowledge of the required curriculum by presenting it to the rest of the class in an engaging way, such as a game. As part of the process, I have them evaluate each other. I use these grades as fifty percent of their actual grades for the project, and I have found students to be honest and non-partisan (or ignorably comical) in their evaluations of each others' work. At the beginning and the end of the project, I give them the rubric. Many students simply will not evaluate their own project. I encourage them to write something about what they think they have learned.

This chapter develops a five part plan for mathematical inquiry. We all do this five part plan in our lives whenever we undertake something important. I myself am immersed in a five part plan as I write this. My first "week" involved reading and rereading Peter's manuscript in an attempt to appropriate the meaning to my own practice. The second, third and fourth "weeks" are figuring out a structure within which to write, then actually writing it down, and finally editing it with a few good friends. And the last "week" will be publication and reflecting on my whole experience. What would I have written differently now that I have more teaching time behind me? What have I done in the last (calendar) week that I would have included had I taught it sooner? What did I leave out but wish I had included? This final stage is critical if students are to make a connection between math and anything other than another requirement. All these constitute valuable starting points for me in the next phase of my journey as a mathematics teacher.

In the music of the Renaissance period, certain sections can begin before the previous one ends. This is called a dovetail cadence. Doesn't life also work this way? We can never just end one thing we do in a neat package and begin the next. I have four children and there literally is never a moment of peace. I would like to be able to simply eat dinner uneventfully, but there is always the diaper to change, the infant to hold, the appointment to make, and the dessert to get. And so many questions! This is what makes life rich. There are seasons of life, but sometimes the waves of living crash upon each other. If you don't target the crest of each wave you may never get anything out of life. Paying special attention to evaluation and celebration with our students is important.

I try to frame my projects to help prepare my students for deciding what interests to pursue in life. How do you decide how to present yourself to an interviewer? How do you put your best foot forward? What exactly is your best foot, anyway? The projects we produce are a bit like art or music. They are a window into our hearts. But students need to be trained well to do all of this. They don't know how to let me observe them as they work because it is such an un-school-like activity. We need to remember to break them in slowly, because their range of comfort with varying strategies of assessment is limited, and their experiences with them have often been negative.

What I try to do the most in the classroom is remain cheerfully balanced about the process. Let students decide for themselves what to think and how to proceed with learning by presenting options. There is exchange in the form of dialogue, but not intellectual bullying toward a particular conclusion. In other words, I try to be nice to everyone. I say this because I have a very hard time doing it sometimes, as I let my emotions get the best of me and it comes out in my vocal inflection. So it can be difficult to interview my students and remain outwardly neutral because I am so curious and passionate about what I do. I ask some good questions, but sometimes people feel like I'm gruff or condescending. Yet ironically I have more success teaching girls than boys.

There truly is a problem of culture in schools, but I'm not convinced it's all teacher–student. At least some, if not all, of it is the splintering of the student culture, especially in schools that are as large as mine. Some students stand out and others fall between the cracks, but it is the students themselves who largely determine who is in what category, even if the adults contribute some of the category labels.

With the restraints of state student and teacher requirements, the hope of our youth is being deflated. We discuss assessment as if it was some benign process, but they know it really amounts to an evaluation of them, followed by putting them in their place. The context and exploration discussed in this book is almost doomed before it begins, because my students don't have any hope of ever getting out from under the weight of the ordinary and mundane. Students feel the irony of society: we live for the weekends. They also feel the irony that even if teachers do something different for a while, it remains something different, not what is really important to their education.

Having painted such a picture I will now attempt to explain the parody I present to my students. I basically tell them that the State Tests and the skills that they are looking for are easy targets to hit, and that my students can hit them easily and peripherally while we focus on other, more interesting and potentially useful (to mind and life) mathematics. The duality of my expectations with my students has come to a head in one of my classes, where I divided them into groups and held them accountable for teaching each other the sections of a unit. They must teach the class, assign homework (or not) and write a quiz for their section. I combine the quizzes to make a test, which I give after each group has its turn to teach. By taking the teacher off the stage almost entirely, the students must own each other's work or fail dismally from lack of attention. While this still leaves the duality of teacher/student intact, it does begin to dissolve the wall between teacher/learner. I always learn from their lessons how they think it should be done, and they learn from teaching the lesson just how hard it is to be on stage and produce. It's easy for them to ask to be entertained, it's harder for them to entertain and teach at the same time. In the process, students will learn all about crowd control and respect and going deeper with the material in order to produce learning. They need to perform my role in order to see things through my eyes. I need to see things from the desks in order to get a feel for their lot in life. I see their potential, yet they only see the task and the need to finish it with a measure of approval from the authority. I set myself up as the non-authority. Their peers are the authority. Their own tests do the evaluation. This is only a small step away from tradition, but it is a step nonetheless.

The next leap for me in doing projects with my students is to have them in on the rubric. If they are permitted to brainstorm and decide the rubric, then they will own the project more. If I have them make a game about something mathematical, then I should ask them what they think makes a good math game. They would respond with an even more critical eye than I, if given the chance to do some group work. A class discussion is too big to focus on, but I can divide and conquer the creation of the rubric. This happened to me quite spontaneously after we were evaluating a game project. I asked for feedback after a project and the students questioned the rubric. The discussion got so in-depth that I finally asked them each to draw up a rubric. I then synthesized them into one and presented the composite to them the next day. They liked the new one much better, and I thought it was more fair and balanced, and less teacherly.

Any rubric will be built on values, whether commonly shared ones or those outwardly imposed. I must engage my students in a rubric for classroom behavior, and teach civility above all things. They need to buy into it as early and as often as I need to enforce it. Communication is becoming my foundational value. Not just to communicate, but to communicate so that others understand you at the most basic level.

My first experiment in teaching a different way came in spring of 2003. While doing research for a class, I stumbled onto a site for Mathematics Awareness Month (MAM), which is April. The theme for 2003 was Math and Art. I had wanted to teach math using art in the past, and decided to teach this unit featuring Escher and tessellations to my seventh grade students. I spent lots of time on the Internet evaluating sites that I could use, and printing helpful activities. I went to the three art teachers in my building for ideas on materials, rubrics for grading, and general organization of an art class. They were very helpful and enthusiastic about the whole idea, and even looked at my class lists to suggest student helpers. I also connected with an art teacher at the high school in my district, in order to get more ideas from a different perspective. I gave a survey at the beginning of March in order to determine the level of interest in a math/art unit among the students.

I boldly decided that I could afford to devote the entire month of April to this seventh grade unit, as we were ahead of the curriculum sequence to which the teachers had agreed the previous summer. I was determined to try to find interesting hands-on activities to get the students involved. I found many good websites. I had to decide what direction to take the unit. Should I hammer away at just tessellations, and see what happens? Should I introduce them to all kinds of mathematical concepts using the artists' work I found? Should I make the unit an "Escher" unit and have them ponder the artist's masterpieces? The hyperbolic geometry in the MAM logo seemed too much for them, at least in my first year of teaching such a unit.

I decided on a middle ground (it is middle school!), which would include all of these to a certain degree. I chose to begin by emphasizing symmetry, beginning with point symmetries (e.g. mandalas, pinwheels, snowflakes), then moving to the seven frieze symmetries (e.g. borders, frames), and finally to the seventeen planar crystallographic groups (e.g. wallpaper, wrapping paper). I would limit this last category to convex geometric shapes, in order to avoid the complications which Escher's lifelike art creates. I wanted students to be able to see and tinker with the symmetries, and then use them to create a final piece of artwork using all that they know. The convex shapes we would use would lend themselves nicely not only to symmetries of form, but also symmetries of color, making the art theme even more worthwhile.

In the end, it was a great experience for all of us. During the process, the art teachers came by to give ideas and encouragement to the students. The students got to grade each other. We invited the entire staff to view our gallery of mathematical artwork and select their favorites. I later invited four students to come after school with one friend each to paint large versions of their work on wood, and now there are four large paintings in the halls of our middle school.

It is very true that the kind of lessons you plan must allow for the proper way to assess what you value. And having a backup plan is crucial, because students forget to bring items to class and simply aren't ready in many situations. They don't put names on their work because they don't value it. Work is not just something you do to get more points for your grade. We need to help them value their work for its own goodness as well as its usefulness as a means to the end of learning.

Action Research 1: Numbers on Trial

Isaiah Manzella

One day I posed a question to my class, "What would you do if you saw a fraction on the street?" My students murmured, "What? Huh? Wha . . .?" I restated the question to them. "What would you do if you were walking down the street one day and you saw a fraction?" "—I'd punch him in the mouth, a voice jumped in." And just as it was said all of us were laughing. But as I laughed briefly I thought, wow this is amazing. Is it that students fear fractions so much if they were personified it would trigger their fight or flight reflexes? It couldn't be, this comment had to have been said purely for laughter. So I asked, "Why would you do that?" "Because," a different student responded, "fractions are stupid, it's not like we need them."

"You're right," I said. "They are stupid. In fact all numbers are stupid, it's not like we need them anyway." (I do this often. Use sarcasm in my classroom. I know I am not supposed to, but I enjoy it so much and I feel it helps the students see the error in their statement or at least it helps me understand why they would make such a statement. Usually their ability or inability to defend their stance helps inform me if they meant it or if it just leapt out of their mouth.) I went on, "I've got a great idea . . . Let's ban all numbers from this classroom. We don't need them anyway." At this point most of the students were looking at me as if I was crazy; after all, it was MATH class we were in. But was I? Did we need numbers in Math class?

We spent the next few minutes trying to decide if we needed numbers or not in this class. Then I decided to take things one step further. Instead of debating the need for all numbers, let's just look at certain ones. I asked all six of my students to pick a number between 1 and 9. Not just any number, but their favorite one. One they felt strongly about. One they felt we needed, or at least they felt they needed. On the spot we began to come up with a very creative and dynamic lesson. I told them it is now their responsibility to defend that number's existence. They must convince the entire class of their number's worth, or else it was going to be banned from our room entirely. We were going to hold a trial for each number that a student picked and that student had the sole responsibility for proving that number's worth.

At first the concept seemed a bit daunting or too ambiguous to actually do, but we talked our way through it. As a class we put some sample questions on the board to help get ourselves started. We picked a number no one had chosen, 9. (*Which seemed an obvious choice for me to ban: who uses 9 anyway? It is just 10's little brother always tagging along trying to part of something it's not; a mere foregone conclusion of what will ultimately be 10. Nine is like the Junior Varsity player—almost there but not the real thing. In my opinion 9 is the fifteen minute warm-up that must take place before you get to actually play in a sporting event of any kind. No one even counts by 9s! Now 10, that is a number you can count by. Regardless of my ramblings and personal views, 9 appeared to be safe for the time being.)*

Q: What does this number mean to you?

A: My little sister is 9.
A: On my 9th birthday I got a bike.
A: My homeroom is C-209!

Q: Are there any important dates throughout history that involve this number?

A: 1991—My Birthday!
A: Mine too!
A: 1892—That's the year Abercrombie and Fitch was founded.

Q: What other numbers is 9 a part of?

A: 1991
A: 19, 29, 39, 49 . . .

Q: Are there any major routes or highways that use this number? Where do they lead you?

A: 295—The Mall
A: 95 Up North

All of the students seemed pretty excited about the trial and their number. It was something new, different, and challenging. It was not a worksheet and I was not telling them what to do. I merely established a framework with their help and they would ultimately decide what their end result would be. I also took a moment to explain to them the magnitude of what we were doing. "This project is only as good or as bad as you make it. If you do no work outside of school that will show, and so will your understanding of your number. If you choose to dive head first into this idea and feverishly work to find data and answers, you will be left with more questions than when you started, but a much broader foundation in which to build your solutions off of."

I also explained to them, "I understand that some of you may think that you have already figured a way around this project; you think that when the day comes for your number to stand trial you will just avoid the trial. You will stay home that day or cut. Well I have news for any of you that may be thinking of doing that. Don't. Just like in real life the world continues to move whether you move with it or if you ignore it. We are holding court and our numbers' existence is at stake. Just like if you were on trial for murder, and your lawyer didn't show up, what do you think will happen to you?" "You're going to jail," a voice yelled out. "Exactly," I responded. "Your number will be banned."

We went back and forth with ideas about the trial for the remainder of the block—some good, some needing work, but all were inventive and fresh. This whole concept was very foreign to us all. I was excited, I couldn't wait to sit down and begin creating the framework of this lesson. I asked the students to begin creating a list for homework; start writing down their own questions about numbers, write down anything they felt important about their number, or questions they felt might help other students defend their numbers better. As the bell rang and they all jumped out of their seats sprinting for their freedom I yelled out my familiar phrase. "Never grow up! Don't get a job! Just stay in school forever." I truly believe this and I tell all of my students this. I am not asking them to be slackers or nonconformists. I am just asking them not to rush into being an adult. We have plenty of time to be adults. Plenty of time to pay bills, deal with other adults, act appropriate all the time, be accountable for all of our actions. Don't waste your opportunity to make mistakes and learn from them, your opportunity to come home at 3 p.m. and watch cartoon network until dinner is made for you. Adulthood will happen—guaranteed—it's childhood that will disappear. In addition to that, if we stay in school forever we are constantly furthering our education. This, in turn, will allow us more career choices when our time does arrive to be an adult.

That day I sat at the computer typing and deleting, trying to find the best, most articulate way of phrasing this lesson. I understood that this whole approach and concept was creative and new, but at the same time had pedagogical roots that go further than I can begin to understand. I asked each student to question what they felt they knew about number sense; to question why we take numbers at face value. I didn't want to ruin it. I wanted it to be something that got them really thinking about Math and numbers, but at the same time engaged them on a level they normally were not. I wanted this lesson to change them. I wanted this one lesson to fill in all the gaps in their learning. To alleviate all of their Math anxieties and overcome any number sense issues they may have. After about an hour of deliberation, this is what I initially came up with.

Why Do We Need Numbers?

- Pick a number 1 – 9 and, prove it!
  My number is _____
- Each student will be able to choose a number they feel is important and defend its existence.
  - o We will have a list of questions and guidelines to help us make our "Case."
  - o We will spend time in class trying to find out as much information as we can about our number.
  - o In addition, you will also be required to do some work on your own at home.
  - o Try and think how the number effects us, what it means to us, how often we use it, and where in our everyday life do we see it, would our life be different without it?

- As a class we will decide which numbers are important based on **only the information provided.**
  - o Students will then present their "case": just like as if you were in a courtroom and a lawyer presented his/her case.
  - o The entire class will be the jury, and I will be the judge. After a person presents their case the jury will vote on their decision in private and all votes and decisions will be revealed after all cases have been presented.

- If it is determined a number is not important, that number will be **banned from our classroom.**
  - o *If you are absent your number is automatically banned!*
    - ■ If your lawyer does not show up, you are probably in trouble too.
  - o If we would like to still use that number or if it comes up during our work, we will have to write it down a different way.
    - ■ For example:

  - o Each time that number comes up you must represent it a different way.
    - ■ For example: If that number is ___, and you already represented it the way we did above, now you must show it as:

"Number Case"

1. What is my number?

2. Do I think my number is needed?
   a. Why or why not?

3. Where do I see it in my everyday life?
   a.
   b.
   c.
   d.
   e.

4. Is the number used for any important dates? For you individually or in history?
   a.
   b.
   c.
   d.
   e.

5. Add any additional information that you feel is necessary.

MAKE SURE YOU INCLUDE AS MUCH INFO AS POSSIBLE

(*Clearly it is not perfect, but for now it will work.*)

The next day, most of my students remembered what the plan was as they walked in the door. It was the first time I had ever assigned homework. It was a rather informal assignment, but none the less, it was homework. Out of the six students, two were absent, three had begun compiling a list of important information about their number, and one simply forgot to do the assignment. Fifty percent of the class had their homework for me. *A fraction just walked into the room and no one attacked it.* Out of the three students who had begun working on their project only one of them had written their work out in a spiral bound notebook. The other two students pulled crumpled pieces of paper from their pockets. I was pleased to see this and I was sure to praise all of them. This was progress and it showed they had an interest in their education.

We began class by me passing out what I had created yesterday. The students began to read it silently on their own. I stood in front of the room and began to go over what I had written. But I noticed something strange as I was doing this. I didn't have to direct anyone back on task, no one was talking, and everyone seemed to be not only following along but answering questions as we went. Taking this cue, I interrupted myself and allowed the class to begin working. I explained to them that I left plenty of room at the bottom of the page for anything they came up with on their own and that they should feel free to ask anyone else in the room for ideas or guidance. For the next twenty minutes I sat back and watched as all four students I had in class that day built their case in order to defend their chosen number. I walked around and peered and prodded, but for the most part that day, I was just in the way. Nothing can please me more as a teacher than days like that. If the class can learn and function without me, then I am doing my job. Those are the days that I enjoy the most: the ones when I am not needed. It shows me that when I am gone or when they have moved on, they learned something from me, the ability to question and a desire to learn.

We spent a total of two full blocks (seventy-two minute periods for a total of one hundred and forty-four minutes) in class preparing for the trial. The students also had three days and evenings to take advantage of work on their own to prepare. What started out as a discussion had become not only a teachable moment, but a great learning experience as well. I asked the students to dress up and fully prepare for the trial: after all, this was no rinky-dink operation we were running here. Presentation is everything. As the trials began I could see the anxiety and tension on all five of their faces. Yes, that's right. Five. Only five students out of a total of six were present that day. Staying true to my word I started the class off by calling for that number's case to be heard. "2? Has anybody seen 2's lawyer? I see 2 in the courtroom today, but I am afraid I do not see 2's lawyer." It turned out no one had seen 2's lawyer and it was all too easy to decide the outcome of that case. Looking around one final time, I exalted, "Number 2, today you are on trial to prove your worth to not only this classroom, but to the entire world. The fact that you do not have representation here today perpetuates the idea that you do not take this courtroom or your existence very seriously. With that being said, I hereby ban you from this classroom on the grounds of you not being important." As my decision came down you could see a look of wonder in the students' eyes. I don't think they were really sure what they were getting themselves into. Nor was I. Looking back, I think it had a great impact going first because it let all of the students know that this was for real.

The next number I called to the stand was the number 1. "Number 1, please have your lawyer come forward. A small-framed, thin build, African American young lady stepped forward. Katrina had her papers in hand. With her shy awkward manner she began to move toward the front of the room. Katrina was one of three freshmen in this class and from the trepidation in her eyes, I could tell she did not want to go first. But Katrina got up in front of a handful of people she did not know very well and turned on a switch. She went from "Katrina—like I'm quiet and shy and like I state all of my answers in the form of a question while I slip in likes and umm . . . to buy time while I look around the room in anticipation of my time in the spotlight to be over" to a confident well-prepared young lady who was going to tell you how it is and you were going to listen.

"1." The number 1 is needed all over. We start counting items at once. If you didn't have the number 1, how could you count? How could you buy things? How would people get paid? How would dollar stores work? Would they? People would be out of jobs. More people would be on welfare. Crime would increase. Taxes would go up. Every number is based off of 1. If 1 does not exist then no number can exist. Really all numbers are just 1 piled into it as many times as needed until it is the number we know. Katrina had such a solid case I didn't have much of a rebuttal. On the spot I could not think of anything to get her on. So all I could say was thank you. I then asked the jury if they had any questions: they had none. I asked Katrina to please leave the courtroom (step out into the hall) so that the jury could decide the fate of her client. While Katrina was in the hall, I listened to the jury discuss her case. I helped them through the process by asking them questions like, "Did she present a good case?" "Was there anything you felt she should have said?" "Did she leave any doubt in your mind if we needed the number or not?" It was unanimous, the number 1 was here to stay. (Good thing too, because I think the class would have been really hard without the number 1.)

Next up was Christina. She had chosen to represent the number 4. Four is a fabulous number. "It is the second number in my address. I guess if 4 did not exist my house number could skip a number—But what about Fourth Street? A lot happens on that street and it's easy to remember because it comes after second and third. Chairs and tables have 4 legs. Our Declaration of Independence was signed on a table. What if we never had tables? Would we have ever had a Declaration of Independence? My lunch is 4d, all students must have a lunch. 4 wheels on a car. Would travel be different? Could we live if we only had bikes?" I noticed many of Christina's facts were actually questions. Her questions forced the jury to think about life without the number 4. While it seemed it could be done, our entire lives would change as a result. In a four to one decision, 4 was not banned.

John led off his argument by appealing to our leisure side. "Prime time television comes on at 8 p.m. My favorite Flyer Mark Recchi wears number 8. CN8 is our local broadcast. 8 dollars is a small amount of money. It seems like a lot less than 9. Nine is almost 10, and 10 is a lot. Cell phone minutes are cheaper after 8 p.m. It is a good round number. It seems to double up well. If I am dividing the number 8, I can get a few different answers. Two cubed is 8. A shape with 8 sides is an octagon. What is the most famous octagon of all?—a stop sign. Without stop signs there would be a lot of chaos or waiting at red lights for no reason. I was born in 1989. 1–800 numbers, what would the world be without telethons and phone numbers to call? Neptune is known as the eighth planet in our solar system. Believe it or not, 8 is everywhere. Please keep 8, we need it." With another strong case and unanimous decision, 8 remained a part of our class.

"Lord of the Rings Trilogy. Need I say more? Without those movies my entire life would be different. Star Wars—all three! Incredible! It's a magic number, that's what School House Rock says." Vincent took an approach that resonated with him—Television and Movie. "MTV is channel 35, Discovery Channel is 31, Sci-Fi is 30; each channel very important to us all. A family can consist of 3. The square root of 9 is 3. The best way to hang out with friends is in 3s. The bus drops me off everyday at 3 p.m.. Detention gets out at 3:30 p.m. We eat 3 meals a day. You would have to add one more or take one away if the number 3 did not exist. It would cost more money to feed people everyday. There are 3 feet in a yard. Football would not be the same. All of the current records in yards would be much larger in feet. Basketball would be very different if there were know 3 pointers. Larry Bird, who was known for his ability to shoot the 3-ball, might have never been known. 3 is not only important to me and my immediate life, but throughout our history as well." It was unanimous, 3 stayed.

Max would have banned his own number given the chance. He knows he did not put in the effort everyone else did. After a short and unprepared presentation; 6 was banned!

After everyone had presented their cases we announced which numbers were banned. The students were not at all pleased that two numbers were taken from them, but at the same time they were relieved it was only two and felt they could manage with what they had. You could also see a sense of pride and ownership for those who successfully defended their numbers. I went on to explain how the numbers would be banned. "Not only are we not allowed to write them, but we are not to speak of them as well. They will now be referred to as, 'Those in which we do not speak of.'" (I stole it from a movie, but it works very well here so we used it.) I also stated that anytime we would be working on a problem, whether in a book, on the board, or on a worksheet we would be required to represent it in another way.

For example, the number 2 was no more. If we were to come across a 2 in our text we had to re-write the problem exactly as it appeared, cross out the 2 and then represent it any other way we could think of just above it in the space provided.

$$\begin{array}{c} (1+1) \\ \cancel{2} + 13 = \underline{\quad} \end{array}$$

Also, every time we represented the number *it had to be a different way*. In this case (1 + 1) was already used on this piece of paper so any other 2s had to be represented a different way each time. This did not excite the class, but it intrigued me. To help everyone understand we did a quick exercise in finding other ways to replace "Those in which we do not speak of." The students really began to get the hang of it and displayed quite an understanding of numbers and basic mathematics. This continued to go on for several weeks. Some days the students tried convincing me to let "Those in which we do not speak of" back in the classroom for the day just to make our lives easier. And I have to admit, some days I wanted to, only because it would have made certain lessons or days much easier. But we stayed strong. Sometimes I would forget and at least one student in the class was always quick to remind me.

After about a month we decided to bring 2 and 6 back in for a re-trial. This time I was not going to give the students any class time to prepare their defense. If they wanted to bring these numbers back into our classroom they were going to have to convince me using their time and their resources. I informed the whole class that the re-trial for 2 and 6 were to be held tomorrow in class at the beginning of the block. All of the students seemed to be excited about this and expressed no concern about the short notice.

Tomorrow came, and each student did an excellent job of explaining why they felt the number 2 was important. Many of them cited how it seemed to come up at least twenty-five times a lesson. They saw it all over the place. You must have a 2nd floor if you are building a city building because there is not enough space to spread out—they must build up. You need 2 people in order to have a couple, to be in love, to be loved, to create a baby. We have 2 hands, 2 feet, 2 eyes, and 2 of several other things to give the human body its symmetry. Each student one by one came up and presented a compelling case on why 2 should be able to stay. But I did not feel 6 was as well represented. At the time, students were so wrapped up in getting 2 back that I think they were willing to sacrifice 6 again if it meant 2 was going to be allowed.

I made the final decision and allowed 2 and 6 back into our classroom. It was quite a relief and at the same time I felt as if we somehow bridged a gap in our learning. Our foundation of Mathematics was now broader and more stable. I felt by taking away a basic symbol they took for granted, they now had a slightly better understanding of the value of numbers or at least how important the number 2 is to us.

Over the course of the seven weeks this lesson took place I can say with absolute satisfaction and honesty, I accomplished my goal. Each one of these students improved their number sense, basic math facts, times tables, logic and reasoning skills, ability to find patterns, problem solving strategies, and most importantly the ability to question what is put in front of them. These students were shown a dynamic concept that they can now apply to their everyday lives.

Show me the information—and I will question it. Answer my questions and I will have more questions, find me more questions that you cannot answer, and we will answer them together.

Isaiah Manzella

I am a Special Education teacher at Rancocas Valley Regional High School. The school and the district are very special to me because not only did I graduate from Rancocas Valley, but I have now been given the opportunity to return there and teach students that I feel connected to in many ways. Over the past five years of teaching here I have taught Resource Center Math, Inclusion Math, Alternative School Math, and now in a self-contained classroom for students with behavioral disorders. In addition to teaching at RVRHS, I also coach soccer, co-run the EA Sports Video Game Club, and the Skateboard Club. With each experience I am able to fully realize how little I actually know. Thank you, everyone.

MathWorlds 1:
Reverse Answer to Questions

1. **Averages**
   a. Find the average of 5, 7, 9, 29, 15, 11, 6, 46
   b. What groups of numbers can you find that have an average of 16?
   c. How many different numbers of number can have an average of 16? Can you characterize the types of groups of numbers that have this average in some way? Come up with more than one way to describe them!
   d. What new questions come to your mind after working on these questions? Choose one of these and begin to explore it.

2. **Shapes**
   a. I'm thinking of a shape that has two sets of parallel sides. What could this shape be?
   b. What is the fewest number of clues you would need to give for someone to be able to figure out that you were thinking of each of these: square, parallelogram, trapezoid, rhombus, kite, isosceles trapezoid?
   c. If I am thinking of a shape with n sides and two sets of sides are parallel, what else do you know about this shape?
   d. What new questions come to your mind after working on these questions? Choose one of these and begin to explore it.

3. **Sewing Shapes Together.** I can make a trapezoid out of a square and a triangle:

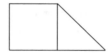

I can also make the same trapezoid by sewing together two triangles:

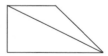

What sorts of shapes can be made by sewing together
   a. two common shapes?
   b. Three common shapes?
   c. Four common shapes?
   d. What new questions come to your mind after working on these questions? Choose one of these and begin to explore it.

4. **Game Dice.** A common game design question asks whether it is better for a given game to use two six-sided cube dice or one twelve-sided dodecahedron die. Since the two options have a range up to 12, but the possible outcomes have different probabilities, the differences can matter for the game.

a. Describe board games where the differences in outcomes would matter, and explain which die/dice choice is best for each game. Describe at least one potential game that you could imagine for each choice being better than the other.

b. Describe alternatives to the die/dice choices that duplicate the same probability outcomes and could be used for these games, using: (i) spinner(s); (ii) coin flips; (iii) something other than dice, spinners or coins.

c. What new questions come to your mind after working on these questions? Choose one of these and begin to explore it.

5. **Story Graphs.**
   a. Make a graph of distance versus time for your trip from home to school.
   b. Tell a story that could be described by this graph if the axes represented weight and and volume instead of distance and time.
   c. Tell a *different* story from the ones in (a.) and (b.), changing the meaning of the variables one more time.
   d. Draw a graph that you think would be *difficult* for someone to come up with a story for. Give the graph to another person (trade with a colleague by scanning or taking a picture of your graph and emailing or faxing the image) so they can take the challenge and tell you a story for it. Discuss whether the attributes of the graph that you thought would make story-telling difficult turned out to be important or not.
   e. What new questions come to your mind after working on these questions? Choose one of these and begin to explore it.

6. **Subtraction Heaven.**

Mersin subtracted like this:	Legza subtracted as follows:	Pabko subtracted like this:

$$
\begin{array}{r}
{\scriptstyle 3\ 1} \\
\not4\not7 \\
-29 \\
\hline
18
\end{array}
$$

$$
\begin{array}{r}
7 \\
40 \\
-30 \\
\hline
9
\end{array} \Big]\ 18
$$

$$
\begin{array}{r}
47 \\
-29 \\
\hline
\end{array}
\longrightarrow
\begin{array}{r}
4\ (17) \\
-(3)\quad 9 \\
\hline
1\quad 8
\end{array}
$$

a. Explain each person's strategy.
b. Come up with at least two other strategies that would work in such a subtraction situation.
c. Which strategies could be generalized for similar three-digit subtraction situations, and which could not? Why or why not?
d. What new questions come to your mind after working on these questions? Choose one of these and begin to explore it.

two
A Psychoanalytic Perspective

What does (mathematics) education ask of us? Let's start off carefully in our answer to this question. First of all, if we are to presume the kind of education that takes place when there are teachers and students working together in some way, then we might say that education demands of all concerned that each recognizes "learning" as something to seek. Or at least as something that exists apart from oneself, so that learning can be an object to which one relates. Next, I want to say that education demands of all concerned that each takes everyone else involved as an object of relation as well, so that a community of learning begins to form out of the relationships that are established. Deborah Britzman (1998), a professor of education at York University outside of Toronto, would say then that education must, after all, address the affects of such relating if teachers and students are to attach to knowledge and each other. But she notes something useful beyond this: the teacher, she writes (1998: 27), can be more like an artist who considers her or his work as crafting the conditions of "libidinality" in learning, as opposed to hardening her or his authority, as might a leader. What she means is that "leaders" retain all of their narcissism, wanting as well to dominate followers in order to use them as instruments for their own purposes. A teacher who acts as a leader is not necessarily leading students to knowledge. Such a teacher may be using the students as objects of her or his own advancement or accomplishments, for example, as tools of self-knowledge and self-promotion, or in order to feel good, perhaps in order to see the effects of her or his actions in the classroom. "Crafting the conditions of libidinality" has to do with establishing routines, rituals, and other features of an infrastructure for classroom "events." The libido is psychic and emotional energy associated with instinctual biological drives; to craft the conditions of libidinality is to set up possibilities for how the individuals in the classroom can make meaning out of what they experience. Britzman gets this idea from the psychoanalyst Hans Sachs, who contrasted the leader with a certain understanding of an artist, as someone who wants her or his work to have an influence on the world and other people, but without ulterior motive. In this conflict between an ulterior motive and the desire for "something," we can find a fundamental binary opposition that is at the heart of mathematics education. We expect our work to lead to "learning," to directly effect changes in our students; yet what, really, are we expecting? We must explore the ways that our own fantasies, dreams, and fears play out in the choices that we make about these routines and rituals. At the same time that we admonish ourselves for "using" our students to satisfy our own desires, it is hardly plausible that we would be involved in this enterprise if we had absolutely no ulterior motives: what could these possibly be? Are we ready to think about them, or to at least try to uncover them? Wouldn't teaching without motives feel like we were not living up to our responsibilities anyway?

So much of mathematics education is about experiencing loss. Let me explain what I mean: our system of education is not designed to notice that learning occurs over time, or even that accidents, chance, and frustration may be as much the source of learning as the affects of learning, and not even be obviously related to any specific actions on the part of an adult. In general, educators focus on what they can observe teachers doing. A book like this one is expected to provide lots of information about things that teachers can and should do in order to create learning as an effect of a teacher's actions. In reviewing Britzman's book for the American Educational Research Association, two Ohio professors of education, Lisa Weems and Patti Lather, note that modern educational theory persists in placing the teacher at the center of education, despite a claim to student-centered learning. For us, this is a significant point to dwell on: to embrace mathematics and to become a teacher of mathematics, we need to "de-center" ourselves; what a tricky business when we are thinking about what it means for us to be a teacher! It seems we need to think about what we do as teachers; we are the experts—on both the mathematics, which the students need to learn, and on teaching, which we are professionals at performing. Yet, as Britzman argues, teacher-as-expert is a metaphor—one of many metaphors we might use to think about our work; this particular one implies that learning is about the mastery of knowledge. Weems and Lather write that the emphasis in contemporary educational reform on "professional development" of teachers, for example, stems from the premise that if we add more classes, more knowledge, more experience, more x, y, z, to the teacher, we have thus refined the "producers" of education and by extension increased the value of the educational product. The move toward "student-centered learning," seductively attached to liberal humanist discourses of unleashing the "natural will" of the student, merely reverses a notion of mastery, identity, and fixity in understanding classroom positions. While this reversal seems an "improvement" over a "teacher-centered" model, Britzman argues that it still leaves the binary of teacher/learner intact. If only we knew more—but we don't; this whole approach to thinking about teaching can only leave us feeling that loss of what we might have known, what we might have been able to do "if only . . ."

Along with Britzman, I want to encourage us to shift our focus away from either the position of the teacher or the learner, toward the relationality of the teaching/learning encounter. To emphasize development as relational is to imply that what happens in learning exceeds the boundaries between inside/outside that are taken for granted in more common psychological theories. Educational psychology privileges cognitive aspects of learning; psychoanalysis highlights among other things affective dimensions of learning such as desires, hopes, and anxieties. We might say learning comes from neither within nor without. Instead we would describe learning as a struggle between two (or more) egos, signaling the uneven movement across and between various positions of knowing. A psychoanalytic inquiry into learning involves the careful tracking of the relations of learning in their various, incomplete, and recursive movements. Furthermore, psychoanalysis ethically obligates teachers to explore the dimensions of one's own otherness (Britzman 1998: 16).

Now, Britzman calls attention to the work of Alice Balint, who writes about "blind spots," where "only the exertions of the educator are considered." Such blind spots shut out the serious work of the learner, ignore the serious mistakes that cannot be admitted or debated, conflicts that are central to the experiences of learning, and so on. Instead of a clear set of understandings about what makes "learning happen," we end up with a lot of defensiveness on the part of the adult, as if the work of learning is only an authentication of or an answer to the question of the adult's capacity to control, predict, and to measure "progress" (Britzman 1998: 26). A psychoanalytic perspective like this raises what Britzman calls "antinomies," paradoxes or contradictions between principles or conclusions that seem equally necessary and reasonable.

> What exactly happens to the work of learning that surprises this defensive logic? First, the educator cannot recognize the learner's logic—and learners know this. From the learner's vantage, the adults cannot understand, cannot remember, their own antinomies, awkwardness, and hesitations in learning even as they may also represent something like what the learner wishes to become. Perhaps, from the learner's vantage, what seems most unfair is that while educators demand that students tolerate the postponing of immediate gratification for the sake of the hard work of learning, educators feed their own immediate gratification by setting the time of learning to the clock of their own efforts!
>
> (Britzman 1998: 26)

So we start the enterprise of education with things that are lost: things that we are blind to and cannot see. Deborah Britzman's colleague at York University, Alice Pitt, uses a metaphor from Edgar Allan Poe's short story, "The Purloined Letter": as we work in educational settings, we are always feeling like the objects of our attention are

Box 2.1

The Imagined Classroom Exercise

This exercise uses mental imagery to lead you through a scene in your imagination. Afterwards, you need to get together in a small group to discuss how the various characteristics of the scene may say reveal "lost" ideas about yourself.

Take your time with this. Pause long enough between each sentence here to make sure you can write down a good description of what you are seeing in your imagination.

> You are walking down a hall . . . As you are walking, you see a classroom. Look at it. A mathematics classroom. Notice its details . . . Walk towards it. What do you notice about it? . . . Look into the room; what do you see happening there? . . . Notice the details of the people and events taking place in this room as you continue to look into it . . . As you observed the classroom, you noticed a way to get in. Now go into the room . . . What do you see? . . . (Nobody there notices you – you are invisible, perhaps you are wearing an invisibility cloak?) Explore this room: what's inside? . . . What do you hear? What does the air feel like? As you explore, you notice a secret door leading to a secret room. Go inside that room. What do you see? . . . Now leave the secret room and go back into the main part of this classroom . . . Now leave the room. As you are walking away from it, you look back at it one more time . . . You are back in the hall once again, walking.
>
> a) Share what you saw, heard and felt during the imaginary exercise in a small discussion group. Feedback from others will be surprisingly insightful.
> b) Based on your discussion, visually "construct" together the one classroom that seems to capture the personality of the group and all of its members.
> c) Comparing classrooms created by different groups can result in some fascinating insights into the unique personality of each group. What do these rooms say in general about how most people feel about mathematics education?

"purloined," that is, lost, displaced, or misplaced (perhaps stolen?). Mathematics pedagogy cannot be simply applied as a set of techniques that you train in toward mastery, no matter how delightful such a fantasy of control over our work might be. Instead, as Alice Pitt writes, our work "relies upon the surprise of discovery in each and every instance of practice" (Pitt 2003: 9).

Antinomies create crises for us because once we seek one member of the paradoxical relationship, we have lost the other, equally reasonable goal that makes up the other half of the antinomy. In a curiously playful and powerfully useful move, Peter Elbow (1986) embraces this as the very pleasure of teaching! Writing in *Embracing Contraries*, Elbow describes the unspeakable that he was forced to confront. First of all, he suddenly found that he had a strong desire not to tell people things they didn't ask him to tell them; second of all, he recognized the pathology of such "holding back" as he did the pathology of "unsolicited telling." On the one hand, he thinks perhaps that he is perversely refusing to tell students things as a petulant backlash; on the other hand, he wonders about this possibly strange hunger to tell people things they didn't ask about. This is indeed a central feature of mathematics teaching and learning. First of all, in most typical mathematics classrooms, students are suddenly confronted with a mathematical idea that they would never have thought of if they had not come to school that day: they can't even begin to desire to be told these things because they do not know of their existence. In such a classroom, the "conditions of libidinality" are crafted in such a way as to make the teacher the one who knows all and the students the ones who do not know. The idea,

I suppose, is that this is what school is all about: surely our students need to be exposed to this stuff so that they can find out that it is interesting, or useful, or relevant, or . . . But students in general do not respond to this kind of a situation with excitement and engaged enthusiasm! Instead, they feel like the knowledge is esoteric or uninteresting or irrelevant to their lives. At best, a few students get good at copying the teacher and receive rewards for this apt mimicry; these students have learned to postpone the gratifications and pleasures of learning, or perhaps to separate themselves from the pleasures of the learning experience, so that they can gain the acceptance of the teacher, please the teacher, please their parents, feel smart, get good grades and go to college, and so on. No student is actually asking the teacher to say what is on her or his mind.

So you walk into the room. You are confronted with eager faces waiting to see what will happen. Do you begin a mathematical conversation? Or, do you think, "Nobody here is really interested; perhaps I shouldn't bother them with this?" A day later, a student is demonstrating a misconception; it is clear that he or she does not understand a mathematical idea. Do you tell them this? Or do you hold back and wait, to see what happens? Britzman suggests that it is impossible to know whether one or another of these sorts of responses is more harmful or useful to learning.

Box 2.2

The Empty-Your-Pockets Exercise

This exercise is pretty simple. Break the class down into small groups. In each group, take turns emptying pockets (or pocketbooks, backpacks, etc.) to show the group what you are carrying with you. No one is required to show anything that they would prefer *not* to show.

The person and the group then discuss what the objects say about the person. Does it reveal something about one's personality and lifestyle? Is there any one object that stands out as a reflection of some aspect of that person? Sometimes what is *missing* says something about the person!

Most mathematics teachers today have settled into an ambivalent acceptance of a rather mundane existence. Students rarely find themselves lost in the thrill of mathematical thinking in these classrooms. Even teachers who try to establish a different kind of environment note how their students come to them with years of expectations for what the conditions of libidinality are in school. Teachers offer the gift of a new, refreshing, and interesting mathematics experience, only to be flustered by students' ennui, blah tedium and resistance to even admitting their own personal interests (see, e.g., Romagnano 1994). Pitt dwells on the ambivalence even more: she asks, "How can we tell the difference between a leap away from and a leap toward learning?" (3). In the moment of the pedagogical encounter, we have lost the ability to know—such abilities are purloined letters. The only thing that is clear is that the standpoint of "teacher" creates a specific desire to be heard:

If I want to be heard at all, I've got to set up a situation in which the options of whether to hear me or tune me out—whether to take me seriously or dismiss me—are more genuine than in a normal classroom field of force. I'm refusing, therefore, to be short-circuited by a role which students react to with the stereotyped responses to authority: either automatic, ungenuine acceptance or else automatic, ungenuine refusal.

(Elbow 1986: 84)

Purloined Purposes

So, on the one hand we enter the mathematics education encounter with a whole lot of stuff that we want to share with youth. We know a great deal. If *they* knew the same things, they would be able to do so much! My new tote bag from the National Council of Teachers of Mathematics annual meeting declares, "Do Math and You can Do Anything." On the other hand, we never really can know what the students are doing, how they learn, precisely, what is happening inside their minds. Is there a chemical change in the brain or heart? A neuro-network restructuring that formed in response to a question from a teacher or other student in the class? We act as if the specific actions of the teacher *cause* precise changes in the student, none of which we can ever really know about. Or we write a history of the event afterwards in which we claim that a specific task or a specific way of asking a question led to a new conceptual understanding on the part of a student, even though there really is no way of knowing at the moment that such a pedagogical move would have that precise effect. This is what comes to be known as a practitioner's wisdom: over time, experience seems to provide so many case studies of what might or might not happen that the teacher seems to develop a repertoire of potential proactive and reactive actions.

Yet we can stop for a moment and disrupt this way of thinking. We do know something about *ourselves* and how *we* learn, what kinds of ways of working and being lead us to make meaning, to consider questions, to take on challenges, and so on. What is the purpose of an activity in the classroom? Whose purpose?

Box 2.3

Purloined Purposes Exercise

Think back to one or two experiences you have had as a student learning mathematics. What was your interpretation of the purpose of what you were doing during this event? Can you describe some purposes that the teacher might have had that would have been different from the ones that you were thinking of at the time?

If education means to let a selection of the world affect a person through the medium of another person, then the one through whom this takes place, rather, who makes it take place through himself [sic], is caught in a strange paradox.

What is otherwise found only as grace, inlaid in the folds of life—the influencing of the lives of others with one's own life—becomes here a function and a law. But since the educator has to such extent replaced the master, the danger has arisen that the new phenomenon, the will to educate, may degenerate into arbitrariness, and that the educator may carry out his selection and his influence from himself and his idea of the pupil, not from the pupil's own reality.

(Buber 1965: 100)

Martin Buber, a philosopher of the twentieth century concerned very much with relationships between and among people, is articulating the work that is going on "behind the scenes" as teachers and students try to communicate, but where, as is commonly the case, the purpose of what is happening is only able to be perceived by the teacher. This can happen even in the most open of classrooms, where the teacher explicitly tells the students the purpose of the activity: from the students' perspective, these purposes are the teacher's purposes, not the students', and so this difference in reality is not dissolved, and instead is pushed further into the realm of the purloined.

Madeline Grumet, who very recently stepped down from the position of Dean of Education at the University of North Carolina in Chapel Hill so that she could return to teaching, used the same quote from Buber when she wrote about the purloined purpose in education (Grumet 1988). Grumet found in Buber the contrast between the apprentice and the student. In some lost, mythical past, the apprentice worked with the master on the task at hand, and it was this work at hand that defined the dimensions of their task. We might imagine classrooms where students are treated as apprentice mathematicians. In such classrooms, the material, essential presence of mathematics in and of the world would draw the one who teaches and the one who learns to each other as they approach it. The mathematical ideas, objects, strategies, interpretations, and so on, become the objects of the students' and teacher's intentionality and the basis for their ability to communicate and understand each other, to be *together* in the world. David Hawkins (1980), too, reaches out to Buber. Hawkins, who was a distinguished emeritus professor of philosophy at the University of Colorado at Boulder, and the co-founder with his wife of the Mountain View Center for Environmental Education, believed mutual respect between adult and child is best developed when both parties focus on something other than themselves or each other. For example, you might want to teach in a way that you are not thinking about what you (the teacher) are doing, what the students are doing, whether or not the students are doing what they are supposed to be doing, and so on. Instead, the group together must find an "it" that piques interest. It is through the relationship that each person creates with "it" that it becomes possible for the people involved in this experience to avoid treating each other like objects to manipulate or react to, and instead to develop a kind of relationship that Buber called "thou." In the chapter "I, Thou, and It," Hawkins urges us to look for *good starting points* rather than for perfect lessons.

The function of the teacher in a classroom based on Hawkins' ideas is to respond to the students, to make what the teacher considers an appropriate response. Such a response is what the student needs to complete the process that she or he is engaged in at a given moment. The adult's function is to provide a kind of external loop, to provide selective feedback from the student's own choice and action. The child's involvement elicits a response from an adult; this response is thus "made available" to the child, it is now another object that the child can relate to. The child learns about himself or herself through joint effects of the non-human and the human world. Hawkins says this process should not go on forever; it terminates at some point in the future, when the adult function is internalized by the child. At this point this child is grown-up, and has become her or his own teacher. This is Hawkins' definition of being educated: no longer needing a teacher, you have learned to play the role of teacher for yourself. At that point you would declare your independence of instruction and become your own teacher.

Psychoanalytic Selves: Learners and Teachers

So what exactly is this role of the adult that Hawkins refers to? It is to be with the student looking at a third "it." This might happen if one looked at what the student is already looking at—by finding what the student is interested in and learning how she or he is looking at it. Or it may take the form of presenting an object for both the teacher and student to look at. In either case, though, it is most easy to slip out of the I–Thou relationship and start to think about what you are doing, what the student is doing, whether or not the "it" is having the effect that the teacher wants, . . . whenever any of this happens, the relationship between teacher and student has changed and is no longer one of I–Thou. Each member of the relation must instead strive to see the "it" as the other sees it, to be together looking at it. As Grumet writes, teachers tend to look out to the world and through the world to the student (1988: 116), instead of with the student to the world, or through the student to the world. She says it is this detour through the world that we call "curriculum," and she says "the look of pedagogy is the sideways glance that watches the student out of the corner of the eye." It is not easy to act like a teacher in the theater of contemporary schools, but we continue to desire this for some— purloined?—reasons, rather than to seek another sort of relation with the world and with our students.

Britzman asks us to consider talk between individuals as the central focus of our work, as in the capacity to respond well and not diminish the contingent work of making relations (1998: 38). "Making relations" might be a psychoanalytic way of saying "living and always becoming who we are." It is the constant making of relations—with things, with people, with ideas; anything can become an object to which we relate—that is the basic way in which we construct who we are and what we are doing. The ongoing making of relations replaces what some people might describe as the self. Notice this is a perpetual, non-ending *process* rather than a unified, fixed, knowable "thing." To see ourselves and our students, indeed to see all human beings as always making relations, and to understand that these object relations are influencing the making of the relations that we are making, is one step way from treating either ourselves and the teacher, or our students, as "its," and to embracing the relations that are part of the "I–Thou" that Hawkins and Buber strive for.

> As our unconscious ego processes are released into objects chosen for the dream
> to evoke a dreaming self by object choice, and as those objects are changed in

the encounter, so too in the waking dream might we choose our objects based on unconscious ego processes and object relations so that a self is evoked. From that encounter, subjectivity may develop. I would that curriculum be understood in this fashion; then what an education that would be!

(Block 1997: 34)

Alan Block, professor of education at the University of Wisconsin-Stout, provides a sort of picture of what is going on in the classroom, whether we think about it this way or not. So, why don't we think about it this way, then? "Being," Block writes, "is the active establishment of object relations" (1997: 22). Psychoanalysis usually goes back to infanthood. The reaching out to the world that is the infant's activity is facilitated by the caregivers; from their responses the child learns a way of being in the world, becomes a self in relation to, and in the use of, objects. This is like Hawkins' idea of the adult as external loop: the responses of the adult mediate the potential object relations, the meaning of the event that has taken place in terms of the development of who this person *is*. Fast-forward several years and find the same child in a classroom. This child, now a student, reaches out to the world; this is the student's way of being. The reaching out to the world that is the student's activity is facilitated by the teacher, and by other students who sometimes take on the caregiver role in this classroom. From their responses the student learns a way of being in the world, becomes a self in relation to, and in the use of objects. Now, in the context of mathematics, many of these objects are mathematical, images of being a mathematician in some way, relations with mathematics and with mathematical objects, with mathematicians, with the teacher of mathematics, and with other students of mathematics. In reaching out to the world in this mathematical context, the student's reaching out to mathematical objects and objects of relation that affect the students' object relations is facilitated by mathematical caregivers, the teacher, the other students, images of mathematicians, and so on; the student becomes a particular mathematical self in relation to, and in the use of, these objects, which we might call objects of relation.

Note, too, that objects may be related to in ways that do not promote a typically "positive" mathematical self. Some attachments—that is, object relations that are part of this self structure of all ongoing object relations—contribute to an alienation from or distanced relation with mathematical ideas, symbols of success in mathematics, relations of mathematical meaning, and so on. The way this constellation of object relations is experienced at any given moment is the sense of mathematical self. (In other words, following Block, the way a student experiences the mathematical self is a reflection of the underlying mathematical self structure, which in turn is made up of the totality of object relations. This is of course constantly in reconstruction as the self is always the totality of the ongoing processes of creating such attachments.) To further complicate matters, the teacher is of course a "mathematical being" in the very same way, so that others in the room are serving as caretakers for the teacher's ongoing construction of attachments. For both teachers and students, the selves consist of object relations and are expressed through the uses of objects. When a teacher uses the student as an object of relation through which to express the self, as in many classrooms, school is serving as a space of being for the teacher. It is very easy to slip into this kind of mode, where

Shadow Exercises

These, too must be done so that you can discuss the experience with a group of others. Do each step in turn without reading further or it won't work for you.

a) Think of teacher you have known whom you don't like very much. Maybe you even hate this person. On a piece of paper, write down a description of that person. Write down what it is about this individual's personality that you don't like. Be as specific as you can.

b) When you've finished writing, draw a box around what they you've written; at the top of the box write "MY SHADOW."

c) Consider this idea: "What you have written down is some hidden part of yourself—some part that you have suppressed or hidden. It is what Jung would call your *Shadow*. Maybe it's a part of you that you fear, can't accept, or hate for some reason. Maybe it's a part of you that needs to be expressed or developed in some way. Maybe you even secretly wish you could be something like that person whom you hate."

d) This exercise might have been done by thinking of a student whom you hate. Now that you "know the trick" it may be too hard to honestly describe the student without reflecting on what your description might mean about yourself. What do you think? Try and see what happens, and discuss this, too, with others.

e) Reactions to the *Shadow* are often mixed. Some folks immediately see a connection; some immediately reject the idea. How many of us have friends or romantic partners who fit the description of the "hated" person? This exercise always leads to interesting discussions about how we project suppressed parts of ourselves onto others, and about why we sometimes choose these "hated" people for our close relationships. What are the implications for teaching mathematics?

everything that is happening in the classroom is about you, the teacher, and who you are. I am a "good teacher." I am "nice teacher." A "smart" teacher, in control, doing my job, *teaching*. But can you say at this point what the students are doing? And with this question I do not mean the objective as listed in the lesson plan. I mean as mathematical beings, as human beings, living in this classroom, how are these individuals doing the important work of constructing object relations, and what is the effect of these attachments on the self? Block gives us a way of thinking about what the conditions of libidinality could nurture: each student's "ego processes released into objects chosen for the dream to evoke a dreaming self by object choice"; activities that do not just allow this to happen, but foster its happening, and joyously respond to its happening, lead to those objects changing in the encounter, and to a (mathematical) self being evoked. The students' selves are present in this kind of classroom.

Transference Exercise #1

As with some other exercises, this only works if you do each step without reading ahead!

a) Think of a significant other, your boyfriend/girlfriend, or husband/wife, or a close friend. Think about some aspect of their personality that you have a strong reaction to, either positive or negative. Now write that down on a piece of paper. Describe what that aspect of their personality is like, and how you react in your thoughts, feelings, and behavior toward that part of their personality.

b) Now draw a box around what you've written, and write at the top of the box, "Is this transference?"

c) Now think about your parents. Is the personality characteristic of the person you wrote about, and your reaction to it . . . is it a kind of replay or recreation of something that went on in your relationship with one (or both) of your parents? For example, does your parent have that same personality trait that you react to so strongly? If so, maybe this reaction to the person you described is a kind of transference from your relationship with your parent.

d) We can apply this idea of transference to relationships between teachers and students. How do strong aspects of our parents' personalities influence the ways we interact with other teachers? With some students? When we react strongly to a student, can we separate what is going on from what might be happening for the student who is enacting transference? When might transference be the defining attribute of a relationship between a teacher and a student? Might the teacher serve as a parent-like figure for a student in later instances of transference?

No matter what we do, this is all happening. This is why the teacher is so important. It is the teacher who creates the conditions under which this occurs, and through which the self is evoked, whether toward a mathematical self that uses mathematical attachments in meaningful ways, or toward a self that constructs objects of attachment that distance mathematical objects in order to maintain a coherent and consistent self. In the majority of classrooms that I have visited and heard about, it is the latter kind of self that is evoked, rather than the kind of self that would readily use mathematical objects to express a mathematical self. The "good students" use mathematical objects to evoke a "good student" self, but rarely to evoke a mathematical self.

Other students use mathematical objects to express another kind of self that is able to weather the storms of mathematical experience; here, too, the relations are not the kind that evoke a mathematical self. Teachers may use mathematical objects to evoke a self of professional success, but rarely do I meet a teacher who uses mathematics to evoke a mathematical self. To do so would be to act as a mathematician, by some definition of that term.

Holding and Listening

Which brings us back to David Hawkins' idea of I–Thou relationships built out of "it," or the introduction of a third term in the pedagogical relationship, and to Deborah Britzman's idea that talking is central to the process of making relations. For Hawkins, the teacher needs to find good starting points for these conversations, and, in the process of looking at something together with the students, to seriously work at responding well to what the students are saying and doing. The purpose of the teacher's response is to make it possible for the students to follow their insights and to support the making of relations. As soon as that sideways glance that Madeleine Grumet refers to makes an appearance, we are no longer in the midst of a pedagogical encounter, and instead we are snagged by curriculum. To teach effectively is to try to recover that lost bit of knowledge, what a child means when she says, as a young girl told Herb Ginsburg, that "two rows of objects both have the mostest" (Ginsburg *et al.* 1998: ix). The curriculum move is to diagnose her misconception and develop a treatment plan. The pedagogical move is to explore with this child how each of these two rows indeed both have the "mostest." Ginsburg is a psychology professor at Columbia University in New York. He has worked for many years to adapt Piaget's idea of the "flexible interview" for teaching in classrooms. His goals are to develop a classroom atmosphere that encourages the expression of thinking, and to help students to learn to express their thinking. Flexible interviewing is one technique that teachers can use as they work towards the I–Thou relationship that Hawkins takes from Martin Buber, in order to support students' making of relations.

The shift to a "thinking classroom" that centers the I–Thou encounter within responsive talk requires that teachers release themselves from a combative stance toward learning. All too often we slip into a mode of being where we are fighting for our students to learn—against the circumstances, against their ignorance, against the seeming unfairness of all that we do not know. It is as if the losses that we feel are being listed in a tally—ten for us, three against? It is tempting to feel like teaching is one whole big mess of conflicts—between the teacher and the students for control of the classroom; between what we know and what we do not know about how our students are thinking; between our desire to tell and our need to hold back so that they can discover for themselves; between our yearning to share what we value for our students' futures and our students' own immediate desires and interests. Most people in conflict are fighting for some kind of victory when what they really want is peace and understanding. As we know from recent world events, peace does not necessarily follow victory. Peace is something that must be created in a context of mutual understanding.

Dan Gottlieb, a psychologist in Philadelphia with a talk-show on National Public Radio, suggests that we can only move from mutual hurt, anger, and buried feelings, toward empathy and compassion, by first experiencing compassion for ourselves. Most of us behave, he says, as though there is something wrong with us. We hide our vulnerable feelings and we overachieve to compensate for what we feel is a defect. We disavow parts of ourselves and yet we long to be understood. If we could only suspend self-criticism and feel compassion for ourselves, it might sound like this: "He threw the book across the room instead of answering my question; that makes me sad and lonely." Or "When she made that joke about me, I felt ashamed and hurt." These are just emotions and don't necessarily mean that you have to do anything. Nor do they mean that a student or colleague is an ogre. They just mean that you are feeling pain. Compassion is noticing that you are hurt and simply caring about it. When you soften to yourself, you can soften to others, and you are ready to listen. If a student in your classroom or another teacher is hurt, that does not mean that you are bad or even responsible. It just means that a person you care about is hurt. Once we get used to examining experience in this way, we are ready to establish a classroom environment where people talk about what they are thinking.

One piece of what Ginsburg calls the "thinking classroom" is that everyone there has the vocabulary with which to express themselves and to describe their thinking. Most students are not used to such a school environment and need to be taught that it is not only o.k. but especially appropriate to use certain ways of talking in this classroom. Thinking, planning, strategies, checking, representing, and proving are all part of this kind of talk.

Box 2.7

The Thinking Classroom

I was *thinking* . . .
I *solved* it by . . .
My *strategy* is to . . .
I *plan* to use a method that . . .
I *checked to make sure that made sense* by . . .
This cube *represents* . . .
I will *prove it* by . . .
(adapted from Ginsburg *et al.* 1998: 47)

Knowledge of the vocabulary of thinking does not develop spontaneously. Ginsburg points to research indicating students typically find it very difficult to talk about their thinking. First of all they need to master the vocabulary—something even teachers have trouble with when they first start to use it; next they have to understand what the expectations are for the kinds of things that count as knowledge in the classroom. Here are some long-term objectives for this type of classroom: 1) develop the ability to think before engaging in problem solving; 2) develop the ability to discuss plans with others; 3) develop the ability to listen to and understand one another's thoughts. When you invite students to begin discussing their problem solving strategies, it is best to use a problem or situation where the question and the answer are not so obvious. More than one correct answer is helpful, as is a problem situation where some information is not given and must be supplied by the students. The problem should need to be more precisely defined by the students. All of this will support an interesting discussion. If a problem is too simple, students are apt to skip over the important process of understanding and clarifying it.

A focus on learning how to plan helps students to shift away from just getting an answer toward thinking about the processes of interpretation. Have students write about their plans, draw them, even record them on audio or video. Have them talk to others. Stop in the middle of processes and ask students to reflect on what they have done so far, and then to plan what they will do next. Have students explain—in writing, speaking, drawing, and so on—*why* they have decided to pursue this path. Establish long-term projects and have students ask the rest of the class to help them plan throughout their process. Here are some issues to consider: will everyone in the class be ready to listen patiently to others? No. They need to learn how to do this. Start off slowly with your expectations and take your cue from the students. Have the class discuss what works well and what does not work well in listening. If someone is taking too long for an explanation to hold the interest of the whole class, ask them to take their time and write down their thoughts; when they are ready to talk, they can raise their hand and the class can then prepare to hear what they have to say.

Psychoanalysts might call this kind of an environment a "holding environment." The term comes from the idea that the primary caretaker of a baby usually "holds" the child so that he or she can do what it needs to do. A holding environment is a secure and comfortable, nurturing atmosphere that enables the children to do the important work of making relations and self-creation. Ginsburg's notion of a thinking classroom is one step toward establishing the mathematics classroom as a holding environment. This involves not just sharing but also learning how to describe what one is thinking, and also developing the tools and skills for listening to others, for it is in representing our ideas and in being heard that the holding environment takes shape.

Learning to record one's thinking is another important piece of representing one's ideas so that someone can listen to you. The teacher must encourage and sometimes push students beyond their initial comfort zones so that they can become skilled at writing about what they are thinking, using pictures to describe their ideas, comparing ideas represented in words and pictures with mathematical symbols for the same concepts, and working with mathematical representations such as equations, numerical sentences, geometric diagrams, and so on. To get started, model what you have in mind:

> After you have thought about how you worked on your question, write down your thoughts on this piece of paper so that somebody else in the room can know everything they need to know about your thinking; they should be able to approach the question in the same way. For example, if I used tangrams, I would say I used tangrams, and I would draw pictures of how I changed the shape I made from one to another, maybe writing some words nearby to explain my picture if I think I need to.

Then I would write my strategy, such as, "I used tangrams to identify potential ways that I could rotate the triangles and see if other polygons were possible, recording in my chart all of the different outcomes that occurred." In other words, the teacher does not have to give step-by-step instructions for how to record thinking, but instead criteria for the product and the work that is done: someone else has to be able to read what is written and be able to replicate what you did. The next step, now, is discussing not just the mathematics that is written down, and not just how accurate or correct it is, but what techniques of representation are effective in communicating ideas. It may very well be the case that a student did not really get anywhere on a mathematical question, or worked with a misconception and thus was not correct. But this same student may have a great idea for how to use pictures and charts in a certain way so that others can understand their thinking. Such a student is perhaps a "better mathematician" than another person who can get an answer but cannot explain the ideas so that another member of the class can reproduce their work. Communicating your thoughts is an important part of acting like a mathematician!

At first students may simply imitate what the teacher does. As long as you continue to model general guidelines without giving detailed or complete routines, you are on the right track. To broaden the kinds of information that students provide in their representations of their work, follow-up the first version, such as the written one, with mini-interviews, either as a whole class or in informal, quick discussions as you walk around

Box 2.8

Follow-up Questions for Representations of Work

How did you figure out that there were . . .?
Why did you use . . .?
Did using unifix cubes help you? How?
What did you do to determine that you could say . . .?

Questions to Extend Group Conversations
What did you learn while you were solving this problem/working on this investigation?
What makes a good problem solver?
Was this hard or easy? Why do you say that?
What new questions do you have based on your work so far?
How will you pursue your new questions?
What did you learn from someone else about this problem?

Questions to Promote Reflection on Communication
How do you think you are doing in mathematics?
What do you like/dislike about mathematics?
What do you do best in mathematics?
What would you like to get better at in mathematics?
Why do you think this is hard for you?
What do you need more help with?
Is it easy for you to write about your strategy? Why?
What parts are you having trouble with that someone else might be able to help with?

the room. If students keep math journals or a mathematician's folder, they can record in these places potential plans for investigations, ideas about recording their work, and examples of the work they are proud of, as well as work that they learned from but may not be proud of. Ideally you will build into the structure of your class ways for students to publish books or pamphlets on how they solved problems, or how they identified interesting mathematical questions that they encourage others to work on. By building this into the classroom structure, you are crafting the conditions of libidinality to include an audience for their work: students must read the books or pamphlets that other mathematicians have written, and be able to discuss what makes a good one, what they learned about communicating ideas from reading another mathematician's pamphlet, and so on. And there must be a reason for doing this. Perhaps these books will be read by others outside of class, and we want them to be good? What would make a good book for *this particular* audience for whom we are writing?

Interviewing

One excellent strategy for helping students to develop the vocabulary of thinking while also helping yourself to practice listening to students is to use Ginsburg's flexible interviewing. You can interview students one-on-one, which gives you the chance to center very carefully on individuals; in small groups, which reduces the individual attention while helping you to observe how each student works within the group; or in the context of whole-class discussions. In any of these situations, the idea is to spend fewer than five minutes in a highly focused listening encounter. Interviews enable your students to demonstrate to themselves and to you their understanding of particular concepts, while also providing a forum for communicating to your students that you value their thinking.

Box 2.9

"Keep in Mind for a Thinking Classroom"

Talk-Show Host Exercise

a) Notice that a facility with words is not always accompanied by accuracy in computing. Some students are very good at sharing incorrect solutions. Since the point of discussion is not to correct students, but instead to understand how *they* are thinking, this presents an interesting pedagogical situation. The point is not to diagnose the misconception, but to appreciate how the student is making meaning.

b) Don't feel frustrated if you find that your students are unable to describe completely the strategy they used. With time and practice, the majority will increasingly be able to do so.

c) Teachers have to learn to question, just as students have to learn how to respond. It's not easy to question students' thinking on the spot. Beginners talk too much and don't let the students explain what they are thinking. Sometimes they ask too many questions at once.

Here is a way to get better at asking questions of students: As suggested in the previous chapter, pretend your classroom is a talk show, either on television or radio. You are the host; your students are the guests.

One approach is to set aside a private space for short mini-interviews with individual students. This requires special classroom management skills. Your class needs to understand the purpose of the interviews and how you expect them to behave while they are happening. For example, the other students need to work independently without your help while you are devoting your attention to a student or a small group of students. It seems best to me that they be busy with other activities that require conversations so that those being interviewed do not have to think about whether others are listening in on what you are talking about. At first a family member or other volunteer may be helpful in monitoring student groups while you are interviewing.

To prepare for an interview, consider possible problems, tasks, or collections of questions that you want to use as the third item (the "it") in the I–Thou–It set of relations. Ask yourself what you want to learn about your students' thinking. Identify a *big idea*, something that is learned over time rather than something that could be memorized and simply recalled, that can be at the heart of what you are talking about. Your goals are to gain insights into the student's conceptual understanding and reasoning, to discover their attitudes towards mathematics, and to assess their ability to communicate their mathematical ideas. The primary focus of this encounter is *not* to instruct or correct, but instead to further the relationship between you and your students.

Box 2.10

Teaching Students How to Reflect and Talk

1. Arrange the seating to foster communication.
2. Tell your students that particular lessons or parts of investigations are specifically going to involve talking about what it is like to work on solving a problem or posing a question, or about identifying a good way to use compare charts and pictures in words, or whatever else will be the focus regarding reflection on and communicating one's mathematical thinking.
3. Motivate your students to talk about what they have done: give them reasons for this, or find out their own goals and help them to see that this discussion will meet some of these goals.
4. Ask your students questions that would help them to think about and evaluate what they have done – not about the answer to a math problem, but about the processes to get to answers or new questions or some other aspect of mathematical work.
5. Show your students that you are interested in what they have to say. Make sure that you use what they say to guide further activity in and out of the classroom so that what they contribute makes a difference.
6. Paraphrase or reflect on what your students say to check that you have understood. This is important to demonstrate that you have heard what they say. But make sure you write down what they say in their own words as much as possible so that you are not changing it into your words.
7. Ask questions to clarify what your students are saying.
8. Involve all students in the discussion.
9. Show unconditional acceptance of your students' thoughts and feelings. Let them know that it is acceptable to be wrong or incorrect (more importantly that misconceptions become correct conceptions when interpreted in the ways that they are likely to be thinking).
10. Give your students feedback. Let them know that they are learning to do something that is difficult but very important.

(adapted from Ginsburg *et al.* 1998: 53)

Remember that the psychoanalytic perspective requires of us that we seek together with our students a meaningful relationship with mathematical ideas and objects, and that it is this attachment to a third object that makes the relationship between you and your students possible. It is easy to slide into the mode of judgment, where you use flexible interviewing to evaluate students' progress on sets of criteria. Moon and Shulman (1995) for example suggest that teachers conduct a series of interviews with the same focus, in order to discover a continuum of student performance, with students clustered at different points. They push teachers to work at identifying these clusters in order to use them to make instructional decisions and to describe the developmental levels associated with them. They assume that a teacher's function is to achieve externally mandated standards. Thus, they imagine that the teacher is using the interview to judge how much progress each student has made toward mastering a particular standard. Such a philosophy of education, which shares some techniques with a psychoanalytic perspective, but differs fundamentally in important ways, suggests that you use interviews to identify the individual strengths of your students upon which you want to build, and to see which behaviors have not yet been demonstrated and perhaps need instructional attention.

Flexible interviewing by teachers in the classroom requires substantial reframing of the teaching/learning encounter. On a surface level, the interview is difficult for both the teacher and the student who are unaccustomed to relating "therapeutically." Teachers want to teach and correct mistakes. Students want to be told if they are correct and if not, how to make themselves correct. The flexible interview, on the other hand, presupposes an open-ended, accepting attitude toward whatever surfaces. Its purpose is not to instruct, but to reveal more than either the interviewer or the interviewee knew about the objects of mathematics before the interview. It looks deeply into the student's thinking and at the same time tells the interviewer how his or her own projections about mathematics affect what the child communicates about what has been learned.

My dear colleague Rochelle Kaplan (who is a professor at William Paterson State University in New Jersey) and I wrote about these differences in an article that we published back in the 1990s (Appelbaum and Kaplan 1998). As in the psychoanalytic process, the interview does not define a clear and objective reality. Rather, just as the psychoanalyst must examine his or her own fears and motivations in the context of the process, so too must a flexible interviewer carry out this task. Therefore, the metacognitive act of a teacher creating his or her own interviewing behavior evokes an encounter with one's own motivations and fears regarding mathematics, children, learning, and teaching. It stirs up issues of one's level of confidence in doing mathematics, what the teacher thinks of his or her own relationship with mathematics, and the extent to which mathematics is seen as "egosyntonic" with one's own self image apart from being a teacher. (Egosyntonic means that mathematical object relations are indeed part of one's concept of self.) The discrepancy between a conscious view of oneself as a mathematical being and the "deeper" unconscious or subconscious feelings that drive behavior is a potentially volatile realm of self-confrontation.

Over the course of time during which Rochelle and I studied the flexible interview process, several interesting patterns revealing teachers' and students' relationships to mathematics emerged. We noted that during the flexible interview, the teacher's questions, responses, his or her every reaction to the child as well as his or her initial selection of an interview task, reveal as much about the interviewer's motivations and object relations as they do about the child's. In essence, what we found is that each level of object attachment that students display toward mathematics is paralleled by a comparable level of active engagement or lack of active engagement with mathematics manifested by teachers. Although these parallel levels do not necessarily imply cause and effect, it is evident to us that teachers certainly filter youths' attachments through their own unconscious investments in mathematics. These investments are defined by the extent to which teachers regard mathematics as essentially egosyntonic or egodystonic (i.e., as part or not as part of a self concept). Observations of interviews suggest that teachers selectively listen to those elements of a youth's encounter with the world of mathematics that can be recognized and acknowledged as consistent with their own perceptions and feelings about mathematics. One result of this is that teachers say they are implementing current professional recommendations even as they interpret the recommendations through their previously constructed notions of what should be happening in the classroom. As Rochelle wrote in another article,

> the constructivist program is filtered through an incompatible lens and what comes out is a distorted version of some seemingly objective curriculum. . . . On a deeper level, . . . her originally stated belief . . . is embedded in a system of other beliefs that defines who [a teacher] is and colors her perception of reality . . . The way in which she communicates to the student(s) defines her real instructional goal.
>
> (Kaplan 1991: 16–17)

Mathematics as Objects

Youth come to the interview with their own emerging relationships with mathematics (Kaplan 1987). In fact, these relationships are quite obvious to the observer who looks beyond the particular content of responses and focuses on the intention of those responses. While Winnicott noted that mathematical ideas begin with the concept of 1 and that this concept derives "in every developing child from the unit self," (Winnicott 1986: 58), it is also true that early object relations involving mathematics come from children's informal senses of quantitative, logical, and spatial qualities attributed to objects in the environment (Piaget 1952). At the early preschool stage, approximately ages 2–5, the child sees mathematics as an object that belongs to the self. All mathematical observations are seen as egosyntonic, a part of the self system. Similar to other objects in the world, mathematical relationships originate in spontaneous concepts rather than imposed concepts from others (Vygotsky 1986/34). The child at this time can be said, in a sense, to be "in love" with mathematics. Numbers, space, and patterns are playful things, things to do something with just to see what happens. It is also a very personal relationship, transforming objective realities to fit the needs and sensibilities of the learner. Let's use very early educational experiences as examples from which to generalize these ideas.

Mathematics as a Student's Personally Relevant Object

For example, when three year old Sal (Ginsburg *et al*. 1992) listens to the interviewer's counting to determine if any mistakes are being made, he can say with complete glee, "You made a mistake!" when the interviewer recites "1, 2, 3, 10." Or he can say, "Another mistake. You have to start with 1!" when the interviewer starts counting with "9, 10," instead of 1, 2, 3. Finally, when the interviewer counts from 1–15 followed by, "17, 19, 30, 31, 32, 100, 200, and 46," the child responds with a smile and tells the interviewer, "That's good." These are all signs of Sal's ownership of the mathematics. The numbers he knows are part of his self concept and may even define that concept to some extent.

Similarly, we have the example of Bruce who stops after counting from 1 to 29 because he does not know what comes next. Yet when pressed by the interviewer to take a guess, he responds, "After 29 comes 10." Subsequently, when the interviewer counts aloud from 1 to 30, Ben first stops her at 25 to exclaim, "You're copying me!" and then when the interviewer gets to 30, she is told by Ben that "No, after 9 comes 10." This child, too, indicates complete ownership of the mathematics and a sense that he has invented it and can tell others how it works.

A final example, of a preschool child's personal connection to mathematics, comes from 4-year-old Stacey, who was asked to compare the amounts of clay in two balls of the substance. After agreeing that the balls have the same amount, the interviewer breaks one of the balls into small pieces, leaving the other intact. When asked if there was still the same amount of clay in the two samples, the child proceeds to count the little pieces accurately to 9 and then points and "counts" the ball of clay as "1, 2, 3, 4." She then concludes that the little pieces "are more because I have 9." We see here that the mathematics used by the child is that of an object internalized to conform to the needs of the learner. Her intuitive mathematical self tells her that the little pieces look like more than the ball. Still she wants to justify that impression and so creates a counting system that works for pieces and wholes and uses her sense of number relations to support her judgment. This child is deeply wedded to her brand of mathematics. Stacey takes it with her and applies it as needed within her personal standards of consistency and as mediated by her senses.

A Student's Attachment to Mathematical Rules

Will this personal attachment and self definition of mathematics endure? Not likely. This is because the preschool mathematics that is internalized is not the mathematics the child will be finding later in school. That mathematics has rules and standards about how it can be interpreted and what it means. No matter how accepting a teacher may try to be about "does anyone have another way to solve this problem?" or however many multiple approaches are entertained, the logic of mathematics cannot be violated in schools. Thus, the child who seriously suggests that the answer to 43 − 17 can be 36 or 24 or 46 depending upon how you do it, will not be allowed to "love" all those answers without bias. Rather, the child will be encouraged to explain all of the ways the answers were derived and then, at best, be gently guided to accept one method as superior to the others.

Stephen Brown's (1993, 2003) work provides a stark contrast in encouraging new interpretations of mathematical operations or searching for cases in which seemingly absurd procedures actually work out sensibly. He is now a professor emeritus of mathematics education and philosophy at the State University of New York at Buffalo. Marion Walter (1996), a professor of mathematics education at the University of Oregon, also suggests such possibilities. Together, Brown and Walter (1983, 1993) have put forth "problem-posing" as the context of such "play" for all ages. Nevertheless, the playfulness of the early childhood period is not part of school mathematics as otherwise conceived, and so the object of mathematics often becomes removed from the self. The self does not internalize the rationale imposed, but just comes to accept the fact that some ways of thinking are considered acceptable and some are not. Depending upon the child's general nature and capacity to remember external facts, this kind of learning about mathematics may proceed toward a successful performance record or to one of increasing failure and misconception as defined by "the rules of mathematics." In school, some children learn that they can now treat mathematics as some object out there, "which has nothing to do with me and which makes no sense." Conversely, a more positive outcome might be that the child still treats mathematics as some object out there, which is not like that mathematics which is mine and that I love, but which I can and must master to survive in school.

In a very real sense, school forms the child in the denial of imagination (Block 1997), in splitting the subject and object and denying the creativity that makes the self possible. As Block writes,

> to deny imagination is to deny the very creativity that makes self possible; it is to perpetuate the hate that results from the inescapable discrepancy between subjective and objective, between the unlimited possibilities of one's dream and what the real world actually offers us.
>
> (Block, 1997: 171)

In other words, because schooling is structured by and about boundaries, it denies creativity and makes us hate ourselves for our thought-dreams (fantasies and ideas); we are taught to split and submerge our thought-dreams. As a result, the student may, according to Block, become a "dictatorial egoist" who actively denies the wishes and needs of the other, and tries to make his or her own wishes alone determine what happens; he or she may become a "passive egoist," retreating from public reality and taking refuge in a world of unexpressed dreams, becoming remote and inaccessible. Or, writes Block, he or she may search to avoid conflict altogether, permitting the outside world to become a dictator, fitting him or herself into that external world and its demands, doing what others want and betraying his or her own wishes and dreams (172). These outcomes are illustrated by two children who were interviewed during their elementary school years, one who sees mathematics as an object to be avoided, and the other who sees mathematics as a fixed, unchanging object.

Mathematics as a Student's Object to be Avoided

Jenn (in Ginsburg *et al.* 1992) represents the kind of blocked approach in a child for whom mathematics is only an object to be avoided. This third grader was asked to recall a basic multiplication number fact, 8 × 4, which she had mistakenly rote memorized as being equal to 35 on one occasion and 28 on another. When asked how she could figure out which answer was correct, she told the interviewer, "Ask the teacher." The interviewer then encouraged her to represent the combination graphically. When the child could not produce anything on her own, the interviewer suggested that she use tally marks and demonstrated how to get started. After a long and laborious process in which Jenn neatly lined up four rows of eight lines, she then proceeded to add the tallies as though they were "1s," coming up with an answer in the millions. When probed further about the correctness of her answers, she indicated that it was definitely not the larger number, but still did not know how to find out the correct answer. We have here an example of a child who is actively engaged, not in the process of connecting to and internalizing some mathematics, but in the process of creating a barrier between herself and the object known as mathematics. Her energies are directed toward preventing the intrusion of mathematical ideas, or the object of mathematics, upon her sense of self. For this child, mathematics is egodystonic (that is, in conflict with the needs and goals of the self) and not something to be manipulated, personalized, adapted, or cared for. The only emotion evoked by the object of mathematics is anxiety, and the child avoids mathematics at all costs in order to diminish anxiety. The process she uses involves providing some answer framed in the language of mathematics in order to "make the teacher leave her alone." Her goal is to rid herself of the mathematics rather than to internalize it.

Mathematics as a Student's Fixed Object

From the practical perspective of school achievement, a more positive resolution of the failure to internalize the object of mathematics as part of the self-image is for the child to be successful at rote memorization. In this case, even though mathematics remains external to the child, the rules of the system and its objective content are competently manipulated in terms of school demands. This child is typically an average achiever, sometimes making silly mistakes because of a faulty memory, but generally able to perform at grade level in a rote manner. In response to a question about why some procedure is performed, this child tends to provide an explanation that does not go beyond reciting rules and describing actions. The relationships within the mathematics are not dealt with nor is the child able to envision a problem in more than one way. In reality, this type of child is also math-avoiding and actively seeking ways to prevent becoming attached to mathematics and mathematical relationships. Rather he or she is engaged in an interaction with rules, authority, and appearances. For this child, both relevant and irrelevant variables in a problem are treated as equally important with as much emphasis placed on form as on content. Third grader Nancy's (Ginsburg *et al.* 1992) interview responses are typical of this kind of surface engagement.

After demonstrating enthusiasm, speedy and accurate recall of multiplication facts, prowess in producing essentially accurate answers to a series of mental arithmetic combinations, and competence with executing the written procedures for multi-digit addition and multiplication procedures, Nancy was asked to solve a simple one-step word problem. The child read and made these comments about the problem: "There were 23 people at each bus stop. So there's 23 people. The bus driver made 3 stops. How many people in all?" When asked what she would do with a problem like that, after some thought Nancy indicated, "You go times . . . 23 three times." When asked how she knew how to do that, the child's response was to restate the facts of the problem before saying, "You read it. You can't add it because they said there was 23 people at each bus stop. You can't subtract it. Like forget it . . . because it says how many people in all. . . ." When pressed further about why it was not possible to use addition in some other way to get the same answer as obtained with multiplication, Nancy responded with, "I don't understand . . . (now showing exaggerated puzzlement and disgust in her facial expression) . . . (shrugs). I don't actually know. . . . uh, uh. . . . it's too hard."

This snippet from an interview typifies the way in which a child who appears to grasp school mathematics, but does not really have a relationship with mathematics, uses standard procedures to automatically apply a known technique to a familiar context. There is no room here for playfulness with the facts, or for interest in the mathematics in the problem. Rather there is a clearly motivated urge to provide answers in order to accrue points or credit for being correct despite, not because of, the mathematics in the situation.

Mathematics as a Student's Flexible Object of Self

At another level, somewhere between the preschooler's complete incorporation and transformation of mathematics as an object of the self and children who avoid active engagement with the mathematics, we have the child who develops a clear image of the self as a mathematical thinker, but does not necessarily make sense of mathematics in conventional ways. This type of child is actively engaged in an interaction with the mathematical world, forcing numbers to make sense and creating his or her own rationale for using mathematics. Whether the conceptions are accurate or inaccurate, this child will always approach mathematics with a positive outlook and an expectation that mathematics can be understood. An illustration of this type of child is a fifth grader, Viola.

Viola was interviewed about her basic conceptions of fractions represented in standard written notational format. When asked how many fifths were in a whole, after some serious thought, she questioningly replied, "Five?" Then she was asked why there were five parts and she responded, "Probably because 5 and 5 is 10 and 10 is an equal number . . . and if it would be anything else, even if it was an equal number like 8, 5 and 3, (smiling now) it doesn't sound right for some reason. I think 10 sounds better." Subsequently the child was asked how many eighths were in a whole. This time Viola responded more quickly and said, "Two?" Then with more confidence she said, "Because 8 plus 2 is 10 . . . That's what you're trying to get at [the 10]." We can see here that Viola takes a different kind of approach from Jenn, who worked with the tallies. Viola ponders the questions and appears to reason them out. Although her responses do not make sense in conventional mathematics terms, she is engaged in trying to force the mathematical object to her perspective. She makes the math conform to her sense of reasonableness and expands upon her judgments by making associations between events as she grapples with them. For this child, mathematics as an object is part of her self image, something that belongs to her, and is, therefore, egosyntonic. Viola is a problem solver rather than a number cruncher. For this type of child, mathematical objects are continually evolving and being adapted along with the internalized sense of self as a mathematical thinker.

Teachers' Object Relations

It is an awkward task to reflect on one's own object relations. Psychoanalysis is usually a very long process that sometimes entails meeting with an analyst several times per week for many years. Slowly, over time, one shifts from resistance to learning about one's own object relations to using the knowledge of this resistance to learn. When Rochelle and I worked with teachers who began using the flexible interview format, we found that we could see some patterns that the teachers themselves might not have been able to see—yet another lost, purloined knowledge that we started to reclaim with these teachers. Together, it was as if we were the analysts and they the clients. While we do not claim psychoanalytic credentials, we did find that their relationships with mathematics paralleled those of the youth they were working with. Some teachers seemed to regard mathematics as a collection of facts and prescribed, procedurally accurate answers. For these teachers, mathematics is something of an alien content, that is, a body of knowledge to which they are minimally connected and which, in general, has nothing to do with definitions of self. Like Jenn, they were essentially engaged with mathematics as a rote activity, the rules of which had been determined by others who have a kind of capacity that they lack. During the flexible interview, such teachers can react in one of two ways.

Mathematics as a Teacher's Object to be Avoided

On the one hand, some teachers tend to approach the interviewing task with awe and an openness to accepting almost anything that the student suggests as indicative of interesting mathematical thinking. For example, we have the teacher-interviewer who asked a fourth grade student, Thomas, to help her figure out how many times she would need to fill an empty quart milk container with water in order to fill up a gallon container. Because the teacher herself actually did not know how to answer the question, she did not bring any preconceived notions about how best to approach the problem other than to just fill the containers. Therefore, she was open to accepting any type of more mathematical answer that the student could offer. During the course of the interview, as Thomas experimented with different ways of combining linear measurements of the two containers, the interviewer expressed repeated approval of his techniques although the efforts did not lead to any clear answer to the original question. In this case, though, the interviewer's own absence of a connection to the task allowed her to "discover" along with the child that they could agree on a variety of interesting observations about volume and dimensions of containers, as well as the use of rulers. Although none of these observations actually made much mathematical sense, this did not necessarily mean that

Thomas' level of problem solving for this task proved to be haphazard and fruitless. Nor did it mean that he was as detached from the mathematics as the interviewer. Rather it suggested that his engagement was at a different level from that of the interviewer, a level somewhere between a procedural rule-based approach and the playful approach of the preschool child who owns that mathematics. Perhaps, though, the teacher-interviewer's own distance from mathematics forced Thomas to shift his own level of object relations vis-à-vis mathematics downward because the interviewer was unable to respond with sharper, more critical questions about his procedural efforts. The question remains open as to whether the interviewer's lack of mathematical engagement actually controlled and diminished the level of his object relations with mathematics during the course of the interview or whether it fostered a more playful problem solving approach for both her and Thomas. It is clear, though, that this teacher entered the interview with no expectations for how the problem should be solved, which is perhaps why Thomas was able to be so creative. Otherwise, the teacher might have been more likely to sway him into coming up with the answer that she believed was correct.

Mathematics as a Teacher's Fixed Object

One the other hand, as interviewers, the same type of mathematically disassociated teachers may listen for accurate recall and statements of fact to questions they pose, questions that for them have only two kinds of answers—correct or incorrect. They do not go beyond the answer given, but tend to greet each response the student makes with a new question, one that is predetermined and not based on the child's response. For example, if such an interviewer heard Jenn say that 8×4 is 28, the follow-up to that response might be, "and how could you check your answer to see if you are right," looking for some kind of procedural explanation. When the child responds with "do it the other way around," the interviewer would be satisfied that the child understands the checking procedure and go no further on this task. Instead he or she would ask the child the answer to a new combination.

To some extent this type of relationship is more a statement of attachment to the interview process than to the object of mathematics. For this teacher, the point of the interview (as of work in school generally) is "task oriented" and a serious business. The teacher is professional about her or his job, serving the student and the system, under the presumption that schooling and learning mathematics is the most serious and impor-tant event in students' social lives. This teacher might emphasize accumulation (or consumption) of skills and knowledge. That is, the more one knows and accumulates skills and knowledge, the "better" one becomes; continuous accumulation becomes the end in itself. This teacher might also compartmentalize time, behavior, or tasks, in the sense that she or he manages the content and ideas of the interview, the tasks, purposes, and organization of ensuing discussion.

Despite the "seriousness" of this approach, there is little evidence in the experience of an awareness of the worth of the tasks imposed. If the seriousness is combined with seeming irrelevance or triviality, then the nature of school mathematics is debased. In James MacDonald's (1995) terms it clouds the development of values in the productive activity of the people involved since all work must be taken seriously whether justified or not. Participants in an interview characterized by this image become alienated from their work (the content of the interview) because the pleasure of worthwhile activity is reduced to satisfaction in the external rewards offered as a substitute for justifiable standards. Inherent is the technically rational planning and organization of work tasks and pupil activity which in the interests of others destroys the spontaneity, creativity, playfulness, and essential risk-taking potentials of everyday living experiences (MacDonald 1995: 122).

A Teacher's Attachment to Mathematical Rules

Other teachers who view mathematics as a set of rules and prescribed procedures also see the logic of these components as well as the contexts in which they can be appropriately applied. For these teachers, mathematics is engaging, but like the case of the student Nancy, the attachment to mathematics is to its authority and predictability rather than to its subtler, intellectually challenging, and even aesthetic characteristics. These procedurally engaged teachers tend to present interesting yet relatively routine problems as the context for the flexible interview. They might ask a student to solve a word problem, pursue a line of questioning intended to reveal the approach(es) used by the student to come to a solution, and even challenge the student's answer in an effort to reveal deeper layers of procedural knowledge. Still, though, this type of teacher-interviewer predominantly notes how students follow or interpret rules and when they apply procedures. The exploration and attachment in this case is confined to mathematical rules and their finite variations rather than to nuances and highly personalized conceptions of mathematical relationships that emerge in more playful and spontaneous interactions with mathematical objects.

These interviewers do not emphasize the work of the interview but instead emphasize power or language. In an interview attending to power, the interviewer disciplines or controls the interview so that it runs smoothly, efficiently, and accountably. A hierarchic relationship is the dominant feature as the interviewer guarantees that his or her goals are met. The student in such an interview cannot be expected to be responsible for the success of the interview in any way since they are constructed as "immature." As MacDonald writes, "The consequences of behavior are sheltered from reality because students learn they are immature, and the imposition of activity is legitimated because adults are mature" (1995: 123). The interviewer teaches a need for rewards that are socially satisfying, thereby constructing a need for the teacher rather than personal pleasure. Students please the teacher for future value, losing their own sense of value in activity and substituting a social satisfaction that makes them dependent on abstract rewards for their sense of worth. This is very much the situation when the interviewer, as described above, focuses on obtaining procedurally correct responses, and the student is left waiting for direction from the interviewer as the power broker. Yet, if approval or disapproval is provided, the student at least knows where he or she stands and does not have to decide on the value or meaning of the response.

Mathematics as a Teacher's Flexible Object of Self

Teacher-interviewers who are searching for a playful manipulation and the creation of mathematical objects are themselves likely to spontaneously engage in exploratory activities on their own. These teachers regard mathematics as part of the soul and view the world quite naturally in mathematical terms. When this kind of interviewer interacts with a student in some mathematical context, if there is a positive outcome there is a kind of synergy that takes place in which both the interviewer and the student travel to a new dimension of knowing beyond that which either might engage alone. In this circumstance, the interviewer becomes part of the student's process and connection to the mathematics. Even when the student is not a spontaneously mathematically connected individual, this kind of interviewer may be able to break down barriers of procedural ritual or illogical random guessing so that in the process of the interview the student is able to see new mathematical meaning in the tasks at hand. Indeed, such a change in relations with mathematics might, over time, be described as more positive; in particular, the student might claim that mathematics is "fun."

For example, Rasheen was asked to compare playgrounds and decide which were "better" than others. Because this youth liked slides and seesaws, she brought the teacher to several playgrounds and showed her which slides she liked best and why. Together, they decided that speed and length of sliding time were important for judging slides, so they used a stopwatch to find ways to measure these, using the student s body length to measure the distance covered in a slide. After comparing data for several slides, they then discussed possible ways to compare the quality of seesaws at playgrounds in the future. To a large extent we see the positive scaffolding effect described by Vygotsky (1986) wherein the more experienced learner is able to act as a catalyst and support for the less experienced one. An interview of this type involves some aspect of discovery of a new way to represent and solve a mathematical problem using more intuitive notions than school-taught procedures, or at least combining intuitive notions with school-taught procedures.

This positive outcome, however, is not always the case when the interviewer approaches the interview with such a close connection to mathematics as an object. Too often, the youth is so removed from the experience of such a relationship to mathematics that the outcome of the interview is either a series of questions to which the youth is unable to make any reasonable response; or more likely, the interviewer is forced to detach him- or herself from the mathematics and pursue a line of questioning that taps more closely into the routine outlook of the youth. In this situation, the engaged teacher in the process of reintegration of self with and through mathematical object relations may in effect be no different in a therapeutic sense for the student than a mathematical dictatorial egoist or passive egoist.

Neither Tough-Love nor Dream of Love

In general then, the flexible interview is a function of the interaction between student and teacher-interviewer in which the individual levels of attachment to mathematics are the main issue. At times the teacher and student may seem well matched and the interview appears to be comfortable for both, whether or not the participants learn anything about mathematics or each other. When the types "clash," the interview is mostly a study in negotiation of purpose for the interview, subsequently moving the discussion to be ostensibly about mathematics but more clearly about interpersonal dynamics. Underlying this interaction are both the student's and the teacher's performance expectations for the mathematics. It is for this reason that "listening" teachers commonly foster an atmosphere about rules and procedures despite their own object-relations with mathematics as an object of the self. Similarly, the student's conception of the role of authority and how to behave in the company of teachers colors the quality of what he or she is willing to put forth regardless of his or her own object-relations with mathematics. What we can say, though, in this context, is that the participants' relationship and expectations for the interview as an object is as important as their relationships with the mathematics. A focus on the teachers, then, raises the experience of the interview as paramount and crucial in the on-going development of object relations with both the experience of learning mathematics in school and mathematics itself.

As we have already noted, flexible interviewing in the classroom context indeed requires substantial reframing of the teaching/learning encounter. A teacher must be able to talk with an individual student while the rest of the class is supposedly "on task" in some "unsupervised" way. Teachers find a variety of options, including: interviewing in the context of a class run on extensive small group work that requires consistent but not persistent monitoring by the teacher; exhaustive use of aides and adult volunteers that work with small groups of students while the teacher interviews a student; interviewing while the rest of the class performs "seatwork" such as "worksheets"; inviting small groups of students to lunch-time or after-school interviews; and interviewing within the context of whole-class discussion, during which the teacher facilitates an interview with the entire class involved in the questioning and elaboration of each others' thoughts. Regardless of the interview context, most of the emerging and experienced teachers I and Rochelle work with find that experiences with individual students are helpful in reflecting on interviewing in general; these teachers typically describe their relationship with mathematics as drastically changed.

In the context of assessment it is evident that the teacher cannot soundly aim to identify and satisfy every need of the student (despite a desire for teachers to view their jobs in heavily medical terms such as diagnosis and prescription). To say that a student "needs" any particular skill or fact is as ludicrous as saying that a child "needs" to have a particular concept readily available for interpreting "reality" mathematically. Equally inappropriate would be to give over the interview as an environment devoted to the child and its need for "holding." The purpose of the flexible interview may be seen as a component of a social space known in practice as a classroom. In this space, it is important for the teacher to continue to reflect on his or her own relationships with the mathematics, perhaps even more so than to "assess" the needs of a particular child. To eroticize the individual child or the mathematics of a conversation as an object of self would be to come too close to a dream-of-love fantasy commonly represented in popular culture media images of teaching and learning. The vision is consistent with fantasies of devotion and self-sacrifice, and, as Judith Robertson (1997) writes, with the rhetoric of scientific efficiency maximized through individualized child-centered learning. Robertson, a professor of education and holder of the 2003 award for excellence in teaching at the University of Calgary, notes, "Child-centered discourses position the . . . teacher as benevolent overseer of a landscape of elemental and natural goodness." She continues,

> What is forgotten in the fiction is that centering children in this way does not in any natural way best develop their potential as learners. Children not only consistently circumvent teachers' intentions in child-centered activities, but the very notion that children can "discover" the truth through "individualized activities" is a denial on many levels. It denies that any discursive system (including those that posture as "individualized") produces its own particular truths, no matter how these truths are veiled or fictionalized. It denies that the "rational" learner who is ostensibly a product of child-centered practices is not in mastery of either knowledge or the self. Finally, it withholds from learners the legitimate right to expect that teachers will intervene in learning in ways that may feel discomfiting, that may not always be easily understood, that may be insistently directive, and that are not always experienced as ego affirming.
>
> (Robertson 1997: 91)

"Dreams of love," writes Robertson, "do not resolve the difficulties of teaching, nor ultimately, do they increase its pleasures" (91). The educational encounter is not the satisfaction of personal drives deriving from particular selves, but an ongoing establishment of relations with objects, within a particular social practice; an educational encounter is an evocation of selves. This is where critics of recent reform efforts in mathematics education may be wrong: by taking a tough-love stance they misinterpret reform-based mathematics as abandoning skill drill whereas reform-based curriculum merely intends to place the meaning of skills and drills in a new context. On the other hand, the "other side" of the math wars debate, advocates of reform in mathematics education, might also be faulted for not paying enough attention to the mathematics itself; they tend to cling to a traditional view of the subject matter while suggesting changes in pedagogy. Where the dream-of-love, child-centered assessment interpretation of reform-based mathematics goes wrong is in its fear of challenging the mathematics itself, and thus failing to enable more substantive changes. It avoids confrontation of the self and its relationship with the mathematics in its desire not to disrupt the dream. In our "ideal classroom," an experientially and reflectively aware aspect of the person is called into existence as the object of his/her own unconscious ego processes; that is, students and teachers become what psychoanalysts call "subjects." Paramount are the perpetually changing object relations that make up a self, and how a teacher's relationship with this theory would change as her or his own object relations alter and coalesce in new ways through work with children.

I am suggesting that we shift our emphasis from skill levels and cover topics toward relationships with mathematical objects and the subsequent uses of objects. As Block writes,

> Our lives may be said to derive from the uses of objects, and this is ultimately a creative process. How objects are used derives from the effects of a facilitating

environment that enables the child to actually find what the child creates, to create and to link up that creation with the Real.

(27)

The facilitating environment he was writing about is Winnicott's "holding environment." A psychoanalytic interpretation of curriculum and the flexible interview dwells on the seriousness of what is at stake: The teacher's relationship with mathematics is crucial to the experience the child has, and therefore affects the child's use of objects. The child's mathematical development is the history of many internal relations, expressed through uses of objects. Assessment, as in for example flexible interviewing, is not necessarily about the child, but is the *teacher's* use of objects, an expression of the *teacher's* self as object relations.

In the end, I urge you to focus on the importance of "listening," as in the work of Julian Weisglass (1990, 1994) and Brent Davis (1996, 1997). For these authors, professors of mathematics education—Weisglass at the University of California at Santa Barbara, and Davis at the University of British Columbia—listening is a form of "embodied action" as opposed to a technique of hearing. Weisglass presents a taxonomy of listening forms that he designates as partially pedagogic; his alternative, dubbed "constructivist," encourages the talker to reflect on the meaning of events and ideas, to express and work through feelings to construct new meanings, and to make decisions. Davis similarly constructs three comparative modes of listening differentiated by their features of attending to the one listened to; beyond evaluative and interpretive listening one finds "hermeneutic listening," which requires a teacher to reach out rather than take in. In hermeneutic listening, listening becomes the development of compassion, increasing the capacity of the listener to be aware of and responsive to the one "listened to"; participants are involved in a project of interrogating taken-for-granted assumptions and prejudices that frame perceptions and actions. For flexible interviewing, such a conception of listening emphasizes the importance of one's own structural dynamic in the evolution of outcomes in interaction with another person, as opposed to functional responses to the other person's actions (as in transmission models of communication and teaching). Interaction is not "instruction"—its effects are not determined by the interaction—rather, changes result from the interaction, determined by the structure of the disturbed system (Davis 1996). Constructivist or hermeneutic listening promotes participation in the unfolding of possibilities through collective action. The talk-show host exercise can help you to think more about listening and the effects it can have on how you yourself interpret what others are saying and doing.

Becoming "Good Enough"

Mathematics education asks of us that we be "good enough" to provide an environment that engages the students' capacities for illusion and disillusion, their capacities to express and understand, and their capacities to tolerate times of being misunderstood and not understanding; mathematics education also asks that we be "good enough" to help *ourselves* in tolerating the results of our own frustrations. I want to leave this chapter with a plea that you not take the main point to be that the strange course of teaching and learning unfolds within and between the teacher's relations to the student, although of course these relations are part of the many stories that might be told about what is going on. Please do not position the teacher as the central moment in the student's learning. In that kind of perspective, something else is purloined.

> What becomes lost is an analysis of one of the central fantasies of omnipotence in education: that there can be a direct link between teaching and learning and that both of these dynamics are a rational outcome of the teacher's conscious efforts.
>
> (Britzman 1998: 41)

We would get caught up in the psychoanalysis of teachers, who we would perhaps say could best profit from intense therapy as preparation for teaching, in order to "work through" their own repressions. This only pathologizes teachers, instead of helping us to see that the pedagogic encounter is always the work on the part of both teachers and students in making relations.

It might nevertheless be a good thing for all educators to pursue analysis. Children learn from an adult's affect toward, and dynamics in, knowledge. Learning is crafted from a curious set of relations, including the self's relation to its own otherness, and the self's relation to the other's otherness. "This is forgotten," Britzman describes, "when the adult's desire for a stable truth, in its insistence upon courage and hope, shuts out the reverberations of losing and being lost" (1998: 134). But Britzman ends with a final, lonely discovery: "teaching, it turns out, is also a psychic event for the teacher" (1998: 134). At once we imagine our work as offering something vitally important to our students, who are provided with "just what they need," yet in our desire for this "rescue fantasy," a wish for stable truth about our work and our knowledge of our students, we also preclude the possibility of us together with our students working through the ambivalence of our own conflicts. The "good enough" teacher insists that learning is hard work, and that somehow pleasure must be made from this very reality. The challenge is to find ways for all in the classroom to tolerate the difficulties of learning, and yet to gain insight from these very difficulties as well as their own conflicted experiments in learning. The key is to move those involved from the fear of losing someone's love to the fear of losing one's own self-respect. The paradox is that self-respect only grows out of relations with others. It was Bruno Bettelheim (1979) who came up with this term, "good enough." Instead of offering love in return for learning, the teacher offers a demand that students learn to make their own demands in learning. Instead of wishing that the students desire to be with the teacher, wrote Bettleheim, the teacher hopes that education can "fortify the child's inner world to serve learning" (1979: 76).

Response to Chapter 2:
Ask Yourself to Change

David Scott Allen

The mindset when I went to school is very different from what it is today. The cultural gap and the modern/post-modern dichotomy are more pronounced. It is accepted that they exist and are in conflict, making them a part of the educational landscape. I must admit that in retrospect, I came to teaching because I love learning and I want to pass that on to the next generation. I want to be a role model to people who need role models. As a trained singer, I'm used to being on stage. Little did I know how differently this generation would view education.

I have experienced firsthand the "refine the producers" technique mentioned in the text. It sadly amuses me that we have categories for how "qualified" a teacher is, when there is no requirement to actually implement any ideas that are recommended, if indeed they are worthy of implementation. A few years ago, my district tried to implement a new curriculum for our seventh graders. The worst part of the experience was that, though the teachers attended training, they were simply not willing to attempt new methods. For them, it was uncomfortable and risky to start all over again after so many years of teaching. I look at it from the point of view that slow change is necessary or I will stagnate and my students will suffer. There are so many interesting ideas and materials out there; I just ignore them, I will be missing out and I won't be doing my students any justice in the classroom. Aren't we asking them to change their thinking when we introduce a new concept or technique? How can we expect them to take us seriously if we do not at least try to change and adapt for new students of new generations?

I confess that I can't really relate to my students' antinomies, or awkwardness, with new mathematical material. I had an easy time with math and always liked it and was good at it. I also loved recreational math, which I found (and still find) much more interesting than textbook math. But I *can* recall almost failing eleventh grade American History because I just couldn't produce. And I had a good teacher! He was very professional, more than helpful, and would not give up on me. I finally made it, but I cannot remember the details of mastery. The only way I can truly experience this again is by learning something with my students. Unfortunately, there is very little like this, if anything, which is on my list for them to know. So, I turn to the problems like those in this book for ideas.

For the most part, school math *is* all about mastering knowledge of process, even if this is just one of the metaphors we can use. Generations of students are programmed to solve equations a certain way, and then to solve more difficult ones, and so on and so on. We do ask students to do the hard work now in order to be better off in some vague future. But in the process we ask them to put off actual learning! They are merely duplicating (or choosing not to duplicate) what we have shown them. They don't need to really think. Our "advanced" students eagerly await the next task, and some of our better thinkers get bogged down in disinterest. We truly work in an atmosphere of performance, and in this ironic way, school does mimic the real world.

When I tried some new things in my classroom, letting them take the direction of the class in hand and asking their own questions, my students got hung up on so much "school" minutiae, that it took a few days before they could see that it wasn't about work, it was about learning. Sustaining that style can be difficult, though I have used it to mix things up on occasion. The upper level students just do what they are told and they do it sadly well.

I understand what it means to say that we can't know how a student learns or what makes them learn anything. Yet I, too, have that reserve of techniques stored up that I use in certain situations. How odd a juxtaposition for me to tell you that I can't know how to do something, describe how I do it anyway, and then wonder if I am actually accomplishing the goal with you as you read this! In mathematics, we can identify this self-reference as recursion. I find that the bag of tricks really only works when a student is attentive to playing that apprentice role, when they are experiencing the math. The teachable moments would be more common if we nurtured them to happen. They only get it when they are thinking about it. The problem is that no two students are going to need the same things in front of them to get them to think about a topic, so it is up to me to encourage learning as the goal as opposed to jumping through teacher hoops.

My students can do amazing things, but addition and multiplication tables are not at the top of that list. They understand them, but they're not automatic. I had one student come to me in tears saying that she could not possibly finish a test I was giving. As we discussed it, I discovered that she had skipped problems because they would require her to know more multiplication facts. The work on her scrap paper was telling. She didn't have her tables memorized. I still believe these to be very important.

All the talk in this chapter about the objects of relation may seem too technical, but we are just putting words on what we do all the time with non-mathematical objects in non-mathematical situations. I indeed have a relationship with my car. I think and act a certain way when it is being used or when it needs repair. My daughter has a relationship with her art materials, my sons with their trains, tools, and pacifier. They take them up and use them because they choose to and it is interesting to note the type and intensity of their interactions with these objects. Mathematical objects, though, are used in the only mathematical setting that students know: school. Then they are tucked away until they are needed again for the next class. Students who are good at manipulating mathematical objects (or worse yet, who actually remember how to do so weeks later) are considered nerds.

I need to foster a more open communicative environment in my classroom; one where all ideas are welcomed, considered, and addressed in a kind of internal democracy. I need to create an atmosphere more than I need to force feed facts. If the student wants to learn, the learning is easier. So I could begin with motivation, but that would be superimposed on them by me. I would rather begin with invitation and exploration. But above all, they must practice communicating clearly and respectfully. That is what is expected and sought in society.

I did the flexible interviewing with my students for one of my classes while I had a student teacher. I haven't really been able to implement it as routine in my classes since, but I am still determined to implement something. If I can focus on thinking in the classroom, and then give a few simple notes on the mechanics of, say, equation solving or adding integers (the "paper" curriculum), I can keep the students talking and engaged with the material while I work with different groups. I think it will be harder for me to do this as well as interview individuals, but I'm going to try it. It would be great to have a student teacher all the time, and in that way we are learning about each other's teaching even as we actually teach. Then we will have our own community, if you will. The results I got when I did the interviews were startling.

The main thing I learned about my students is that the best behaving students ("good students") are for the most part the ones who are just going through the motions of school. The ones who are the worst behaviorally ("problem" students) are the ones who actually have an alternative vision of reality. They have insights into their work and math class (as opposed to mathematics itself) that are truly unique and creative. They also produce work that is respected by their peers and in many ways are superior to those who are just "doing school." They simply do not produce work that is what the teacher expects or desires.

One of the reasons that I believe this to be true is that I interviewed all of my seventh grade students about their mathart projects which they created for my classroom. Some of them are in an above average and some of them are in an average class. Most of the (relative) behavior problems are in the average class, yet some of these students created works of mathart which were voted by all the students to be the best. I left it up to the individual student to decide what "best" meant to them, and then I interviewed them as to the reasons for their choices. To my surprise, the problem students showed more purpose and reason for their choices than did the good students. They used a wider vocabulary when justifying their choices, and were more intelligently critical (positive and negative) of all the works presented.

Another justification for my observations is that I could see no pattern in the voting that would lead me to believe it was along clique lines. In other words, students who weren't friends with the problem students voted for them just as much as the problem students' friends voted for them. Many of the problem students readily admitted that their project wasn't the best in the group, and could point to one or more specific reasons why. Many of the good students either picked their own as the best, or one of the problem students' pieces.

As a result of these observations, I adjusted many of the projects' grades upward to reflect the respect of their peers as well as my deepened insight into the meaning of their works. These students forced me to look at their product differently, as if using a new paradigm. This was legitimate and worthy of the reward of adjusting the rubric.

As students get older, the playfulness of life and math go by the wayside as our sponge-like students learn bodies of knowledge which they spit back at the establishment. But they will still find ways to explore and play. Why not in the classroom? What good is all that technique without sense? The things students want to learn quickly diverge from what they must learn, while what they can learn is somewhere in the middle, and is much deeper than either extreme. The teacher holds the actual direction of the class in hand. But change the students, and you can change the whole learning curve.

As for me, math is a flexible object of self, except as it intersects my teaching. I feel the rigidity of my testing structures every day. Here's to the potential to rethink our teaching and restart, renew, and refresh our careers in this way. It all starts with experimenting.

Action Research 2:
Flexible Interview Project

Karen Cipriano

Since I interviewed my entire class, I will speak generally about my students and give specific examples of why I feel the way I do. First of all, I questioned my students on their feelings about the interviews. We discussed whether they liked being interviewed in groups or individually. The class was split right in half. Then I asked the students to choose from these four answers to the question "How does the interviewing make you feel?": 1) Never nervous; 2) Always nervous; 3) Nervous in the beginning but not anymore; 4) Was not nervous in the beginning but it makes me nervous now. The majority of the class said that they were never nervous or they were nervous in the beginning, but now it really does not make them nervous. Only about a fifth of the class is still really nervous about the interviewing.

Naturally the next questions I asked were, "What makes you nervous?" "How could you feel less nervous?" The students came up with honest answers and good suggestions. They were afraid of getting the wrong answer, getting in trouble for not knowing something, and afraid it would affect their grades. I began to cringe when they told me they were afraid of getting the wrong answer because I always tell them that the process is what I want to see. Even when they do their homework I collect all of the work and tell them the work is important. I do not want to see just answers on the paper because the answers do not tell me the whole story. Obviously they have had years of teaching that makes the answer important, or I am not doing a very good job at modeling to them that the process is important. When I asked them about the interviews, I held back any comments because I wanted their feedback and did not want to sway their responses.

They really did not like when I was writing while they were talking. They said it made them nervous because they did not know what I was writing down. I explained that I am just writing what they are saying because I cannot remember. They suggested that they could write what they are thinking (although that actually defeats the whole purpose of having them speak during the interview). Other suggestions were that I write after the interview is finished or watch the video and then write what I need to when they are not there.

I am glad I asked the students about the interviewing because it really did give me insights and suggestions for next year. Also, the article, "Listening for Differences: An Evolving Conception of Mathematics Teaching" (Davis 1997) is awesome. I really wish I had read that article before starting the interview process. It was like reading an article about myself. I know it is not a great habit to put oneself down, but I really am one of the worst listeners. As I read this article, I began to realize that I did not listen to my students; I was listening instead for the answer I wanted. I actually applied this idea in the last interview I did with my students. I was looking for the students to realize that square numbers were the open lockers in the problem we were talking about, and this meant 2 × 2, 3 × 3 . . . Then, when Charmaze said to me, "I noticed the pattern + 3, + 5, + 7, + . . ." I almost said, no!!! I began to listen and let her finish what she was saying. Then I asked her to continue the pattern for a while and see what numbers came up. She found what I was looking for in a different way from how I was thinking they would.

Interviewing is something I definitely want to practice next year. In a graduate course I was taking at the time I was trying this interviewing, others kept asking me how I measured their progress. Progress was *my* word. I kept on using that word, and yet I never explained what I meant. Revealing what I meant is going to make me look like a hypocrite . . . I guess I meant by progress that they could come up with the right answer or the process *I* had in mind. I have come to realize that I have notes on what they said and have organized those notes, but what was I writing? I began to think about what the students said about writing *after* they spoke. I began to realize that is a great idea. I think to *really* listen to what they are saying I should listen intently and then, when the interview is over, I should write the important things down while they are still fresh in my memory. If the students were not giving me the answer I wanted, I would think they did not know what to do. I never *really* listened. This is something that I would like to try differently next year.

I also want to consider better ways of keeping records. I decided to make a spreadsheet, using the computer blocks with each student's name in them. This worked to help me keep track of who I interviewed and who I did not. I also had the students in the same order, so I could reflect quickly to get a single view of a student instead of just how the class was doing. Then I realized it would be very helpful to write down examples of what the students were saying; as I introduced this, I would take those notes and write individual notes for the student. This was rather time consuming. I find interviewing beneficial, so I want to make it useful, but not burdensome. Next year I will try using a folder for each student. I will have two sections. One section will be notes on word problem solving and the other will contain notes on other math concepts. I could make a checklist for myself to keep track of who I interviewed and who I did not. I will listen as the student speaks, and observe what the student is doing if I have a question that involves manipulatives. Then, after the interview, I will take my notes and put them in the folder.

In addition to working toward my goals of being a better listener, collecting anecdotal records in the folders, and employing a check-off list, I would really like to meet with my fifth grade team and share my experience, results, and ideas for improvement with them. We worked very separately this year and I do not think that it hurts just teachers to do that; it hurts the students. I worked on a rubric for a research paper with a former teammate. I have revised it every year for the last three years; the students now use the packet for their research as I continue to make it better. I also work with the librarian. She took our packet to a meeting of all elementary schools and guess what? Our packet is now shared with all of the elementary schools in the district—they did not yet have anything in place like that. I felt it was quite a compliment, but the important thing is that a TEAM of three teachers created a packet and rubric.

That teacher I worked with on the packet is on maternity leave and I do not know if she is coming back. Unfortunately, the other teachers I have on my team are not as excited about being team players. I have always gained a tremendous amount of ideas through collaboration, and I really want to improve on being the "cheerleader" for working together. The hardest part for me is to listen to others' ideas. I have many of my own, but listening does come into play. I want to take my journals and my experiences with interviewing to my team in the hope that they could add their input too and make the process even better.

The best mathematical experience that my students could have next year is the opportunity to "explore" mathematics and "speak" mathematically to each other. I like to let my students work in groups, but sometimes they do not because they would rather work at their own pace. I feel my students have learned not to fear mathematics and to communicate better. By not fearing mathematics, I mean that they are willing to take on challenging problems and not give up. Take the locker question for instance. I know that if I gave my students that problem in the beginning of the year, they would not have spent the time they did on it; they would have simply given up. I had students who had free time and took out the problem on their own because they were trying to figure it out. Not all of my students are at this point yet, but they are moving in this direction; they need the practice and confidence to motivate themselves to be independent learners.

I have noticed the communication improving. By now, it is very rare for my students to just sit when I ask them questions in an interview. If I question them, they are able to respond. Some even try to use mathematical terms in their responses, and I encourage that. Also, I think they have learned to talk to each other. Just two weeks ago when they were working in groups on the locker problem before I interviewed them, Charmaze and Isaiah were disagreeing on how to solve the problem. They began the highly mature, "Yeah, huh." "Uh uh." I went over to them and they said they disagreed. I simply told them to just take turns explaining why they think the problem should be solved in the way each thinks and then determine which method would be best. They said, "Oh." Isaiah started to speak and the next time I looked over they were able to come to some type of agreement. My students were not as vocal as that in the beginning of the year. They have begun to trust their own mathematical thinking and know it is okay to share with others without making fun.

If my students have the opportunity to be continually challenged and to speak to each other, they will be getting ready for the future. Most jobs from CEO to McDonald's you have to work with other people, and it is best to learn communication skills early. The students need to speak, listen, and question next year.

I am able to make progress reports on each interview I did and that is what I have recorded, but my record keeping did not allow me the opportunity to make any long-term evaluations. Like I mentioned earlier, I do have a plan in place to try and improve that next year. Overall, the interview process has opened my eyes to another form of evaluating my students. The practice this year allows me to have a plan set in place next year. Then every year I can continue to make improvements.

Karen Cipriano

I am a fifth grade teacher at Stewart Middle School in Norristown, Pennsylvania. I have been teaching at-risk students for the past twelve years in a self-contained classroom. It is such a pleasure when my enthusiasm in math carries over to my students and we see an increase in their standardized test scores. Just watching my students work cooperatively to solve challenging word problems by questioning, conferring, and evaluating their results is exciting. I am very family oriented and I enjoy outdoor activities such as walking, biking, and a variety of sports.

MathWorlds 2:
Multiple Answers

1. **Grass Seed.** "Covers two square feet," is written on the packaging for a bag of grass seed.
   a. Your plot of lawn is rectangular, measuring about three feet by five feet. How many bags of seed should you buy? Come up with at least three different answers to the question based on your calculations
   b. Find at least three different ways to work on this problem as a mathematician.
   c. Change the original question so that it is more realistic, by changing the dimensions of the lawn, the amount of area a bag of seed covers, the shape of the lawn, to anything else you imagine might be relevant.
   d. Make a chart that includes the variables in parts (a.), (b.), and (c.), e.g., coverage of a bag, dimensions of plot, area of plot, numerical calculation, and number of bags purchased, along with others that you came up with. Fill in as much of the chart as you need to answer the following questions: Which variables make the most difference in the decision of how many bags to buy? What kinds of differences do they make?
   e. Change the original question so that it requires analogous mathematical work with coverage of unit amounts in predetermined quantities, but so that it is about a situation in the world that you or someone else would actually be interested in.
   f. What new questions come to your mind after working on these questions? Choose one of these and begin to explore it.

2. **Division & Decimals**
   a. Use long division to divide 7 by 8 in order to find the decimal representation of the number sometimes written as $\frac{7}{8}$.
   b. How do you know when you've gone far enough with the division in (a.)?
   c. Divide 2 by 7 to do the same thing for $\frac{2}{7}$. How do you decide when you've gone far enough with the division?
   d. Do all fractions have the same kinds of results in this situation? If eighths and sevenths are special cases, characterize the types of decimal expansion representations you would get for all "rational numbers," that is, any number that can be written $\frac{a}{b}$ for any integers a and b (b not zero, of course).
   e. Does this imply any decimal number could be written in the form $\frac{a}{b}$? Consider 3.7962 and 0.123123123123 . . .
   f. Have you considered the situation where decimals either don't stop or don't repeat?
   g. What situations in everyday life are analogous to the relationship between fractions and decimals?
   h. What new questions come to your mind after working on these questions? Choose one of these and begin to explore it.

3. **Iterative Functions.** Create an "iterative" function by repeatedly replacing x by $f(x)$ in the following function definition: $f(x) = 3x + 7$, starting number for $x = 2$.

a. Describe the behavior of the results of this iterative process.
b. Put the following functions into categories that you think would be useful in describing how their iterative processes behave:

$$g(x) = \frac{x^2 + 1}{x^2 + 3} \qquad h(x) = \frac{(x-2)(x+5)}{4500 - x^2} \qquad J(x) = \frac{x-7}{x+5}$$

$$K(X) = SINX) \qquad M(x) = 3x^3 + 2x - 2 \qquad n(x) = \frac{x^2 - x}{x^3 - x^2}$$

$$p(x) = \frac{\text{Volume of a sphere}}{\text{Surface area of this sphere}} \qquad q(x) = \begin{cases} 100 & \text{If } x > 0 \\ -50 & \text{If } x < 0 \\ 0 & \text{If } x = 0 \end{cases}$$

c. Describe two more functions that would require new categories to be added to those you came up with in (b.).
d. Have you thought about varying the starting value of x? If not, go back to (a.), (b.), and (c.) and see if your responses would change if the starting value could be something other than 2.
e. What else could we be interested in regarding these functions? Make a list of five other things, and choose which one of the five you think is the most mathematically significant thing on your list. Explain your reasoning in terms of potential connections to further exploration of functions.

4. **Fermi Problems.** Named after the Italian physicist Enrico Fermi, who liked to make up questions that seem at first unsolvable, but for which someone could come up with a reasonable estimate through creative thinking, these questions can promote inventive problem posing through imaginative play with numbers.
   a. Research Fermi problems with an Internet search.
   b. Compose three different Fermi problems of your own.
   c. Work on one of the Fermi problems you composed.

5. **Chairs and Tables.** If two points determine a plane, why do chairs and tables usually have four legs?

6. **Angle Bisectors**
   a. What is important about angle bisectors in triangles?
   b. Describe at least four different things you can say about various triangles once you know some information about one or more of the angle bisectors.
   c. What new questions come to your mind after working on these questions? Choose one of these and begin to explore it.

three
You Are a Mathematician

How does a mathematician work? The answer to this question depends on who we consider a mathematician. In this chapter, we approach the question from the standpoint that everyone is a mathematician. In particular, *you* are a mathematician. This may seem strange if you have not thought this way before. But this approach has powerful implications for teaching. If we say that everyone is a mathematician, then it is the teacher's job to figure out *how* each student is a mathematician. As opposed to showing the students how to be a mathematician, we have to assume that they already are mathematicians. When and where do they exhibit the behaviors that we would associate with a mathematician? How is what they are doing at any given moment an example of a mathematical way of being in the world? How do our students' actions change our assumptions about what it means to do mathematics? The title of this chapter is taken from a book by David Wells, whose work I always find particularly insightful in communicating the nature of mathematics. In this book, David Wells, a British author of numerous collections of mathematical puzzles and popular journalistic articles in mathematics, suggests that there are certain especially mathematical ways of working that we all do sometimes; mathematicians, he might say, do these things a *lot* of the time. The idea is to go beyond the ordinary way that we might describe an object that we see or think about, to play and explore in order to find "hidden worlds" of connections that are not always obvious.

Everyone is familiar with the simplest mathematical objects, such as straight lines and squares and circles, and the counting numbers. To be a mathematician all you have to do is learn to look at these objects with some insight and imagination, maybe do a few experiments, too, and be able to draw reasonable conclusions . . . The results of these activities – which are also familiar to you from everyday life – is that you soon see the square as more than something with four equal sides and four right angles; a circle as much more than, well, just a plain circle; and the number 8 as much more than merely the next number after 7.

(Wells 1995: 1–2)

The main idea here is that most anything can be thought of in a mathematical way, but that we must go beyond, "Oh, I see that," to probe enough—with imagination, experiments, and just trying out new ideas. Also, we need to draw conclusions on our own from our investigations, rather than sit around and wait for someone else to tell us what to look at, or what to say about what we see. This mathematical attitude has been studied by mathematicians and educators for many years, so we have some models that you can try out for yourself in thinking about when and how you are a mathematician. Everyone does these things. Everyone is a mathematician. Once you start looking at what you do, you may find that you even *feel* like a mathematician. You will be curious about the most mundane phenomena, and you will cherish a good surprise along the way.

Take out your mathematician's notebook to use as you go through this chapter. On your first blank page, start by drawing a triangle—any triangle. Leave room for some other stuff. Somewhere else nearby, draw a *different* triangle. Write down next to this triangle how or why it is different from the first one. Now comes the challenge: draw yet another triangle, different from the first two; nearby, describe how or why it is different. Keep going, until you have four to six different triangles, with brief descriptions of how or why they are different from the others that came first. Now compare your "triangle data" with several friends or classmates who have done this at the same time as you. How have each of you thought about triangles as you did this simple task? *Do this before you read any further.*

People usually differ in some ways when they are given the triangle task, because the directions are so vague, and because it is designed so that it is impossible to do it incorrectly. There seem to be several common languages that people employ to informally describe shapes that they draw:

1. *Size*. One triangle may be larger or smaller than another.
2. *Orientation*. Perhaps one triangle is pointing up, while another is pointing down? Slanted to the right as compared with exactly to the left?
3. *Names*. Many of us use shape names to refer to "right triangles," "isosceles triangles," or "scalene." Here we classify shapes as belonging to families or categories.
4. *Analysis*. We may note how particular parts of shapes compare, such as sides being longer or shorter in one triangle or another, changing an angle so that it is smaller or larger. Here we are understanding shapes as made out of parts that are related to each other in certain ways.
5. *Synthesis*. At other times we may note how our shape is a piece of a larger whole, for example one of our triangles might be half of a square, while another is one-third of a trapezoid.

I believe I have covered all the ways people can talk about triangles. At least, these are the various kinds of discussions that have taken place with people with whom I myself have done this. If you have another idea, please contact me and let me know. My point with this exercise can be made with just these at any rate: When we think about or communicate with each other about shapes, we employ one of several possible languages. In some ways, the words we use influence what we can say or think about. In other ways, no matter what language we use, we can say *something* about shapes, and this something is the start of possible mathematical explorations.

I bring up this idea about language because it is different from another popular way of understanding mathematical thinking. Some people have suggested that there are levels of sophistication in geometry, where the higher levels are in some sense more mathematical than those that are lower. For me, these are not levels but different languages. And each language is appropriate for specific times in our lives. Choosing the appropriate language would be for me a sign of sophistication, rather than a propensity to apply one language over the others regardless of the context. In our triangle task, we had no context, so each of us was free to arbitrarily discover how we think about triangles when we are not given a purpose for doing so—a rather silly situation.

One set of levels for geometry was introduced by Dina and Pierre Maria van Hiele. Dina, a student of the psychologist Piaget, approached developmental psychology from the perspective of a teacher. Imagining that learners are located at developmental levels, she decided her job is to move the learner to the next level, instead of to wait for children to arrive at the next level on their own, which some developmentalists think is necessary. She created a taxonomy for geometry as a test area for her instructional theory that a teacher could help students progress from one level to the next. Later, after her premature death, her husband continued her work. Together the van Hieles came up with the following:

Table 3.1: Van Hiele Levels

Level	Description
0	*Shape Names*. The learner refers to shapes as circles, squares, triangles, etc. The learner can differentiate shapes.
1	*Parts of Shapes*. The learner can discuss how a square is made up of four equal sides, or four equal angles; can say that a certain hexagon must have an angle greater than 90°.
2	*Compare across shapes*. The learner can compare shapes, such as a rectangle could be a square if all four sides were equal, or that parallelograms and trapezoids both have at least one pair of parallel sides.
3	*Make and argument*. The learner can explain a deductive claim, such as why knowing that an isosceles trapezoid has two equal sides would have to guarantee that it also has two equal angles.
4	*Alternative geometries*. The learner can work in a variety of different geometric systems; for example, not only in Euclidean geometry, but in geometry on a sphere, where triangles do not always have 180° total in their angles.

Languages and Levels

Go back to your page of triangles.

a) Try to use the five different languages to re-describe how your triangles differ from each other. Which of the languages are easier for you than others for your particular shapes? Did your first set of ideas limit the kinds of differences you might have thought of? How does thinking about the other languages suggest kinds of triangles that you might add to your personal data? Draw as many new kinds of triangles as you can, using these other languages.

b) Try to organize your triangles into categories that you believe capture the range of different triangles that can be drawn. Use any names for these categories that you want, but try to use as few as possible yet as many as you need to describe what is important.

c) Repeat your processes for (a) and (b), but using van Hiele levels instead of the languages.

d) How do the van Hiele levels change the way you are thinking in comparison with how you thought when you used the idea of languages instead of levels?

The van Hieles' goal was to be able to provide specific tasks that help a learner at one level to move to the next. My own experience, however, is that learners use whichever language I use in my interactions with them. Once, while working with 3- to 5-year-olds in a preschool, I casually discussed the sides and corners of the shapes that the children were experimenting with. Naturally, the children also began to use such language as their own. Because we were organizing pattern blocks and other materials using this language, I was amused when an observer was astounded that these young children seemed to be operating at van Hiele levels of 1 and 2, whereas they were expected to reach level zero by the time they entered kindergarten. The lesson I took from this experience is that you talk about what you look at. And you look at what you talk about.

Surely it would not be more sophisticated, but strange, to make deductive claims about shapes, or to describe them in alternative geometries, if you are in a tile store shopping for a new bathroom floor. It would perhaps make more sense to ask about unusually shaped tiles either by name, or by particular properties of the shapes. For example, I might ask whether or not I could purchase an unusually shaped tile, such as triangles or hexagons instead of squares, or whether I might design a pattern using more than one shape of tile. Far from being less sophisticated mathematically, I would be employing the language that makes the most sense in this context.

will this work?

Box 3.2

Tile Designs

a) Experiment enough with an interesting tile design until you can ask questions that you do not readily know how to answer. Write down these questions to compare with others. Consider how difficult it might be to use this pattern in corner areas of a bathroom floor, and along the edges of the room.

b) What shapes are impossible to use alone, by themselves, because they leave spaces that can't be filled? What can you say about such shapes?

I am sitting at a table and I notice a square shadow. What is making this shadow? Does the object making this shadow have to be square itself? I decide to hold up a few objects in the light to see what kinds of shadows they make. Can I tilt any of them to make a square shadow? My notebook? A can of soda? A pen? What shapes can I make out of paper that could be used to make a square shadow?

Box 3.3

Square Shadows

a) After you have explored the square shadow situation, think about the approach that you used. Did you cut out shapes from paper and hold them between a light source and a table or wall, experimenting with physical objects? Did you collect data to help you to reach a conclusion? Did you pursue this until you had a good idea of what properties a shape must have if it can produce a square shadow in the right circumstances? Or did you lose interest as soon as you were not sure what to do?

b) Did you work with diagrams on paper instead of, or in addition to, physical experimentation?

c) Write down the questions you have *at this point* and think about how you might go about trying to figure out answers to each question on your own.

Dear Reader,

Our class is setting up a fun, interactive math museum. We're going to invite other classes in the school to visit our museum. Each group of us is making an exhibit. My group is going to have large cardboard shapes for visitors to try to make square shadows with. Can you advise us on what would make a good collection of shapes to have available? Also, we want visitors to be really interested in our exhibit. What questions should we ask them to think about so that they will spend enough time playing at our exhibit to learn something cool about shapes? Thanks for whatever help you can offer,

Anya, Kendra, Lincoln, and Isaiah

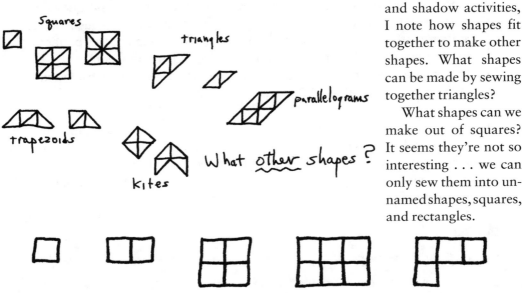

Working with tiles and shadow activities, I note how shapes fit together to make other shapes. What shapes can be made by sewing together triangles?

What shapes can we make out of squares? It seems they're not so interesting . . . we can only sew them into un-named shapes, squares, and rectangles.

Yet, I am distracted from my initial pursuit of shapes that can be made from simple shapes, and I am now thinking that these shapes without names are indeed interesting themselves: How many might there be? As a mathematician I try special cases. For example, I think of just one square. What can I do with it? Not much. It is just itself. Two squares give me two possibilities, , and , but they really look like the same shape to me, if I do not dwell on orientation—I can move one (glide or flip it) onto the other, so I think they are the "same." Three squares is a more interesting case: . These *are* really different from each other!

How about four squares?

Is this all of them? No, I can think of one more. What if I use *five* squares? Six? One hundred? Will it be easy to find all of the possible arrangements?

Having done this for five squares, are you ready to do it for one hundred? Can we get an idea of how many there would be to find? It really seems like a daunting task. Let's create a chart to collect what we know; do we see a pattern?

Box 3.6

Number of Squares	Number of shapes made from multiple copies of the square
1	1
2	1
3	2
4	5
5	?
...	
100	?

Do I need to figure out six to predict one hundred?

Suddenly, I can't stop thinking about squares! At first they were so boring, but now they raise so many questions! What if I arrange them into "staircases" . . . How many do I need for each staircase?

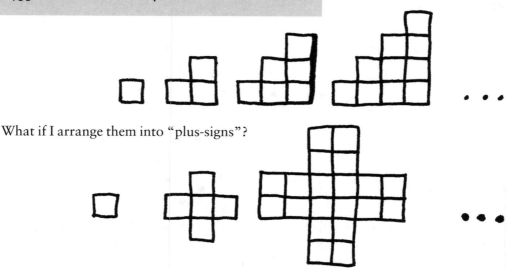

What if I arrange them into "plus-signs"?

But the plus-sign pattern is different from the pentagram question and the staircase investigation—because I am just replacing each square with a four-square copy of a square each time, I now realize I am just multiplying by four to make each new plus-sign. The other patterns were not as obvious, so I find them more interesting.

The staircases remind me of triangles again. It is time consuming and uses a lot of space in my notebook to draw the squares, so I am going to use dots instead:

The dots lead me to arrangements of any objects, to where I no longer think about a floor being tiled, and no longer worry about the spaces between arrangements of these objects. I have these "triangular" arrangements that come in 1, 3, 5, 10, 15, . . . And I have square arrangements that come in 1, 4, 9, 16, . . .

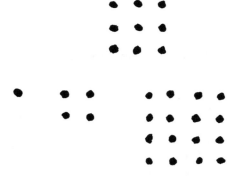

(Note that square numbers are, indeed, the "square" of a number—the number of rows and columns in my arrangements, $1 = 1^2$, $4 = 2^2$, $9 = 3^2$, $16 = 4^2$, . . .) Can I make rectangular numbers? . . . Sure:

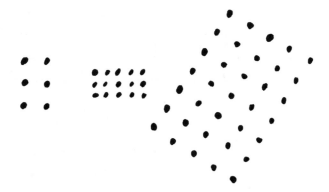

I suppose every square number is a special kind of rectangular number that just happens to have the same length and width, that is, the same number of columns and rows. Rectangles feel a lot more plentiful than squares, though. Is there any number that *can't* be made as a rectangle of dots? Here I must decide, for myself, what I mean by "rectangle." Are 5 or 7 rectangular numbers? I can't find any way to make 5 or 7 other than as a single row or column. These single-row numbers seem in some way to be *special*, or different from the others. I want to say that I should have a special name for them.

And now I realize these numbers do have a special name that I once learned in school. They are called "prime numbers," which means that they can only be obtained by multiplying themselves by one; they have no factors, no numbers other than themselves and one that could be multiplied together to make them, which, of course, is very much the same thing as saying that they cannot be arranged into rows and columns other than as one long row or column.

Another kind of special number can be arranged in rows and columns in more than one way. For example, 12 can be a three by four rectangle or a two by six rectangle (as well as the single row or column of twelve); 24 can be a three by four rectangle, a three by eight, or a four by six rectangle. I do not think there is a special name for such numbers that I learned in school, but my dots investigation leads me to want to make up my own name for them. How about "factor-rich"? I also want to try other shapes! How about pentagonal numbers?

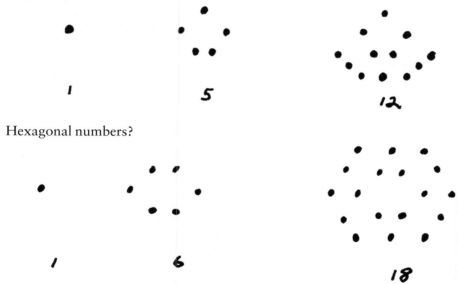

Hexagonal numbers?

And, I can organize a lot of what I have found out in a chart:

Box 3.7			
Triangles	Squares	Pentagons	Hexagons
1	1	1	1
3	4	5	7
6	9	12	
10	16		

But this chart makes me want ask *new* questions. Are there relationships between different kinds of numbers? Are triangular numbers related to square numbers? . . . Yes, since each triangle is close to half of a square, but not quite. I play around a bit and figure out that the triangles next to each other in my list fit together to make squares: On my chart, this means that I can add the previous triangular number to my current number to obtain the corresponding square number! I feel *certain*, I have a strong *hunch*, that I can find even more of these relationships in my chart. And so begins another investigation.

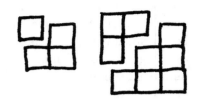

So far, this chapter has a stream-of-consciousness quality, as I attempt to share with you my own path into mathematics from seemingly nothing. My main idea is that anything can be "mathematized" if we think in a mathematical way. So, for example, we did the same thing in the prologue with a game of cat's cradle. (I'm sorry I led us away for the moment from shadows because I have so much more I want to think about in three dimensions and how they might be reduced to two, and even so much more with buildings and shapes off the paper with triangles and squares and rectangles and other shapes, and . . . and . . .) You, too, are a mathematician, so I can dream of what you may share with *me* as well that I have yet to think about. But what we can do now, at this point, is look back over the ways of working that we have entered into with these initial investigations, and develop a language *about* the kinds of processes that we go through when we act as mathematicians.

Box 3.8

Shaped Numbers

Find at least two other relationships between shaped numbers in our chart. Identify what seems to be the key idea to important techniques that you use to arrive at these invented relationships.

Hidden Worlds

In a way, what we are doing is uncovering what David Wells calls "hidden worlds" of mathematics. Sometimes this means we simply need to look at something we have not looked at before, like the way triangles fit together to make parallelograms and trapezoids. At other times, we add things to what we first notice, such as lines inside squares that make triangles inside; we name little pieces of what we see (the sides and angles of a hexagon, for example). Then the very things we name or recognize as an object to look at suggest more general questions: How do these angles fit together with other angles in the shapes that are touching each other? If we think about a similar image, where are the new corresponding parts to be found? We manufacture rules for combining the things we name, and then we cannot help but ask questions about the effects of our rules: If we make trapezoids out of one square and one triangle, where must they meet? What sort of trapezoid do we produce?

"Needless to say," writes David Wells, "the search for these connections is another way of finding out yet more about the object" (15). He suggests the simplest way to look for connections is to do a physical experiment and to draw pictures whenever you can.

Add things to your picture, name parts of the picture, and write down further the relationships that you are beginning to recognize. How does a mathematician *feel* upon discovering that these already-special objects are connected among themselves? She or he naturally accepts these relationships as a sign of progress. "Mathematicians are continually discovering odd and idiosyncratic facts about mathematical objects, but they are also continually finding connections between them, and they get a kick from the weird facts and the connections between them" (15). But doesn't there have to be more to it than merely collecting novel objects and the surprising relationships among them? If they were all connected together in obvious and trivial ways, nobody would ever care. So you must explain to yourself, and eventually to others, why these connections are interesting, worth knowing, more than simple trivia. Are they so subtle and difficult to find that they are fascinating in themselves? Is there an unusually striking pattern of connection? Does this connection help you to understand or demonstrate another, already interesting relationship that you or somebody else has been working on? Mathematicians produce a purpose for their work as they work.

One key idea is to examine more than one type of the object you are looking at. For example, if we look at a bunch of random triangles, it is not impossible for us to reveal to ourselves some simple relationships among them. However, if we take some triangles that already have a connection, we may be more likely to find some worthwhile connections. The other day I was thinking about three very different triangles, each of which has a *right angle*. "What might I learn about right triangles by looking at such different versions of them?," I asked myself. In this way, I created a purpose to my work, because I examined a range of types of the same thing, which naturally led me to be able to make important observations about right triangles in general.

Using what they have in common:
Things I know:

- Each is half of a rectangle, and if both of the sides making the right angle are the same length, then we get a special rectangle, a square. (I still needed to ask myself how I knew that each of these were really half of a rectangle; but I put this concern aside for the moment, to come back to later.)
- Together, the other two angles of the triangle always fit together to be another corner of a rectangle, suggesting that the other two angles add up to 90° and together make a right angle. The total of all three angles in the triangle would then seem to be 180°.
- So, what happens if we arrange the corners of three copies of the same right triangle in such a way that all three angles come together? We should get that they make a straight line:

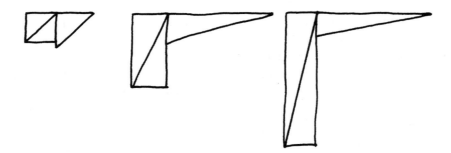

- But, of course, we already sort of knew that we could arrange two of them to make a rectangle. We are just placing this rectangle next to the original right angle. (But I wouldn't have seen it in this way without all of this experimenting!)
- Did we need to use *right* triangles to do any of this?
- Could we find relationships like this for other shapes, such as trapezoid or pentagons?

"Just as a sculptor taking a hammer and chisel to a piece of marble can see a vast number of potential statues within the marble," writes Wells, "so [a] mathematician expects to find an endless variety of theorems in the simple shape of a triangle" (16).

With number patterns, the issue of interest and value is especially important. Anyone who spends any time with numbers will start to gather useful facts, but how can we say whether they are valuable or worth remembering? To use another example from Wells (30), anyone might notice that the last single digit number in our common number system is 9, which is a perfect square ($9 = 3 \times 3 = 3^2$). Is this deeply significant? Perhaps, he says, it is more of an accident. But right here, I would say, is a perfectly interesting question worthy of investigation. *Is* the fact that the last single digit number is 9 something we would want to talk about with others? Compare the fact that 9 is a square number with Wells' fact that numbers that are one less than perfect squares happen to be the product of two numbers which themselves differ by two: for example, $16 - 1 = 15$, and $15 = 3 \times 5$; $9 - 1 = 8$ and $8 = 2 \times 4$; $36 - 1 = 35$ and $35 = 5 \times 7$. Wells finds this kind of fact more appealing because it seems to reveal an underlying *pattern*. "It is just so tempting to suppose that this property applies to any square at all, and the moment we start to wonder whether it does, then we are thinking mathematically" (31). The next step is then to try to figure out whether it really *does* work in this way for any square at all.

Science

One way to discover mathematical facts and theorems is to treat mathematical objects as if they were objects in the real world, and then to make observations and do experiments. Working like a scientist, we make models of the objects we cannot actually make in the real world; by drawing pictures, we measure, calculate with numbers, and collect data to be analyzed. We organize our data and through examining it we make a *conjecture*, that is, we come up with a theory about "what is the case." Now we test our theory with experiments, maintaining those "variables" that we believe are important in our conjecture; we change everything else that we can change without changing what our conjecture is about. When we are very confident that our conjecture is true, we *prove* it by organizing a persuasive argument or explanation that others will be able to understand. Now it is the others' turn to see if they can understand our argument or explanation, and if they can reproduce the same results.

Suppose somebody, maybe even ourselves, comes up with a situation where our conjecture does not seem to work out as we thought. Does this mean we have to give up? Not at all! We just need to modify our conjecture to say that it does not work in this case. We need to go back and better explain what situations it *does* work for. Imré Lakatos (1976) called this "monster-barring," when we re-state what we now know to make it clear that possible situations are *not* part of what we can say in our conjecture or theorem. Those weird cases where the conjecture does not work are "monsters."

This is very much the way we have worked so far in this chapter. Our mathematician's notebook is similar to a lab notebook where we collect our observations, organize our data into charts for analysis, record our conjectures, design experiments, collect new data for further analysis as we test our conjectures, and then organize what we have learned into a draft of what we will say or write about our work when we share it with others. The next step will be to present our results publicly, either in an oral presentation or as a published paper to be read by others (in our class, or in a more formal setting). One thing to keep in mind when we work at science is that scientific knowledge is never certain. It is always tentative and open to future examples of where we have misunderstood a phenomenon. We tend to forget this when we treat mathematical objects like they are real things in the world; but as long as we are prepared for such future questioning of our mathematical knowledge, it is possible to experience such a vital and exciting way of being a mathematician.

School mathematics does not always embrace mathematics as science, because we are sometimes worried about the uncertainty that accompanies science. Yet, what mathematics as science allows is for the living nature of mathematics to be part of school mathematics. Instead of "dead math" we are breathing life into the mathematics that we explore. In this kind of work, we are identifying areas of interest, articulating questions that we ourselves have within these areas of interest, and then pursuing what these questions might mean, as well as trying to answer our questions. "Do we have the right question?" might be a common question in our classroom.

When we act like scientists, our questions tend to take a particular form, in which we ask whether or something is "true," or whether certain circumstances lead to others. But sometimes we work in a different way, when we hear of a question that somebody else has formulated. In this case, we need to understand what the other people are asking, and then try to answer *their* question. Too much of school mathematics expects us to work in this way, so that we rarely are able to comprehend why the question was asked, or why the question is interesting, and thus we often find it a challenge to care about the answer. Mathematics as science carries with it the natural motivation to care about our results and to experience the reactions of others to our work, since we asked the questions ourselves, and because we know why the answers matter. Nevertheless, both traditional school mathematics and mathematics as science require that we get better at solving problems and answering questions, so it is worth our time to examine further how we might help ourselves become more skilled at the solving of problems. The next section turns to problem solving skills. We will come back to the issue of who poses the questions later, toward the end of the chapter.

Polya and Problem Solving Phases

Georg Polya, a mathematician, wrote a lovely book in the 1940s based on his analysis of how mathematicians solve problems. He thought this might help him train his students to do the same things. Reflecting on his own thought processes, and through conversations with other mathematicians, he came up with four phases during which mathematicians ask themselves questions about the thinking processes that they are using. Each phase uses different kinds of questions. He realized that he could "keep going" and not get "stuck" in a paralysis of not knowing what to do if he told himself to ask these questions as he moved through four phases of work. As a teacher, he would model this guided questioning for students, working through examples himself. Then he would guide them through their own problem solving experiences, asking these questions for them. Slowly, he would do less and less of the questioning so that the students could internalize the teacher's monitoring of their movement through the problem solving phases themselves, asking themselves the same questions that the teacher would have asked.

When Polya begins to work on a problem, he tries to *understand the problem* by asking what he knows and what he does not know. He asks himself if he has ever done a problem like this before, and if so, what does he remember about how he solved that problem? What exactly does he want to know in this problem? What does he already know that might be helpful in thinking about this problem? Suppose I am asked, or want to find, further relationships among triangular, square, and pentagonal numbers. I can think about our previous work with just the triangular and square numbers. The triangles fit together to make squares. Do they fit together to make pentagons? I was able to combine numbers in one column of my chart to create numbers in another column of my chart. Can I do more of this, in order to use triangular and/or square numbers to make pentagonal or hexagonal numbers? Already there are things I can do to work on this problem. I have identified parts of my chart, potential connections to explore, and experiments I can begin to perform.

Box 3.9

Triangles	Squares	Pentagons	Hexagons
1	1	1	1
3	4	5	6
6	9	12	15
10	16	22	28
15	25	?	?

Once Polya has begun to understand the problem he moves into the next phase of work: *Plan What You Will Do*. Here are some questions that help him to plan. Can you draw pictures? Can you make a chart? What are the relationships between what you know and what you want to know? Can you name things? This last question can help you to define variables, those things that have different value in different places in your problem, or at different times in a situation. You may be able to write about the relationships among variables using equations. Soon you will find yourself, like Polya, shifting into the third phase of work, *Carry Out Your Plan.* Can you add things to your picture?

If so, you can search for ways that these new things may lead to connections, or new variables. Can you find a pattern in your chart? Can you solve your equations? These questions will help you start.

As you are working you must be ready to go through many starts and stops. You do not know whether or not you are making progress toward a solution. Nevertheless, every connection you make is something that you have discovered for yourself. What I do is periodically look back over what I have amassed, crossed out, circled as important, put giant question marks next to, and so on. I then write down for myself in

Box 3.10

Triangles	Squares	Pentagons	Hexagons
1	1	1	1
3	4	5	6
6	9	12	15
10	16	22	28
15	25	?	?

some kind of organized way what I now know and do not know, along with how I led myself to what I know. I record why I am curious about what I still do not know, so that I can keep track of the purpose of my work. I write all of this carefully and with a lot of details, as if it will be read by somebody else, because I may not remember all of these details when I return to my work later on or on another day. What this means is, even as I am going back and forth between Polya's Phase 2 and Phase 3, I am also experiencing his Phase 4, *Look Back.* In this phase, which Polya describes as the most important part of the work, I ask: What have I learned? Can I find an easier way than I have done? What is the *meaning* of my answer? What I have learned may or may not be obviously related to what I originally hoped to find out. Nevertheless, it may be important in the future, perhaps *more* important. I can begin to move further along as well, as long as I do not skip the step of thinking about the meaning of my answers to the questions. And, in my personal experience, this always moves a mathematician into *new* questions. The new questions are almost always more interesting, more important, and more central to my learning than those that came before. And so I start the cycle of phases all over again.

Mason: Specializing and Generalizing

John Mason and his colleagues provide another useful lens for thinking about the problem solving process (Mason *et al.* 1995). In their book *Thinking Mathematically* they describe mathematicians as searching for an initial understanding through "special cases." If I do not know what to do next, I can always see what happens when I test the number 1, or 0, or 5, or 10, or 100; I can look at a triangle, or, specifically, an isosceles triangle and a non-isosceles triangle, or perhaps a square and a rectangle that is not a square. I can collect my knowledge of these special cases, organize them into a chart, and then look for patterns or relationships. When I change from 1 to 2 to 3, what happens? When I add sides to my shape, what happens? If I make the angle larger, does this change anything else? This collection of special cases will help me to risk a statement that seems reasonable, a conjecture or generalization. Most of my conjectures will turn out to be false, or not quite right; but they get me going, they give me something to think about, they move my work along so that I am not "stuck" or paralyzed. Besides, I tell myself, if I make a conjecture and then figure out that I wasn't quite right, that is just as interesting as if I had come up with something that I could be confident was right: knowing that something that seems to be the case is not really the case is an important kind of thing to know. Conjectures based on special cases keep me thinking about the mathematics so that my mind does not wander into all the other things that are going on in my life.

When I am making a conjecture, I act, as Mason suggests, as if I really believe it is a good conjecture. I test it with special cases. I think about my variables, those things that can vary within the context that I am working, and try the range of how the variables may vary— odd and even numbers;

Box 3.11

The Conjecture Cycle

For each of these questions, adapted from John Mason and his colleagues' book, try out numerous special cases until you have enough of them that you can organize them by categories or patterns. Use the categories or patterns to formulate a conjecture. Next, test out your conjecture with new, artfully chosen, special cases. Do you need to do any "monster-barring"? Finally, present an argument in support of your conjecture to another mathematician; your goal is to convince them that your conjecture makes sense.

(a) John Mason's Painted Tyres (Mason *et al.* 1995: 65). Once while riding a bike along a path, J.M. crossed a strip of wet paint about 6 inches wide. After riding a short time in a straight line, he looked back at the marks on the pavement left by the wet paint picked up by his tyres (he's British, so that's how he spells tires). What did he see?

(b) Furniture (66). A very heavy armchair needs to be moved, but the only possible movement is to rotate it up to 90° about any of its corners. Can it be moved so that it is exactly beside its starting position and facing the same way?

prime and not prime; right angles, smaller angles, and really big angles; number of sides of my shape; parallel and non-parallel sides; and so on. Next I challenge myself: I play a game with myself and distrust the conjecture, trying to refute it. I look for an especially "nasty" special case to see if it will demonstrate a weakness in my conjecture. But after a while, I feel like I am ready to say that I can get a good sense of why the conjecture is true, or how I should modify it, through the use of more examples. These examples lead me into a position that allows me to make new conjectures, starting the cycle all over again.

On Reading a Book: How Mathematicians Read

Mathematicians look at something and "mathematize" it. They go beyond what they already know, experiment with what they see, and try to see more. They celebrate the new connections they find and work hard at putting together a way to communicate what they have discovered with others. This sounds very approachable. It is something that all of us can do. Yet, when most of us pick up a mathematics book, we are intimidated. It is too hard to read! So many of these books appear impenetrable, so we do not believe we ourselves could or would even want to write one. What we do not see is the mathematician's notebooks full of all of the experiments, errors, confusions, lack of confidence, and surprises. We do not see the several days when she or he may have had no idea what to do next. What we see instead is what the mathematician Reuben Hersh (1997) calls the "front" as opposed to what is going on in the "back." He uses the analogy of a restaurant, where we sit in the dining room and are served a beautifully prepared meal. What is going on behind those closed doors in the kitchen? Spilled sauce, burnt cheese, a cut finger, the exclamation, "What'll I do? We're all out of salt!?" The perfectly prepared dinner we are served can only exist if the kitchen exists. Our notebooks are our kitchens. We also need dining rooms in school, but we cannot forget to experiment in the kitchen. And when we read a math book, we sometimes need to peek into the kitchen to understand how we could make the same meal. Unfortunately, the book does not come with the notebook. This is a shame. It would be so valuable if we shared more of what we do in the kitchen with our mathematical guests, because being a mathematician is more like being a cook than like being the one who eats the perfectly prepared meal.

Luckily for us, there are ways to find out how mathematicians do their kitchen work. We can invite mathematicians into our classrooms to talk about how they work—local mathematicians who work in industry, or at a nearby college; family members who work mathematically as a beautician, carpet installer, financial analyst, or botanist; older students from other classes in our school. And we can talk with each other: in a mathematics department at a university, there is always a coffee lounge where mathematicians go to take a break; there may be a daily "afternoon tea," say at 4 p.m.. They share their ideas and help each other think through their work, helping them get "unstuck" so they can return to their office ready to work. We can establish similar experiences in our classrooms, but we also read mathematics books with the kitchen in mind, which means we keep a reading notebook by our side, and rebuild that kitchen work for ourselves. We run down false paths, make a mound of trash that can't be used (but which may serve as compost in the future). Now, John Mason and his colleagues have generously provided a peek into their kitchen. In the chapter on conjectures, which includes the bicycle and furniture problems, they also work through another problem, sharing their ideas every step of the way. As I read this book, I *still* needed my kitchen notebook next to me, because I would only really understand a mathematics book if I tried it out in my own kitchen. I share some of my reading with you here.

They start with a situation that most of us are aware of, but which few of us have pursued mathematically. This is the first important step in working mathematically, looking at something beyond our initial awareness of what it is. Have you noticed how some numbers are the sum of a string of consecutive positive (counting) numbers? Sure, I had noticed this before. $7 = 3 + 4$; $5 = 2 + 3$; $6 = 1 + 2 + 3$; $11 = 5 + 6$. But I had never gone past just noticing that this is possible. It is easy to create such numbers for example by just adding different collections of numbers together—$21 + 22 + 23 = 66$; $1042 + 1043 = 2085$. The *interesting* question, it seems, is, "Can any number be written this way?" or, "Exactly which numbers have this property?"

I got hooked on this part of their book because I felt a connection to my earlier work with triangular and square numbers. Triangular numbers may seem like magical, special numbers. But it dawned on me that other numbers are "slices" of these triangles. For example, $7 = 3 + 4$, which is part of a larger triangle:

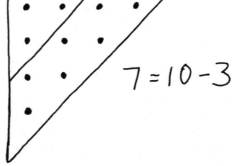

$7 = 3 + 4$

$14 = 25 - 1$. Another way to say this is that many numbers can be made from differences of *triangular numbers*: $7 = 10 - 3$:

$7 = 10 - 3$

I wonder: if *every* number could be thought of as a difference of triangular numbers, then everything we know about triangular numbers could be used to talk about numbers in general. I would only have to study *some* numbers to be able to think about *all* numbers! These differences turn out to be a way to look at sums of consecutive numbers, since each new triangular number is made from the previous one by adding the next counting number: $10 = 1 + 2 + 3 + 4$; $15 = 1 + 2 + 3 + 4 + 5$, and so on. The new question becomes, "Can every number be the sum of consecutive numbers? If so, then every number is a difference of triangular numbers."

So, I was reading about this in a book by John Mason and his colleagues and it had a section about whether any number might be able to be thought of as the sum of strings of consecutive numbers. Do you have any hunches? Try lots of examples. Try changing the questions, extending their scope in some way. Be systematic in the special cases you select, so that you can organize your work into charts and then look for patterns. I too wanted to be systematic as I was reading this book. I asked myself how. Should I start with 1 and go through every number for a while, trying to see if I could make these numbers by adding consecutive numbers? Or, should I try every combination of numbers added together in strings, and see if these sums would in the end produce every possible number? I tried the first idea because it was suggested in the book, but also because there would be one very clear system, going through the numbers in order, and this seemed nice to me at the time.

Already I had to make some decisions. $1 = 0 + 1$. Is this allowed? It could be, but I was really thinking we were talking about sums of numbers that start with 1 or more. I'll keep $1 = 0 + 1$ in mind, I thought, but maybe 1 is a *very* special case as the first positive number. Two is impossible, 3 is easy. 4 is impossible. Wow! Here's a possible conjecture: Even numbers are not the sum of consecutive numbers. I think of this because 2 and 4 are not possible. I'll keep going. $5 = 2 + 3$; $6 = 1 + 2 + 3$.

Box 3.12

$11 = 5 + 6$
$12 = 4 + 5 + 6$
$18 = 5 + 6 + 7$
$10 = 1 + 2 + 3 + 4$
$15 = 1 + 2 + 3 + 4 + 5$
$20 = 2 + 3 + 4 + 5 + 6$
$14 = 2 + 3 + 4 + 5$
$18 = 3 + 4 + 5 + 6$

Box 3.13

$1 = ?$
$2 = ?$
$3 = 1 + 2$
$4 = ?$

Oops! My first conjecture is already something for the trash. But hold on; maybe there's something that 2 and 4 have in common that 6 does not. There may be another conjecture in the making. What do 2, 4, and 8 have in common? They are all powers of 2, or multiples of just 2. I am ready to leap into another conjecture: Powers of 2 cannot be the sum of consecutive numbers. I *predict* 16 will not work either. Can I test this?

Box 3.14

$5 = 2 + 3$
$6 = 1 + 2 + 3$
$7 = 3 + 4$
$8 = ?$

I need to be more systematic! I am losing track of all of the combinations of two, three, four, and five numbers. I still have not found a way to make 16, but I haven't found 13, 17, or 19 either (conjecture? Some prime numbers are impossible?) yet there are so many possible combinations that I am increasingly convinced that I could get virtually all numbers by adding other consecutive numbers.

Time to collect my thoughts. Just by trying special cases, I quickly leap into generalizing through what some of the special cases have in common. Specializing gets me doing something, but it also gives me a feel for what is going on. The "feel" is some kind of pattern underlying the surface facts that I try to express as a generalization. I write down the generalization as a conjecture. Now I can examine my conjecture, test it, challenge it, and confirm it. So far my specializing confirms my conjecture about powers of 2, and it also introduces a conjecture about primes greater than 11. I need to organize my data:

Box 3.15

Starting with 1:	**Starting with 2:**	**Starting with 3:**
$3 = 1 + 2$	$5 = 2 + 3$	$7 = 3 + 4$
$6 = 1 + 2 + 3$	$9 = 2 + 3 + 4$	$12 = 3 + 4 + 5$
$10 = 1 + 2 + 3 + 4$	$14 = 2 + 3 + 4 + 5$	$18 = 3 + 4 + 5 + 6$
$15 = 1 + 2 + 3 + 4 + 5$	$20 = 2 + 3 + 4 + 5 + 6$	$25 = 3 + 4 + 5 + 6 + 7$
$21 = 1 + 2 + 3 + 4 + 5 + 6$	$27 = 2 + 3 + 4 + 5 + 6 + 7$	$33 = 3 + 4 + 5 + 6 + 7 + 8$
Starting with 4:	**Starting with 5:**	**Starting with 6:**
$9 = 4 + 5$	$11 = 5 + 6$	$\boxed{13 = 6 + 7}$ ***
$15 = 4 + 5 + 6$	$18 = 5 + 6 + 7$	$21 = 6 + 7 + 8$
$22 = 4 + 5 + 6 + 7$	$26 = 5 + 6 + 7 + 8$	$30 = 6 + 7 + 8 + 9$
$30 = 4 + 5 + 6 + 7 + 8$	$35 = 5 + 6 + 7 + 8 + 9$	$40 = 6 + 7 + 8 + 9 + 10$
$39 = 4 + 5 + 6 + 7 + 8 + 9$		
Starting with 7:	**Starting with 8:**	**Starting with 9:**
$15 = 7 + 8$	$\boxed{17 = 8 + 9}$ ***	$\boxed{19 = 9 + 10}$ ***
$24 = 7 + 8 + 9$	$27 = 8 + 9 + 10$	$30 = 9 + 10 + 11$
$34 = 7 + 8 + 9 + 10$	$38 = 8 + 9 + 10 + 11$	$42 = 9 + 10 + 11 + 12$

I have quickly ruled out my prime number conjecture, but I still believe the power-of-2 conjecture, as 16 and 32 still do not appear in my tables. So it is time to get a sense of why this conjecture might be right. What is it really saying? I believe: (a) Powers of 2 can never be written as the sum of consecutive numbers; (b) All other numbers *can* be written as the sum of consecutive numbers. What would this *mean*? (a) Why would being a power of 2 prevent a number from being the sum of consecutive numbers? And, (b) If a number is *not* a power of 2, how does this *allow* or *enable* it to be the sum of consecutive numbers? Central to my inquiry is what is so special about powers of 2.

Following Polya, what do I know about powers of 2 that I can try to use in a connection with what I still do not know? All I know is what it means to be a power of 2: a power of 2 is some number of 2s (and *only* 2s) multiplied together. In other words, the only factors of numbers that are powers of 2, other than the number 1, are themselves powers of 2; specifically, the only factors of these numbers that are powers of 2 are even numbers. Powers of 2 have only even factors. They can only be made by multiplying even numbers. For example, the factors of 16 are 16, 8, 4, and 2 (other than 1). Another number, like, say, 22, has the factors 22, 11, and 2 (as well as 1), and 11 is *odd*. 18 has the factors 18, 9, and 2 (other than 1), and 9 is *odd*.

I feel ready to make another conjecture. Every number that is *not* a power of 2 has an odd factor other than 1. Now, I can spend a lot of time on *this* conjecture: do more special cases confirm it? But, before I do, is it worth pursuing this conjecture about odd factors? At this point I only care about it if it helps me to figure out which numbers can be the sum of consecutive numbers. So, I only really care about this conjecture, the one about factors being odd, if it makes a difference. For this reason, I want to hold off on testing it. I'll come *back* to this conjecture if it is an important one for my work. Right now, I need to reflect on whether or not odd factors matter for consecutive sums.

How might odd factors matter for consecutive sums? I need to reorganize my data according to odd factors rather than according to how I added the numbers, to see if the odd factors lead to any connections.

Numbers that have odd factors: Start with a small odd factor bigger than 1, that is, 3.

Box 3.16

3 as a factor:	5 as a factor:	7 as a factor:
$3 = 1 + 2$	$5 = 2 + 3$	$7 = 3 + 4$
$6 = 1 + 2 + 3$	$10 = 1 + 2 + 3 + 4$	$14 = 2 + 3 + 4 + 5$
$9 = 2 + 3 + 4$	$15 = 1 + 2 + 3 + 4 + 5$	$21 = 1 + 2 + 3 + 4 + 5 + 6$
$12 = 3 + 4 + 5$	$20 = 2 + 3 + 4 + 5 + 6$	
$15 = 4 + 5 + 6$	$25 = 3 + 4 + 5 + 6 + 7$	

It looks like I had an easy time reading this book, doesn't it? But it was at this point that I got stuck. Suddenly, the authors were writing about a new conjecture that I could not understand, I read the next two pages dozens of times. Then it hit me: They are saying that it seems almost to be the case that a pattern is emerging, having to do with the number of consecutive numbers that are being added together to make these multiples of odd numbers. For multiples of 3, three consecutive numbers are added after the "basic" 3 is made from only $1 + 2$. For multiples of 5, five consecutive numbers are added after 5 and 10 are made from fewer numbers. Perhaps if I continued my 7 table, I would find that seven consecutive numbers are needed to make the multiples of 7? (The book only showed 3s and 5s; I need to try more kitchen experiments to get their idea more firmly in my own mind.)

Box 3.17

$7 =$	$3 + 4$
$14 =$	$2 + 3 + 4 + 5$
$21 =$	$1 + 2 + 3 + 4 + 5 + 6$
$28 =$	$1 + 2 + 3 + 4 + 5 + 6 + 7$
$35 =$	$2 + 3 + 4 + 5 + 6 + 7 + 8$
$42 =$	$3 + 4 + 5 + 6 + 7 + 8 + 9$
$49 =$	$4 + 5 + 6 + 7 + 8 + 9 + 10$

I found a pattern to use in my sums: Once I have seven terms, I can remove the first one and add the next consecutive number to make the next multiple of 7. Indeed, I see now, and only now, that this *makes sense*, because, I have numbers seven apart, so I will get another multiple of 7 when I remove the first in my sum and add the next, just *because* there are seven consecutive numbers between them (1 and 8, 2 and 9, 3 and 10, 4 and 11; subtracting the first and adding the other is just like adding 7!).

I need to *name* what is going on if I am to go further, so I can study this phenomenon. First of all, how do I describe the idea that a number is odd, or that a number has an odd factor? Odd means, for mathematicians, *not* even, that is, *not* a multiple of 2; in other words, an odd number is always one more than an even number, which *is* a multiple of 2.

I will write an odd number as 2k+1, which means, k can vary but 2k is always even, as a multiple of 2, and 2k+1 is always odd because it is one more than an even number. And I can use this to *write down* the next conjecture: A number, call it N for Number, which has an odd factor, which I will name 2k+1, can often be written as the sum of 2k+1 consecutive numbers (once we get a big enough number).

O.K., this fits with the numbers in my data so far, after the smaller numbers on the lists. Once the first number in the list is written with 2k+1 consecutive numbers, it appears the rest of the multiples of the same odd factor continue to be the sums of that many consecutive numbers. As a mathematician, I want to test this out by varying k, my variable. When k = 1, I have the situation with 2k+1 = 3, which is already part of my data. My data also already includes when k = 2 (2k+1 then equals 5) and when k = 3 (when 2k + 1 = 7).

Again, I had trouble reading this book. The next paragraph had an *F* in it! What is this F? Fifteen minutes later, I realized that the author used F to name another part of the charts. Nowhere was this written, but what the author apparently noticed is another property of an odd number. If you list an odd number of things, there is another thing going on: there is one number in the middle of the list. So we can name the *middle* number "F." Thus, when k = 1, we have multiples of 2k + 1 = 3:

Box 3.18

$3 \times 2 = 1 + 2 + 3$

$3 \times 3 = 2 + 3 + 4$

$3 \times 4 = 3 + 4 + 5$

$3 \times F = (F - 1) + F + (F+1)$

When $F = 2, 3 \times F = 1 + F + 3$

$F = 3, 3 \times F = 2 + F + 4$

etc.

And, when k = 2, we get multiples of 5, once we go down the list to the first one with five consecutive numbers in the sum:

Box 3.19

$5 \times \boxed{3} = 1 + 2 + \boxed{3} + 4 + 5$

$5 \times \boxed{4} = 2 + 3 + \boxed{4} + 5 + 6$

$5 \times \boxed{5} = 3 + 4 + \boxed{5} + 6 + 7$

$5 \times \boxed{F} = (F - 2) + (F - 1) + \boxed{F} + (F + 1) + (F + 2)$

and we can say, if N = (2k + 1) × F, then when the N is a big enough multiple so that there are 2k + 1 consecutive numbers in the sum, then the middle number is F. That is, the middle number is the number that determines the multiple of the odd factor that was used to make N. That F is both the multiple of the odd factor and the middle of the sum; that odd factor, meanwhile, tells us how many numbers are in the sum.

This feels like a "result," so it is time once again to pause and reflect on the *meaning* of this result. Again, this pausing was not done for me by the author as I read. I needed to do this myself, even though this author is very accessible and understandable. If a number is a multiple of an odd factor, what do these ks and Fs do to help with finding a response to the original question? So I have to look back over the last several pages in my kitchen notebook. Powers of 2 seem to be a problem for finding sums of consecutive numbers. O.K., I got that. Next, all other numbers have an odd factor. So far, so good. The following point was that this means that we can write these non-powers of 2 as an odd number times some *other* number. We wrote the odd number as 2k+1 because every odd number is one more than some even number. We named the even number 2k, and the odd factor 2k+1. And we named what we multiply the odd factor by as F. So we have a new language for describing all the non-powers of 2. Any number, named N, which is *not* a power of 2, can be thought of as having parts that we can look at. One part is the odd factor that we know it has, and which we now call 2k+1; the other part is what we multiply the odd factor by to make N, which is named F. And we have found a relationship between N and F: N can be written a lot of the time as k consecutive numbers, centered around F. It was actually at this point that I went back and boxed those numbers in my list of multiples of 5 (see Box 3.19), to remind me that F is the center of the list of consecutive numbers.

What does this mean, though, again in relation to the question I am working on? It shows that I can always write a non-power of 2 as 2k+1 consecutive numbers (where 2k+1 is the odd factor of my number). Indeed, except for the first few, smaller numbers in my lists, this pretty much "proves" the conjecture. I just need to get a handle on how to describe these smaller multiples of odd factors.

It was then, at this point, that I realized I had a new "thing" that could be understood better: these sums of consecutive numbers centered around Fs. I'll name them "C" for *c*onsecutive, *c*entered sums. Does every consecutive, centered sum make a number with an odd factor? There are always 2k+1 terms, and the sum looks like:

Box 3.20

$$
\begin{array}{lcl}
F & & = F \\
(F-1) + (F+1) & & = 2F \\
(F-2) \quad + \quad (F+2) & & = 2F \\
\cdots & & \cdots \\
(F-k) \quad\quad + \quad\quad (F+k) & & = 2F
\end{array}
$$

Add these up: = . . . well, there are k+1 equations here, with k of them adding up to 2F, and there's that first one that's just F, so we get $(F + k \times 2F)$ on the left side, and $(2k+1) \times F$ on the right side. This confirms for me that every C, every consecutive centered sum, has an odd factor, namely $(2k+1)$. I have ended up demonstrating something to myself that is not exactly the same thing as what I am seeking, but it is close, and it is very cool: Every non-power of 2 can be written as a consecutive centered sum, in *some* way. This sum, though: what do I know about every number in it? Not too much really, except I do know that these sums always indeed have an odd factor, which is reassuring. But I do not know about these sums beyond this. For example, an arbitrarily chosen consecutive centered sum might have negative numbers in it, or a zero. For example, I might be describing a way to write 7 as $-2 + -1 + 0 + 1 + 2 + 3 + 4$. This adds up to 7, but does not have the form we have been studying, where the numbers in the sum are all positive. I am *so* close to finding out if every non-power of 2 can be the sum of consecutive counting numbers greater than one; in the meantime, I *do* know such numbers can be written as consecutive centered sums if I allow negative numbers and zero.

> I was struck . . . by the quick flow of conjectures and the pleasure they gave me. Being aware of the spiral nature of conjecturing prompts me to give more careful attention to conjectures than I once did. When I am stuck I can go back and look for where I departed from the spiral by failing to test and disbelieve my conjecture.
>
> (Mason *et al.* 1985: 76)

Suffice it to say, there is more to this yet! I encourage you to find Mason and colleagues' book, not only to read the punch line for how and why he can say, for sure, that non-powers of 2 can always be written as sums of consecutive, positive numbers. And not only to find out more about the conjecture regarding odd factors being something that every non-power of 2 has. This book also introduces work on the "next" questions we might have: can we predict how many different ways a single number can be written as the sum of consecutive positive numbers? What numbers can be written as the sum of differences of *squares* of numbers? And so on. Mostly, though, this book introduces ways of thinking mathematically that we all would profit from. For our purposes right now, I wish to highlight only two more points from the book. First, "conjectures are like butterflies. When one flutters by, there are usually many more close behind" (76). Like butterflies, conjectures flutter by and are not easy to capture; and each one distracts you from the others. The wise skill is to pin them down with words before you lose them so that you can study them later. Also, it is important to distinguish between a conjecture and a strongly confirmed conjecture. How much evidence is necessary before you are certain that a conjecture is true? It is always better to treat every statement as a conjecture that needs to be challenged than to assume it is safe to accept it. If you always mark conjectures carefully in your notebook, you can return to them at another time to test them and play the game of distrusting them. This will help you to strengthen the arguments that justify your important conjectures. (In my reading notebook, for example, I still need to go back and understand that conjecture about non-powers of 2 always having odd factors, because it is, in the end, the basis of everything that follows.) Conjectures rely on a search for patterns. If no conjecture flits before you, try further special cases, or reorganize your data using a different system. (In my reading notebook, I switched from listing the consecutive sums by counting numbers to listing them by multiples of odd factors.)

The Art of Problem Posing

One final topic for this chapter, and I believe it to be the most important one, has to do with where the questions come from in the first place. We can *pretend* to be mathematicians as much as we want as long as we work on other people's questions. But mathematicians ask *their own* questions. It is when we pose the questions ourselves that we move toward working as a mathematician works. It is worth dwelling on this, and examining the processes of problem *posing* so that we can demystify what is involved. Once we understand that posing, too, is a matter of skills to be applied, rather than inborn talent, we will be closer to understanding how we are—how every person *is*—a mathematician. Through this realization, we can lay the foundation for classrooms where students and teachers live and work together, thinking mathematically, and sharing the joys and frustrations (which themselves can be shared as another sort of joy) of mathematics.

Where do mathematics questions come from? Here are three places we can go when we pose questions: things, situations, and problems we have or somebody else has already posed. First, we can make mathematics questions from a thing. Name parts of the thing and ask how they are related to each other. Already you are asking questions! Much of this chapter has already presented this approach: I have a right triangle; how are the other angles related to each other? I have as many squares as I want; how are two, three, four, five, or any number of copies of the same square related to each other? I have triangular numbers and square numbers in my chart; are they related, and if so, how? But there is another way of working with things. If we note specific attributes, we can ask what happens when we change one or more of these attributes. For example, I have a triangle with three sides and three angles. What happens if I require one angle to be a right angle, or two sides to be equal? I have triangular numbers; what if I make them square numbers? I see how I can make triangular and square numbers. Is there a way I could invent to make pentagonal and hexagonal numbers? If so, is there a pattern to them like with the triangular numbers? I know that I can make square numbers by adding certain triangular numbers; can I make pentagonal numbers by adding certain numbers in my table?

Situations also generate mathematics questions. I am playing the African stone game mancala; is there a best first move? Is it always good to go again? Is there a way to minimize my opponent's opportunities to steal? I am walking to school and the road is curved; is there a side of the street that I should walk on if I want the time it takes to be the shortest, because my backpack is heavy?!? I am not so anxious to get there as soon as possible, so can I chose a side that makes the walk just a bit longer? (Think about what runners try to do around a racetrack.) I have shaped tiles that could be used to make a design on a bathroom floor; now I ask many questions about possible combinations of shapes. I could, if I wanted, add together consecutive numbers; among the many potential questions I might ask hides the one about which numbers could be the result of such sums.

The most interesting situations for me are those from everyday life. Should I buy a car or use public transportation? How many months do I need to work before I could afford to travel across the country? How many vans or buses should we hire to take us to the anti-war protest? If we raise the price of pencils at the school store, will we simply make more profit, or will we lose money when kids think the pencils are too expensive? What shaped cardboard will make the best box for mailing this present to my friend? How much fabric do I need to make this costume for the play?

A rich source of questions is those very questions we have already worked on. This works well in school where we feel the pressure to use the textbooks that we have been given. They certainly contain many given problems, none of which are very interesting, but all of which have the potential to be turned into something interesting. Having worked on a question which someone else thinks is important for us to think about, what questions do *we* have now? The question talked about even numbers; would it matter if the numbers were odd? The question asked about squares; how would the question change if we used parallelograms instead? The question said we were walking at three miles per hour; what if we walked faster or rode a bicycle? Or crawled? Are there extreme limits to the speed that make the problem silly, or more interesting? If these triangles were on a sphere, like on a globe, instead of on a flat piece of paper, would the question still make sense? The question asked us to find the average of certain numbers; what if we started with an average and wanted to find numbers that might have this average?

The basic strategy of problem posing is to list attributes of an object, or of the original question, and ask "What-if-not?" about any of the attributes. This is a technique made popular by Stephen Brown and Marion Walter in their classic book, *The Art of Problem Posing* (1983). Here is a fundamental attribute that we can *always* use: The question assumes we want to find the answer. What if the answer were not our goal? What else might we work toward? Perhaps we want an answer, but to another, more interesting question? Maybe a collection of all of the strategies we might use to obtain the answer? What are all of the things we could do with the answer once we have found it? We might reflect on what sorts of strategies we personally favor over others, the kind of problem solver we think we are, or the sort of mathematician we would like to become. We might identify skills we need to practice because our sloppiness with them is distracting us from

Box 3.21

What-if-Not Problem Posing

Attribute	New Question
The question uses even numbers.	Is the same thing possible for odd numbers?
The question is about triangles.	Can we find anything special about specific triangles, such as isosceles or right triangles?
The question states 3 mi/hr.	What if we changed the speed? Are there any limits where the question no longer makes sense, or where it becomes more interesting?
The question asks us to calculate the average.	What if we were given the average and then wanted find numbers that could have this average?
The question asks us to make a graph.	What if we were given the graph, and were asked to come up with questions that would lead to this graph?
The question has an answer.	Can we come up with at least three *different* answers?

what we really wish to be able to do. Or, we might organize questions into "types" to help us understand how to recognize good approaches to problems that follow common patterns.

You probably realize I was listening to some of the comments you made to each other a few minutes ago. You have to admit I have come a long way in using strategies to encourage kids to think that they are seeing patterns or even solving problems that look to them as if they are the first people in the world to do so, but aren't you pushing the limit a bit? If you start to encourage them to pose problems of their own and to listen to all that talk about the value of errors and metaphors . . . Then you'll never end up covering what you need to get through . . . It is possible [too] that they will pose problems that neither they nor you know how to solve.

(Brown 1996: 290)

This interjection is spoken by Eloise Farrong, a character in a novelette about problem posing written by Stephen Brown. The spirit of her ideas, as Brown sees them, can be found in her name, which sounds so much like "far wrong." Yet she is the chair of the mathematics department at Cutting Ej High School, in a position of power over teachers who are exploring exciting and meaningful approaches to teaching mathematics. She is both "the enemy" and the source of support for our efforts. She is also that secret voice inside each of us, expressing doubt. Can it be the case that school is not a good place for teachers and students to act like the mathematicians they are? Even Eloise agrees that problem solving by students is important, and that we should not be spoon feeding them material. As a good chair, she wants each teacher to embody their most cherished beliefs in the classroom. But at heart she *knows* that "you are the one in charge of what gets done in the classroom, what problems students choose, and which answers are correct" (290).

"Wouldn't you agree," asks a teacher named Sy in the story, "that students would be better motivated to *solve* problems that they have posed on their own?" "Maybe," responds Eloise, "and you can sometimes give them the illusion that they are posing the problems to solve. You can even sometimes actually let them do it. But you can't make that a steady diet. You would never cover anything in the curriculum" (290).

Some people wonder with Eloise if it is possible for youth to pose their own problems and to learn mathematics through problem posing experiences. They might ask how a student is supposed to ask a question about something that he or she does not yet know. Would the teacher not need to guide them so much that it would look like ordinary, teacher-directed "guided discovery" anyway? The wonderful thing about school, as I see it, is that students do not come to the classroom as "blank slates," even if so much of the typical curriculum treats them as if they are. They bring previous school experiences as well as life experiences with them. I have found that some students in every class know *something* about most every idea that we are expected to study. My job is to find out how they have informally experienced the concepts and skills of our assigned curriculum so that I can encourage them to

Box 3.23

Continuing the Conversation

Here is a bit more of the dialogue between Sy and Eloise from Stephen Brown's Novelette. Both of these people needed to continue the conversation at another time, so it had to be left at this point. You should know, however, that they are not "fighting." They are good colleagues who appreciate each other's perspective. Suppose *you* are the teacher or department head in this conversation. What would you want to say the next time you both have a chance to talk?

"As a start, don't you think we should admit that students are always posing problems on their own?" Sy began. "Even if we don't acknowledge it, don't you think they are asking questions like, 'Why did the teacher put an *x* on this side of the equation and not that side?' Sometimes the problems are about the math we try to teach; sometimes about the purpose of doing it at all; sometimes about their daydreams to escape from it all."

"But it's one thing to be aware that they are doing that. It's another thing to acknowledge and honor that kind of activity as part of the curriculum," said Eloise.

(Brown 1996: 292)

explore these informal sets of prior knowledge mathematically—and eventually to name them as such. A good teacher will still worry whether his/her students will develop their own, idiosyncratic and peculiar versions of the school curriculum. Will they be able to do well on standardized tests? Will they be ready for a new teacher in the future, who may not share our terminology or strange procedures? I, too, am concerned about this. So I take seriously the idea that mathematicians communicate their ideas to each other. We must also read the work of other mathematicians! "Has anyone already explored what I am investigating?" is a valuable question in the classroom. What does our textbook have to say about this? How can we find it in our textbook, if it is in there? Can any other students, family members, or other people in the community help us with this inquiry? My students are exposed to traditional mathematics, because it is even more relevant to their investigations than the drill and practice curriculum that may be occurring in another classroom.

Questions have different lives at different times in the mathematical community of a classroom. What is at one moment an exercise for practice may feel at another moment like a genuine problem of interest. At still other times, the same question may seem like a puzzle. David Kirshner, a mathematics education professor at Louisiana State University in Baton Rouge, explains that this feeling has something to do with how much a student has been acculturated into the particular mathematical community (Kirshner 2000). Early on, as one is at first beginning to investigate some mathematical phenomena, most questions are still unclear and not quite understandable. They are part of a collection of "problems." Once one has worked with the ideas a bit, things start to act like "exercises" because they follow certain patterns and fit into types of questions for which procedures have been established. For example, in our investigation of shaped numbers, we worked with a collection of problems such as "Can there be square, pentagonal, and hexagonal numbers?" or, "What would these numbers look like?" and "Can they be combined in a way that is related to combining shapes to make other shapes?" Then, when we started to fill in our charts, we completed a series of exercises, such as "What is the next pentagonal number?" It was only after going through these exercises that we were able to consider some new problems, such as "Can square numbers be made out of triangular numbers?" Earlier, this question would have made no sense to us; we would not have been acculturated to the point where the question would be of interest. At this point, after figuring out that we can indeed use triangular numbers to make square numbers, we are "hooked"; we are ready to receive challenges that push us beyond the simple understanding of the material toward important mathematical work. We can ask questions such as "Can we make the pentagonal numbers from triangular numbers?" While the question is difficult to answer, we appreciate it as an obviously related question. Such questions are "puzzles," questions that signal membership in the mathematical community. Puzzles will only be interesting to someone who has become "part of the club"; the interest in the puzzle and its solution is cherished by the mathematician as a symbol that he or she already knows a lot, even if he or she can't yet figure out an answer to the puzzle. The fact that I can understand the question, framed as it is, in specialized language and assuming motivation on the part of the mathematician, referring as it does to special objects that I know about, can only be appreciated by those who "belong."

Continuing the Conversation

The art of problem posing is built on the premise that *everyone* is a mathematician. But mathematicians actually pose exercises, problems, and puzzles for themselves and each other as part of their work in order to establish these special communities that they can be a part of. We can use this idea in our classrooms as well, so that class members can enter into the communities that we create together and create a common "classroom culture." We agree on a name for a part of a mathematical object that we are working with, and only we can claim to have been there at its birth. Even if I tell a friend in another class about it, my story acts so as to demonstrate my part in the community of my classroom, and my friend's vicarious participation will never be the same as my own role in my community. Specialized languages, too, work to support the creation of community. When we refer to a result by the name of the mathematician who introduced it—for example, by calling the fact that all squares are also rectangles "Kevin's Rectangle rule," or by naming the accepted practice of solving an equation by adding the same amount to both sides as "Britney's Equation technique" because Kevin and Britney were the first people to suggest these ideas—we solidify the membership in our community. Both those whose names are used, and also everyone else in the class who was present when the name was established, is made a central member of the community by this act of naming.

Just as important are those questions for which we still have no complete solution. When Eloise worries about youth posing questions that neither they nor their teachers can solve, I expect Sy to be thinking in the back of his mind that such problems can remain alive throughout the school year. Perhaps it is March and a member of the class notes that "We still haven't figured out Tasha's question about the differences of two square numbers." Or, perhaps we have set aside ten minutes each Thursday to discuss any further results that anybody has come up with outside of class time on those unanswered questions. We may have them posted on a large sheet of newsprint on the side wall, each associated with the names of those who posed them. Tahss, Ashley, Samir, and Jen are treated like Fermat, whose last theorem was not proven for centuries, or Goldbach, whose conjecture remains unproven. Unanswered questions are part of the life of mathematicians, who must always decide whether they should spend their time trying to understand these questions that others have not been able to solve for years, or work on something else. In this way, too, posing a problem that nobody can solve is another sign of membership in the community, as would be not being able to solve it. Unanswered questions are especially challenging puzzles.

However, not being able to solve an exercise is not the same thing as not being able to solve a puzzle. Everyone on a football team should be able to catch a toss from five feet away, and we practice this every day after school. Every knitter must be able to turn a skein of yarn into a ball, know the difference between a knit and a purl, and be able to knit a simple scarf. Likewise, everyone in our classroom needs to be able to add fractions like $\frac{1}{3}$ and $\frac{1}{2}$ or we will never be able to talk about patterns with fractions. So we understand the need to practice adding fractions every day. The strength of the five-part unit discussed in Chapter 1 is that it provides opportunities for both inquiry, the pursuit of problems and the posing of problems, through projects, and exercises, the practice of techniques that the teacher says are necessary, through mini lessons and clinics. What this structure of assessment and instruction does, however, is replace the traditional focus of classrooms on drill and practice with an alternative emphasis on inquiry. Nothing is lost. There is just a shift in the ways of working so that practice is valued for what it leads to rather than merely for itself. Similarly, the five-part structure begins with an introductory set of experiences that orient the class community toward a potential set of inquiries. In this beginning part of the unit, students are introduced to basic ideas out of which they may develop their own questions. So we no longer need to worry about whether students are ready to pose questions about something they do not yet know about. The five-part structure supports the creations of a community so that even the practice is part of what makes the interesting work in the classroom possible. Such ways of working ensure that everyone in the classroom *is* a mathematician.

> The "math" of mathematics class, those bits of pre-fabricated knowledge that are identified in curriculum documents, aren't the important things. Rather, they are the visible consequences of an invisible backdrop of mathematized sensibilities.
>
> (Brent Davis 1998: 36)

It is only now that we are ready to admit this deep, dark secret, which returns us to the psychoanalytic perspective of Chapter 2. To become a mathematician is to embrace those mathematical ways of relating to our world. And to teach mathematics is to embrace the frustration that we cannot simply push a button or do the right kind of dance in front of our students in order to make them learn. Teaching does not itself cause learning. Many teachers pretend that telling is an efficient, tried, and tested method of imparting knowledge, and that students learn best when new material is broken down into discrete chunks that are carefully explained. In teaching this way, teachers are honestly trying to help their students. But students do not think mathematically when they copy the dance their teacher is performing in front of them. They only learn to cook by cooking, not by eating fine meals prepared for them. What makes this tricky to understand is that we only learn to appreciate a good dance or a fine meal when we experience them prepared by highly skilled others. In the community of the classroom we can make this happen by sharing our own investigation and those of other mathematicians. This is partly the role of the key skills in the curriculum: to tell us what finely prepared mathematical meals we might serve. Another, more important role for the teacher, possibly invisible to non-teachers, is to serve as examples of exquisitely prepared models of what we are all striving to become. This last role requires us each to see ourselves as mathematicians, for only mathematicians are able to teach mathematics. It is a certain kind of mathematician, however, who has thought deeply about the kinds of examples they can provide, in addition to establishing an infrastructure for working together *as mathematicians*.

Response to Chapter 3:
Explore the Vastness of Mathematics

David Scott Allen

A professional mathematician must spend much of his or her time trying to go beyond the limits which previous mathematicians have constructed. While they appreciate the ground that has been broken by their predecessors, they may not think about mathematics in the same way that it is presented to them. They still need to reconstruct the ideas in their own minds and figure out a way of thinking about it that makes sense to them, as well as to others.

I want to know what my students' passions are, and then figure out ways to get them to mathematize those interests. All mathematicians do is look at a certain body of knowledge more carefully than others, because they are more interested in it than others. I don't buy the whole "I'm never going to need this" posture for many reasons. Life is not only about needs, but also about wants. When am I ever going to need being a high school varsity pitcher again? When am I ever going to need my high school German classes again? When am I ever going to need the social studies or science from junior high? No, I'm not going to include reading or writing here, but I will include novels and sentence diagramming. When am I ever going to need my musical training?

The answer is nearly always: when I really want it. And even then I will go back and relearn (an oxymoron?) the material. I taught Pre-Calculus to a class of very bright ninth grade students last year, and I had to really reach back and relearn just a step ahead of them. But I could do this because I knew how, and I had been there before. Mathematical topics are means to ends. The problem is that when you are twelve years old, you don't know what that end is. All of the ideas are metaphors from which to draw in the future. Then one day you put a couple of ideas together and you know who you are and what you (think) you want to do. This will change many times in your life, but all the more reason to have options from which to choose.

Imagine a large piece of graph paper with many large squares. Then imagine each square having smaller squares inside it, and each smaller square having even smaller squares inside it. Such is the knowledge of the human race and the attention of any individual human. Because our days are numbered (that is, we have finite time to spend), we must decide how to spend our energies. We can choose to examine one very miniscule square in detail for a very long time. We can choose to look at the relationships among some nearby squares. We can imagine what it would be like to find the material from Square A suddenly appearing in Square B. Life is much about scale and detail. Mathematicians choose to pay more attention to the details of properties of and relationships between numbers and figures. My wife spends most of her time paying more attention to the properties of and relationships between her friends. We clash when we attempt to get the other to mathematize the things we mathematize as much as we mathematize them! A person's knowledge in a given area will outstrip anyone else who has not taken the time to look at those matters in detail. There are experts and geniuses in every field, but in many cases school does not reward student experts in certain topics.

The mathematician's notebook is a wonderful thing. I love paper, and I love to write and draw on it in exploration of "esoteric" mathematical minutiae. The triangle experiment is an interesting one when you are working with twelve and thirteen year olds willing to participate. Who gets to decide what relationships are worth cataloguing and pursuing and which ones aren't as important? Not you. But there are times when something strikes you as being so intriguing that you follow it up even though it's not "important." To name and label the new triangles and discuss their qualities is the beginning of mathematics.

I did the triangle experiment with two of my classes, and they enjoyed challenging themselves to find different (self-invented) categories for new triangles. The list of informal shape languages did not include many of my students' discrepancies. The overarching one is "redefinition" of a triangle. One of my twelve year olds differentiated between happy and sad triangles by drawing faces inside them. Should we call this Personification of Triangles? One of them constructed a triangle by drawing a straight line from the top of the paper's edge diagonally to the right edge, making two of the "sides" be the edges of the paper itself. Is this Difference in Material Construction? Somebody drew a triangle as the end of a triangular prism. Could this be called Triangular Function? My students broke out of Euclidean geometry to surreal definitions of triangle. Three points connected by three "squiggly" lines (like the sine wave). They connected vertices with repeated text, small circles, and double lines. Some of them did so without referring to the fact! Students labeled their triangles social and antisocial depending on whether or not they lived in a colony (common sides) with other triangles. Students drew triangles into real objects and labeled them "cheesy" and such. Some triangles had appendages. One of them created a work of art with the arrangement of the triangles on the page. Another student created a triangle which was too big for the page. We will write more on this in Chapter 6 under the name of space-off.

Let's take the case of the sides being serrated (saw toothed) lines to illustrate this issue. In a usual classroom situation, the serration of the lines would be dismissed as not related to the topic at hand. However, in the world of tessellations, the serration and its variations are actually a technique for creating and classifying the tiling patterns that cover the plane. They help define the enantiomorphic (left- and right-handed) patterns displaying "similar differences." So why are they not legitimate in the usual classroom setting? Perhaps because the teacher is unaware of the aforementioned mathematics, but perhaps because the teacher is passing along to the student what they are supposed to be looking at. This will do nothing but perpetuate a narrow band of mathematics, which will keep us figuratively (and perhaps literally) well within Euclidean space. There are infinite worlds to be discovered. I believe that not all of them will be found by adults or professionals.

One triangle was colored in as opposed to just drawn. The coloring schemes of tessellations is also a field in its own right. In related geometry, the four-color problem has only recently been solved (among some controversy because of the types of issues which we are now discussing) using a proof which includes a computer program and its results. Perhaps the first pioneer in the field of color tilings is M.C. Escher, the Dutch graphic artist of optical illusions and visual paradoxes. He was not a mathematician, yet his notebooks are full of analyses of this type of situation. He was nearly obsessed with the regular division of the plane, but took it one step further as an artist to include how he could take a certain tessellation and color it differently. And so, he was indeed a mathematician after all.

While I believe that there are helpful rudiments of any field to remember and revisit, I also believe that many such constructed hierarchies are merely circles of connection, rather than pyramids built on a foundation. My modern born and bred mind likes the organization of the Van Heile levels. But upon further review, it's not quite that cut and dry. My own children use certain turns of phrase and vocabulary simply because I do. After hearing them enough, my two year old even uses them correctly! They are merely one more useful grid through which to look at geometric thinking, but not the progression once thought by the professional experimenters who adapted it from Van Heile.

Nothing illustrates the idea that you talk about what you look at and vice versa more than teenagers! They are social animals, and so they examine every nuance in their relationships internally (well, the girls do, anyway), and then discuss it using its own, sometimes invented, vocabulary. The online communication system has given rise to hundreds of codes for brevity, such as LOL for "laugh out loud." Why invent them? It's better than repeating far more lines of text.

The fact that many of the experiments in this book are borrowed from geometry encourages me that geometry is making a comeback. The square combinatorics problems are difficult to do, rewarding to pursue, and a bear to calculate. Yet they are self-determining sets, and the answers are there for the discovery. I have done a similar experiment myself as a teacher and as a person who loves geometric beauty and design.

You may have heard of the knight's tour problem in chess. Place a knight on any square and then jump it around the board until it touches every square exactly once before touching any square twice. I explored this same idea from the perspective of making polygonal figures out of the jumps connecting squares (now converted to vertices). The number of different figures using four or six jumps is manageable, but with eight jumps the number rises sharply. It was an eight-year journey (off and on) to finding exactly how many eight-jump polygons existed. My next personal mathematical investigation project will be to do something similar on the hexagonal grid.

These projects are somewhat analogous to a baseball game. When is a baseball game over? When certain activities (outs) have taken place. Or take tennis. Win a certain number of points within a certain conceptual framework, and you are the victor. The game itself, and human indeterminacy, determines its own limits. It ain't over till it's over. Other sports and games have timers. The game ends when the clock says zero. But the clock is arbitrary and a third party regulates it. Baseball and tennis determine themselves, so in this way they are criterion-referenced, rather than performance-based.

We must name things, define the term we invented, and then ask questions about it. This is mathematics. Mathematicians produce a purpose for their work as they work indeed. Many mathematicians die not knowing their contributions or their implications. So they need each other to connect the chain of significance and understanding to keep their ideas alive. Often, uses for mathematics are found only after the mathematics is discovered. Categories arise out of the naming as they become larger in some significant way.

One set of digital discoveries I just happened to notice is a set of coincidences with my anniversary and the birthdates of my four children:

		Years since previous		Age at next occurrence		
Wedding	10/15/89	–		15	Birth Month/Day	Birth Year
1st child	11/27/94	5 years, 1 month		10	Odd/odd	Even
2nd child	1/10/99	4 years, 2 months		6	Odd/even	Odd
3rd child	4/19/02	3 years, 3 months		3	Even/odd	Even
4th child	8/4/04	2 years, 4 months		1	Even/even	Even

Do you see the patterns? The third column is fairly obvious, but the fourth column is a list of the first five triangular numbers in descending order, and represented our ages on 8/4/05. The fifth column represents every permutation of odd and even dates. Finally we have three even and one odd birth year, and we have three boys and one girl. According to the first pattern, child #5 should have come along on 1/10/06. This did not happen. I would like to state emphatically that I do not believe in numerology. If you can stomach the psychological aspects of the movie, there is a great monologue in the motion picture *Pi* (π) which explains the difference between numerology and mathematics. How did I figure this all out? I was just playing around with the numbers and examining them closely one day, because I chose at that moment in time to mathematize the dates.

So is language communication or is communication language? They are one and the same in mathematics, where without the language there is no communication, and without communication, there is no need for the language. It may seem a bit artificially constructed in textbooks, but it really can be helpful to speak about the same things the same way with as many people as possible.

The point brought up in the text about the tilings of bathroom floors may seem an odd example, but that belies a problem we face: why do some people mathematize some things and not others? I definitely lean toward certain types of flooring because the patterns give me hours of endless enjoyment when I'm stuck in a given room trying to get my infant to sleep. But other people seem to ignore the fact that they are in almost every case making a geometric decision, not an aesthetic one. The geometry of perpendicularity is so intuitive in us that we do not know how to function without it. The fact that we stand at right angles to the ground in nature imbues us with a sense of perpendicularity. Any departure from this disorients most of us. It's not coincidental that most people choose patterns resembling a square grid with some relatively minor adornment or variation on it. This point is lost on most because they just go in to the store, see what's available, and buy what they like. So why don't people care more and consciously think more about the issues raised here?

Approaching mathematics as science is the way I do many things. Does this mean I approach everything as mathematics, or that I mathematize everything, or that I approach everything as approaching mathematics scientifically? My aforementioned work on knight move polygons is a process of making sure (in my own mind anyway) how many figures of a certain type exist. I also used a computer program. I still don't have a handle on the general case (i.e. a pattern) but I have the specific cases I want for now, at least until I learn more powerful computing languages than the archaic one I used to assist me in my endeavors.

Polya's book is one of those ironic icons of mathematics education. His work was important and groundbreaking at the time, but apparently misunderstood in the realm of education; so much so that there are sections in some of my own textbooks related to his problem solving techniques. His book has come to be used as a catalogue of all the ways that problems can be solved, and turned into a textbook for how to teach problem solving. We are indeed our own laboratory rats in this scenario! The point of the book was to help students learn how to solve problems, not to problematize problem solving as something to be learned and taught for its own sake! This is a bit analogous to knowing how to play the violin or swim vs. being able to play the violin or swim. Knowledge is not the issue. What he writes in his book seems like such common sense until you start to dissect it for pedagogical purposes. Understand the problem, Plan what you will do, Carry out the plan, Look back. This is the seed of the archaeological digs discussed in the text. It is very important to examine your journey and conclusions, and ask new questions.

It is Math Team that has taught me that you are not really problem solving if you know what to do. It is sad that even the bright students in my voluntary Math Team are sometimes too afraid to give an answer (even one that is later proved to be correct) because they are afraid they might be wrong. Yet they are short on words when I ask questions to the effect of, "What's the worst thing that would happen if you did get it wrong," or, "So you get it wrong, and then what?" We do hundreds of problems over the course of the year practicing for our School and Regional competitions. This year, there were thirty-six problems still unsolved after we had gone through all of them. It occurred to me that this was the best time to do problem solving: when you don't know the answer after having tried to get it and failed. Students are a bit bewildered by this, but they have lots of perseverance built up after our challenging two-hour sessions once a week.

Mathematicians do ask their own questions, but students are expected to have the answers to types of questions which other people ask. This is where the rubber meets the road. Do I get them to want to ask the questions that the test-givers want them to answer, or do I teach to the test in a short amount of time and teach them all of the rest alongside? In other words, I could acknowledge the testing structure only as an aside to the curriculum, or as a core of the curriculum upon which we can add our own ideas. Right now I'm somewhere in between. If asking their own questions is truly the essence of mathematicianhood, then the students will never truly experience it, because they have the tests looming in the distance. But what if they learned what they had to learn and still had time? So we have another set of poles—the questions they would ask and pursue if we let them go, and the ones which we expect them to know the answers to. Textbooks are notorious for their streamlined sterility and lack of deep problems. Maybe if we published the kitchen work it would be more attractive to the reader. Use text squares for conclusions, but after the false starts and tangents have been worked through. This sounds like an idea for a novel!

I also have the issue looming of the fact that my students are at such varied levels of maturity, motivation, and knowledge of mathematics and life in general that they amount to a bunch of people thrown into a room to be force fed mathematical objects which they would rather not look at or know about. They would rather harass each other than deal with a mathematical proposition. For now, I debate with myself how to turn this tide around in their hearts and minds, not so much in their classroom behavior and test performance. Maybe ideally everyone needs to know how to add fractions so that we can discuss patterns with fractions, but if the person does not care to become a part of the community, they are still part of my class, and I can't ignore their presence. I can encourage, but I can't get them to be interested in the process of discovery. For some, practice may be valued for what it leads to, but if it will not lead anywhere interesting for others, it creates a have vs. have-not situation in one room. Ironically, it is when it works that it is exhilarating, and when it does not that it feels forced.

Action Research 3: Mathematics Journals

Karen Cipriano

I feel the students came a long way with the journal project in my math class. I first introduced the students to the journals in January. I chose a "thought provoking" question that pertained to the lesson taught that day and had the students write or draw a picture about it. They had five types of worksheets to choose from in organizing their work, and they began choosing two out of the five consistently, so I chose to develop only those two types of papers. The students would write their journals at the end of class, take their homework and leave for the day. They were a little apprehensive at first, but then some of the students began to blossom, while others would write less and less.

I began to change a few things. First, I made an expectation sheet called *Math Journal Writing Guide*. The students and I reviewed the guide and discussed it. Most began to put more effort into the project, but some would only give me their minimum. I checked the journals each day and wrote comments and feedback that I hoped would encourage them. Then when I read Chapter 4 and 5 in *Finding the Connections*, by Jean Moon and Linda Shulman (1995), I felt I had to do more documentation. I began reading about documentation and rubrics and felt this would help my students focus on the parts of the problem they were doing correctly and incorrectly. I read carefully through the rubrics in the Moon and Schulman book and used the rubric from the Pennsylvania State PSSA test in mathematics to create a rubric that fit into my own journal project.

I introduced the students to the Math Journal Writing Rubric and the descriptions on how they would be scored. I made the rubric from the expectation sheet I had given the students at the beginning, so it was really an extension of what I did then. The students responded well and were more cautious on meeting each criterion, which was what I was hoping for.

I then changed the time the students would write their journal. At first I had them write at the end of the class, which always seemed rushed. The students leave my class at 11:40 a.m. and lunch is at 11:45 a.m., so they rushed to write in order to leave and go to lunch. I decided later on to try a question from the previous day's lesson and complete the journal writing first before moving on to the new day's lesson. It has worked very well. I combine some students from my homeroom with the next class, so as we switch, I have one student who is in charge of distribution and collection of folders. When I say it is time to switch, I do not even have to remind her. The journal is on the board, so those students who are already in my room begin their journal, while the other students file in with their folder already on their desk. They take out their homework for me to check as they write in their journals. The transitional noise has dramatically decreased and students are working productively, while I am checking to accept or reject homework. (Work must be shown!) There is a purpose to the question asked, so it is not busy paperwork to keep students occupied.

New Directions for Equity in Mathematics Education (Secada and Adajian 1995) has been great because I feel it fits my situation. I want to make sure that I am useful to my students and giving them the best education possible. I feel that is a teacher's job and I take it very seriously. The *New Directions* book made me feel that I was not alone in striving for equity, and made me feel determined to succeed in being equitable. I feel no matter where you teach, students need to be challenged, and I think the book made me think about *my* students. With those thoughts in mind and reading yet another book, *Mathematics, Pedagogy and Secondary Teacher Education* (Cooney *et al.* 1996), I was able to focus more on what types of questions I should use for journal prompts, or what types of problems would be challenging to the students. I wanted them to "think" beyond the daily lessons. I felt both of these books enriched my thinking and challenged me to do just that with my project.

In the last phase of my project I adjusted my rubric. I changed the order of the criteria because a picture or graph did not always have a place in the questions I asked, so I put that last. I also changed some of the wording for rubric criteria. I did not have a score of zero because I thought it would scare the students away, but I did not feel comfortable giving a student a "1" when they did not even attempt an answer. So, I decided to make the "1" category an attempt category and a "0" if the student left it blank. I want my students to get into the habit of restating the question in their answer, so I decided to add "use complete sentences" to earn a score of "5." I then adjusted the wording of the sentences for the rest of the rubric. I also changed some of the wording for the numerical answers too, so it would be more specific in what I was looking for. Finally, I put the four criteria at the end of the journal worksheet paper, so all I need to do is fill in the numbers and give my written comments.

Interviewing is something I need to work on for next year and I think that is an extension I will add onto the journals. I spoke with the students more informally during this last half of the year, but I really was unsuccessful in the formal process of keeping to an exact schedule. I printed out a schedule, but honestly, I need to tack that on to next year. I am really disappointed, because I think the interview part is extremely important, and if I keep to the schedule the students will become accustomed to it. I believe it will become easier as time goes on. The students seem to enjoy talking to me and discussing their journal and they would even remind me that "today is their day" to talk to me. I feel once I field-test the interview, it will make the journal process more meaningful to the students and myself. I hope to take what I have learned so far and add on to it for next year.

Another project that I have already thought about for next year is to compose weekly problem solving journals that the students do at home. I will hand them out once a week, say on a Wednesday, an involved problem the student needs to work on for the week, and it will be due the following week. I hope to work over the summer in finding resources to use for the problems. I do not want it to be a "regular" word problem. We discussed and read in a graduate course I took about developing skills in problem solving and I hope this will help my students. I focus so much on their basic skills sometimes that I leave the problem solving last. My goal is to change that and I am hoping that next year with the journal writing, interviews, and problem solving, I can move my students to be mathematical thinkers instead of mathematical robots.

Karen Cipriano

I am a fifth grade teacher at Stewart Middle School in Norristown, Pennsylvania. I have been teaching at-risk students for the past twelve years in a self-contained classroom. It is such a pleasure when my enthusiasm in math carries over to my students and we see an increase in their standardized test scores. Just watching my students work cooperatively to solve challenging word problems by questioning, conferring, and evaluating their results is exciting. I am very family oriented and I enjoy outdoor activities such as walking, biking, and a variety of sports.

MathWorlds 3:
Reading and Writing Mathematics

1. **How did you read this last chapter?** Share with at least two other people a sample of how you read this last chapter. Did you write notes in the margin? Highlight things? Work out some of the math on a separate piece of paper? What ways of reading mathematics texts work well for you? Based on your discussion with others, what new ways of reading will you try the next time you read a mathematics text?

2. **Trying new ways of reading.** Use a new strategy of reading that you mention in question #1 on a mathematics text of your own choice. Don't just try it once! Keep using this method until it works for you. What did it take to make it work for you? Based on this experience, what ideas do you have for supporting students who are learning new methods of reading, and using writing to support reading?

four
Critical Thinkers Thinking Critically

How am I to think of "critical thinking" in my classroom? I find this an overwhelming task at the current juncture in mathematics education. Once finding solace in the National Council of Teachers of Mathematics *Standards'* (NCTM 1989, 1991, 1995, 2000) support for problem solving, reasoning, communication, and assessment that features these general goals, I now find myself caught in the cross-fire of a turf war among students' expectations for a skill-based lecture format, parents' desires that range from a delight in the "reform" movement to horror at the open-ended assignments coming home, standardized tests that have not yet caught up with reform, and a media-amplified backlash reminiscent of the seventies' Back-to-Basics. I cherish the central place of mathematics in the school curriculum, and note the legacy of mathematical associations with critical thinking, having heard numerous clichés throughout my life that refer to mathematics as contributing clarity of thought, an appreciation for logic, and a propensity to analyze and generalize arguments and presumptions. Yet I also note with concern the historical inconsistency of any sort of "transfer of learning" of these skills to non-school experiences. And I further recognize the role of mathematics in perpetuating an ideology of "reason" that can contribute to regimes of truth and power rather than to a project of social justice (Appelbaum 1995, 1998; Mellin-Olsen 1987; Swetz 1987; Walkerdine 1987).

> There has probably never been a time in the history of American education when the development of critical and reflective thought was not recognized as a desirable outcome of . . . school. Within recent years, however, this outcome has assumed increasing importance and has had a far-reaching effect on the nature of the curriculum.
>
> (Fawcett 1938: 1)

130

Thank *goodness* such attention was paid to critical thinking back in the thirties when Harold Fawcett's dissertation on his geometry class was printed as the NCTM yearbook. More than seventy years later, we surely should be able to recognize innumerable examples of Fawcett's attributes of a student "using critical thinking well"; such a student:

1. selects the significant words and phrases in any statement that is important, and asks that they be carefully defined;
2. requires evidence supporting conclusions he or she is pressed to accept;
3. analyzes that evidence and distinguishes fact from assumption;
4. recognizes stated and unstated assumptions essential to the conclusion;
5. evaluates these assumptions, accepting some and rejecting others;
6. evaluates the argument, accepting or rejecting the conclusion;
7. constantly reexamines the assumptions which are behind his or her beliefs and actions.

(Fawcett 1938: 11–12, paraphrased)

The fact that this list is referenced as vital to our understanding as recently as 1993 (O'Daffer & Thornquist) would presumably attest to over a half-century of accumulated wisdom regarding how students studying mathematics involve themselves in a process described by Costa (1985) and Ennis (1985) as effectively using thinking skills to help one make, evaluate, and apply decisions about what to believe or do.

One problem, though, is the continued under-theorizing of critical thinking in an individualized, or egocentric and antisocial, politics of education that echoes another early twentieth-century formulation often quoted: Robert Hutchins, once the president of the University of Chicago.

It must be remembered that the purpose of education is not to fill the minds of students with facts . . . it is to teach them to think, if that is possible, and always to think for themselves.

(quoted in O'Daffer & Thornquist 1993: 39)

An observation I want to stress in this chapter is that I no longer construct "critical thinking" for myself as thinking "skills," and find the notion that I "teach" critical thinking a barrier to successful experiences in my classroom. I prefer to take my students *as critical thinkers* who enrich their abilities and deepen their conceptions of themselves as thinkers through our efforts in class. This is an extrapolation from recent transformations of "problem solving" in mathematics. In 1980, the National Council launched Problem Solving as the number one "basic skill" for the eighties (a clever response to the seventies' Back-to-Basics movement). Horror of horrors: we found problem solving curricula sequencing the various problem solving skills and strategies added on to an already crowded smattering of mathematics, a whole new realm of opportunities for some students to feel good about themselves and others to "learn" that they are no good at solving problems. By the nineties we had the *Standards'* presentation of problem solving not as a skill to be explicitly taught, but rather as the context through which all mathematics should be learned. Having tried this out I personally moved on to "critical thinking" as the context rather than the objective of my classroom curriculum.

A claim to be made again and again is the haphazard association of mathematics with any sort of rationality or clarity of thought. The links we make are based on a cultural convention of mathematical activity in schools as opposed to any universal quality of mathematics that the subject exemplifies (Appelbaum 1995; Hersh 1997; Pinxten *et al.* 1983; Rotman 1993). This claim has been well made in different terms recently by Heinrich Bauersfeld (1995). As Bauersfeld tries to research children's thinking, "certain problems arise with the structuring of the internal process":

> How does a child learn to correct an inadequate habit of constructing meaning? Teachers can easily correct the products, but there is no direct access to the individual (internal) processes of constructing. Thus, on the surface of the official classroom communication, everything can be said and presented acceptably, but the hidden strategies of constructing may lead the child astray in other strategies or in the face of even minor variations.
>
> (285)

When it comes down to it, writes Bauersfeld, "there is no help from mathematics itself, that is, through rational thinking or logical constraints, as teachers often assume. Mathematics does not have self-explaining power, nor does it have compelling inference; for the learner there are only conventions" (287). (As a cognitive psychologist, Bauersfeld is tip-toeing towards the psychoanalytic framework discussed in Chapter 2. In this chapter, we explore some concrete approaches that might bridge the cognitive and psychoanalytic perspectives.) For me, any claim to critical thinking is not unleashed by the mathematics; it is an attribute of the classroom activity, or a description of pedagogical dynamics.

Another driving feature of my current understanding of critical thinking is the dangerous pleasure it affords as an "objective" toward which I steer my students, and for which I reward them. In the words of psychologist and educational critic Alfie Kohn (1993: 31), "a brief smile and nod are just as controlling as a dollar bill – more so, perhaps, since social rewards may have a more enduring effect than tangible rewards." I now work at reminding myself that critical thinking is not something my students need to be tricked into performing, but rather a process they will go through as human beings as long as my organization of classroom activity does not stifle it, reward it, or distort it.

It is necessary, according to Erna Yackel (1995: 158), "for teachers to understand that students' activity is reflexively related to their individual contexts, and that the teacher contributes, as do the children, to the interactive constitution of the immediate situation as a social event." Alan Schoenfeld (1989), meanwhile, collects research to support the notion that mathematics can be taught in a problem based way so that students experience the subject as a discipline of reason developed because of the need to solve problems and for intellectual curiosity. Jack Lochhead (1987), on the other hand, stresses attitudes over methods, encouraging us to have students choose or construct their own problem solving techniques, rather than follow a specific method; students are forced through a structure to choose and evaluate their method. And the ever-quoted "bible" of nineties mathematics education, the NCTM's *Curriculum and Evaluation Standards* (1989) chimes in:

> A climate should be established in the classroom that places *critical thinking* at the heart of instruction . . . To give students access to mathematics as a powerful way of making sense of the world, it is essential that an emphasis on *reasoning* pervades all mathematical activity.
>
> (25)

Critical Points in Pedagogy

So how am I to think of critical thinking in mathematics? *I imagine curves of motion through space, each of which is a trace of the above approaches, and each of which has particular moments of critical change.* In geometry we sometimes speak of "critical points" of a curve, and as I look back over my shifting pedagogy I can identify such points in the flow of my classroom life, *points at which the flow has a sudden shift in acceleration toward a critical thinking classroom.* What follows is a list of eight critical points in the historical trace of my teaching/learning strategies, each of which invite considerations of how to enrich critical thinking in the teaching and learning of mathematics in schools. What I share here is a collection of "before" and "after" images: what I used to do, which was pretty good, and then what I now do, which is even better.

Box 4.1

Thinking Skills Explorations

a) Find a chapter on problem solving or critical thinking in a mathematics textbook. How does this book suggest that students think about thinking? Compare how students are supposed to behave according to the book with your own observations of when and how a group of students demonstrates these same skills on their own when they are outside of the classroom—in the lunchroom, during recess, or at another time in their lives.

b) Find a page in your mathematician's notebook that you are proud of. Would you say that it is easy to find evidence that you have used each of the skills of critical thinking that are in Fawcett's list? Why or why not? What does this make you think about in relation to critical thinking in mathematics?

c) Experiment with a small group of students, where you assume that they each already are capable of the skills in Fawcett's list; tell them that you know they already know how to do each skill when it is relevant to your group discussion.

Treat Mathematical Actors as Mathematical Critics

I used to have my students invent their own procedures and algorithms, and to always search for another way to do a problem. The result was presentation of multiple perspectives on a single situation. This helped to establish mathematics as a humanly constructed technology of meaning which I hoped would lead to two results: (a) a view of oneself as a maker of meaning; and (b) a view of mathematics as made by people, and thus subject to the same critique as other human endeavors, according to criteria of value.

Now I recognize that it is not enough to provide a forum of presentation or solipsism. Dewey admonished that a democracy provides not just access but the opportunity to be *heard*. Students in my class not only explain their strategy or procedures. They now have to use another person's strategy or procedure in a similar problem/situation, and participate in a discussion that notes the strengths and weaknesses of each. Students are asked to identify a situation in which they would use each strategy offered (for example, to explain to a younger child, to impress a town council member during a presentation on a local issue, to calculate most quickly, to be most sure of their result . . .). Now my students perceive mathematical thinking through multiple perspectives, and can articulate a plausible reason for selecting each perspective over others in particular contexts of use.

For example, we once needed to determine how many bags of concrete mix to purchase if one bag would fill two square feet of area (three inches thick was the recommendation), and the area was a rectangle measuring three feet by five feet. Kudan suggested eight bags, because $3 \times 5 = 15$ and $15/2 = 7\frac{1}{2}$, and he figured he would need to have to buy whole bags. Marlee came to the same conclusion by reasoning in terms of a ratio: 1:2 is equivalent to $7\frac{1}{2}$:15. Xandie drew a picture of a three by five rectangle, drew lines at every foot to create a fifteen square grid inside the rectangle, and proceeded to color in two squares at a time, counting up to seven, which left an empty square; the empty square called for another bag, making the total eight bags. Pearline skip counted by twos on her fingers until she got to fourteen, and then figured another bag would make eight.

Students discussed their solution strategies. The group noted that Xandie's and Pearline's methods were similar in that they counted the bags needed, one visually and one numerically. The students liked Marlee's ratio approach the best because it seemed the simplest to do, even though it was the most challenging to understand why it worked. Kudan's strategy seemed fastest and the most reasonable one to use in a situation where the numbers involved might be cumbersome. In another problem, if the numbers were something like 53 feet by 294 feet and 3 inches, and a bag covers $7\frac{1}{2}$ square feet, they felt that the counting strategies would be too confusing, and the ratio too difficult to solve. This group of students went on to work through another similar problem using all four strategies each, including in journals their thoughts on what audience would most appreciate each strategy as an explanation.

Make a Choice, Pursue it, and Consider the Consequences

I used to structure my activities to include a collection of critical thinking skills, in order to facilitate my students' development and refinement of these skills: comparing, contrasting, conjecturing, inducing, generalizing, specializing, classifying, categorizing, deducing, visualizing, sequencing, ordering, predicting, validating, proving, relating, analyzing, evaluating, and patterning (O'Daffer & Thornquist 1993).

Now I no longer view my job as a trainer in skills. I instead recognize that my students come to me with varying inclinations to use skills of critical thinking in school contexts. I presume that critical thinking is a trait of human experience. I can take advantage of this trait and make critical thinking the context through which mathematics is learned. I provide open-ended situations in which students use the above skills of critical thinking to draw a mathematical conclusion or accomplish a mathematical task. I ask the students to design the questions and investigations themselves. Students then must *choose* one question or investigation and work together based on their selections. They then report to the classroom community about their group's ideas, new questions, findings, or frustrations when they believe this new information could make an impact on the community's evolving understanding. Here they must choose as well. We discuss whether their choices of question, investigation, timing, and format of reporting were good ones, and how they made these decisions.

My class once was investigating calculator patterns, as we discussed in the Prologue. In our class, we selected a starting number from 0 to 9, and an adding constant from 0 to 9. Students would enter some starting number, and add to it the constant number; then repeatedly adding the constant number, patterns emerged in the 1s digits on the calculator screen. After initial explorations, ideas for investigation were collected in a class discussion: odd versus even constant or starting numbers; the relationship between the constant and the length of the resulting pattern; the effect of a starting number on a particular constant chosen; and so on. Students chose to work in groups based on which investigation seemed most interesting or promising to them.

Another day we explored which four-sided shapes could make a square shadow when held up to a light source, as discussed in Chapter 3. Groups investigated the relative importance of angles, parallel sides, lines of symmetry, and distance from the light source. Several groups split into two research teams that either worked abstractly in their notebooks or preferred an experimental approach, cutting out shapes and tilting them against a light source to see the shadows produced. This was a great activity that led to many insights in geometry. But most fascinating was the class' conclusion that teams working abstractly were able to understand the significant issues more readily than the people who had worked with the actual shadows.

Obsess About Functional Relationships

I used to collect data in explorations and support students' identification of patterns in the data, encouraging them to search for more than one pattern or to articulate more than one rule or description of the same pattern.

Now my students are pressed to go further by recognizing how changes in one or more categories of data are related to changes in other categories. In an investigation of the behavior of bouncing balls, students studied the fact that the ratio of the height a ball is dropped from to its return height is consistent and a special property of a ball (this ratio is called a "coefficient of restitution"). Research groups collected data on weight, circumference of the ball's equator, density, and heights of bounce, comparing data across these different categories of measurement. Another group explored heights at which the ratio no longer held, depending on the different characteristics that had been measured.

On another day this same class was studying water drops. After measuring the rate of absorption of a drop of water for different materials, one group switched to maximum number of drops a material could absorb. By changing their variables, they were able to convince the class of the importance of their research for athletic clothing, sanitary napkins, and Band-Aids.

Analyzing a survey of interest in new bike racks versus new stall doors for the second floor bathrooms, my class noted confusion over how race was defined in their survey. These students suggested that affirmative action forms and surveys unwittingly perpetuate an image of "minority" by lumping together some groups into one big category while dividing other groups into specific categories. (For example, Dominican-American and Haitian-American were important distinctions in this school, whereas Italian-Americans, Polish-Americans, recent Russian immigrants, and some Hispanic students would all identify themselves as "white".) Class members felt that the "minority" status of some groups should be questioned.

Strategy games also offer an opportunity to understand the relationships among variables. Mancala is a game I often use: it involves moving "stones" in and out of "pots." Usually there are four stones in each pot to start, and a typical board has six pots on each player's side. Playing the game with standard rules offers numerous opportunities for strategy and decision discussion, especially when we expand the conversation by shifting the goal of play: to win, to lose, to keep the game lasting as long as possible, to "tie." But we can also study how changes in variables—the number of stones in each pot to start, the number of pots, the direction of move on each turn—affect the strategies for a game "well played."

Problematize the "Answer"

I used to have students share the different ways they obtained an answer, having them offer their strategy for arriving at an answer as an argument for why people should accept their answer. Later, they would express how they got their answer without using any numbers or shape names in their story. Then they would explain why their answer was important in terms of the question asked.

Now I encourage students to offer several possible answers based on their calculations. They explain to each other how they might come to an answer despite the seeming universality of their computations. They write up or present in play form a decision involving a choice of three or more answers or actions based on their calculation that dramatizes the complexity of choosing with conflicting criteria.

Returning to the concrete problem discussed above, each strategy came to the same numerical calculation, seven and a half. Students suggested eight bags. Until they were encouraged to problematize their answer. Perhaps they should buy fifteen bags and share with a friend who also needs concrete. Perhaps the company cleverly packages bags for two square feet knowing that anytime people have an odd number of square feet they round up and buy extra mix; let's buy only seven bags and sacrifice a tiny bit of thickness.

Sometimes my class becomes members of a game design team. The Property management people ask them to consider using one twelve-sided dodecahedron die instead of the usual two six-sided cubic dice, since there is an overstock on the dodecahedron dice. In their game, players who roll a six or higher move forward, while those rolling less than six move backward. Students can use a variety of methods to determine that players will move forward more often with the two cubic dice. Yet the decision still remains: will it be a better game with players moving forward more often, or just a shorter game?

Problematizing the answer can also be done by making the answer into the question, as described in Chapter 3: I start with the answer and ask what might lead to it. Instead of asking for the average of a list of numbers, I give the average and ask for several sets of numbers that could have that average. Instead of asking for the graph, start with the graph and ask for the story. The square shadows activity is like this as well: given a square shadow, what might be the shape?

Problematize the Pedagogy

I used to offer options for learning a particular topic or skill. Sometimes I would arrange centers based on multiple intelligences; sometimes I would use jigsaw cooperative learning techniques around key components of a unit.

Now I present the teaching of a particular topic or skill as a controversy among educators. I have students experience at least three different ways that people have thought of (I avoid a binary continuum/happy medium situation) and have students evaluate the approaches (a) for themselves, (b) according to what they perceived as the logic of mathematics, and (c) how it helps them connect to things in their life.

For a unit on percent, students were asked to critique an area model of percent, base-ten blocks described as fractions of the "whole" square flat, money as percent of a dollar, and a ratio model during their bouncing ball investigation. A follow-up activity studying meat labeled "low fat" led students to suggest such labels were deceptive marketing ploys because even "low fat" meat by weight is extremely high in fat by volume. It just happens that fat takes up lots of space and weighs hardly anything when compared with meat. Class members prepared explanations for parents and family members according to the model of percent they thought would best communicate their ideas to the particular audience.

Another study of percent involved some students interviewing a professional (a carpenter, a baker, a rug layer, a taxi driver), while others researched non-fiction materials in the school library, and still others read books written for them by former students. Class discussion focused on how the pedagogies differed, and how the pedagogies influenced interpretations of the meaning of percent. Students then created their own books, puppet shows for younger children, and rap songs and manuals for older children, on percent; they discussed presentation as effecting the representation of percent as an idea. (For a discussion of such curriculum as "post-modern," see Gough 1998.)

Understand Mathematics as Rhetoric

In the past I would collect examples of the use of mathematics in newspapers and on television. These would prompt investigations in my class, based on graphs, charts, and other statistics. Students would gain an appreciation for the role of mathematics in everyday life, and pursue their own studies of the use of mathematics in different sections of newspapers, and in magazines they read themselves (videogame magazines, *American Girl*, etc.).

Now I ask my students to study how the mathematics is represented, and to consider why the author chose a mathematical representation over other possibilities. We consider the difference in scales on a graph, choices among graph and chart types, units of measure, and the placement of mathematics in the argument.

Recently we were discussing an economic plan presented by a candidate running for reelection. Obtaining platform papers from the local party office, some students noted the prominence of mathematical language and models used for prediction based on equations. Their report to the class facilitated a discussion regarding the need to stuff an economic plan with mathematical rhetoric and imagery just to impress the audience with its seriousness. The group offered two alternative summaries of the plan that did not use equations or numerical facts, but instead generalized the main points of argument in the plan. One summary seemed lacking in content, but the other seemed to them and the rest of the class to be potentially as convincing as the mathematical jargon.

Realize Apprentice Mathematicians and Citizens as Objects of Mathematics

I used to make sure my students experienced decision making as authentic creators of mathematics as well as in the role of mathematically literate citizens. In the first context I imagined my students as apprentice mathematicians learning a craft. In the second context they behaved more like discriminating members of a democratic community, forming opinions about environmental, consumer, health, and political issues with the help of mathematical thinking.

I now add a third position toward mathematics (Weinstein, 1996). We study the ways in which mathematics is used to turn ourselves into objects of study. Last year we examined district standardized test scores and reports on them in the local newspapers. Students interviewed adults in their neighborhoods about their impressions of the scores and the local schools. The principal agreed to be interviewed about the school's scores and how she felt they affected discussions of the curriculum among the teachers.

Recently the connections between theoretical and experimental probability were introduced through a project on market research. Groups of students chose a product and surveyed schoolmates regarding their preferences. A major component of insight during this project was when students noted that they could shift a large number of people toward a second or third choice by varying the relative price for each. Links were drawn between the appearance of "choice" in stores and the likelihood that consumers are led to buy products in order to clear out stock or for other marketing needs. Conjectures about ways in which marketing decisions could be manipulated were tested through further market research surveys.

In these projects we also talk about issues of representative samples. I often bring in examples of notorious mispredictions based on non-representative samples, such as the effects of heart medications on women being misunderstood for many years when studies only included male patients, the inaccuracies of political polls, and the magnifications of error when one uses a sample to estimate a quantity for a larger population.

Perform Celebratory Archaeology

I used to assess student understanding and performance with a variety of assessment strategies, including tests, performance tasks scored with rubrics, portfolios of work selected as exemplary by students, and clinical interviews. I still use these, but find it crucial for students to see for themselves that they have learned particular skills and concepts, and that they can apply these in new contexts. We carefully look over a period of time and collect lists of what students find important, facts they can organize in more than two ways, and conjectures or arguments they believe are central to a summary of the material learned. Finally, we collect a "good list" of investigations they wish they could pursue if they had more time (and perhaps will as part of optional future work), conjectures they think they could "prove" to

> **Box 4.2**
>
> **Eight Critical Points**
>
> Each of these critical points provides ideas for adapting existing lessons and activities into more dynamic classroom experiences.
> a) Find several good lessons that you have experienced yourself or that sound good in teacher-materials, and modify them in eight different ways so as to emphasize the points in this chapter.
> b) Organize your ideas as classroom plans with five parts, as suggested in Chapter 1.

someone else, and questions for further discussion. Important here is our discussion of criteria for including something on these lists, and the need to keep the list to a meaningful length.

Critical Thinkers Thinking

This list of critical points is my way of organizing what I have learned about critical thinking in my classroom, both conceptual and skill-based critical points of pedagogy for mathematics education. I interpret my teaching practice as making the political more pedagogical, because my practice embodies political interests that are emancipatory in nature. I treat students as critical agents, make knowledge problematic, utilize critical and affirmative dialogue, and make the case for struggling for a qualitatively better world for all people (Shapiro 1993).

> In part this suggests [taking] seriously the need to give students an active voice in their learning experiences. It also means developing a critical vernacular that is attentive to problems experienced at the level of everyday life, particularly as they are related to pedagogical experiences connected to classroom practice.
>
> (Shapiro 1993: 277)

I do not expect all readers to agree with me. Indeed, many people want to clutch the (false) certainty of mathematics as a scaffold to critical thinking, while others believe that strong basic skills in mathematical calculations are a prerequisite to their *later* application in critical thinking. I am proposing that we turn this around: starting with the critical thinking that our students bring with them, we create experiences through which we can recognize occasional instances of apparent certainty, and through which we can develop conceptual understanding that leads to a cultivated collection of calculation skills. But more than this is the *presence* of a kind of *critical insight* (Shapiro 1993), an awareness that pervades the ideology of surface description (in which our world is named in particular and distorting ways).

My pedagogical starting point is not the individual critical thinking student, but the critical insight of individuals and groups in their various cultural, class, racial, historical, gendered, and other settings, in which diverse problems, hopes, dreams, and fears become particular to individual students. In my practice, I do not search for the perfect critical thinking lesson. What I do is rethink my current lessons and units (Paul *et al.* 1990) in terms of the critical points of practice I have listed above.

Are my students "less gullible, more logical and more critical in their thinking," as Harold Fawcett asked of himself back in the 1930s? And if so will my list of critical points be packaged, sealed, and marketed in your local teacher store (Tanner 1985)? I certainly hope not. I write with the thought that you too will write, and that I will read what you have written. You will describe what you do, in order to make meaning of your teaching, confronting how you came to be like this, and reconstructing how you might do things differently (Kincheloe 1993; Smyth 1989).

In my own experience, I was able to construct a meaning by performing a standard move in mathematics: widen the context. I found that I could teach a critical thinking classroom by listening to the critical insight of my critical thinking students. I stopped narrowing my attention to what *I* should do, and placed what I do in the context of how everybody in that room is thinking. What and how *are* they thinking? I have stopped searching for lessons that teach critical thinking and instead document for myself the ways in which my students exhibit critical insight in a mathematics class. Am I *allowed* to give up certainty, indubitability, timelessness, or tenseness? Am I unfair? Hersh (1997: 249) ends his work, *What is Mathematics, Really?*, by raising the same issue: when we drop restrictions like certainty or timelessness it is like breaking out of the restrictions of the real number line in algebra.

> Dropping the insistence on certainty and indubitability is like moving off the line into the complex plane. . . . We don't throw away all sound sense. The guiding principles remain: intelligibility, consistency with experience, computability with philosophy of science and general philosophy.

A humanistic philosophy of critical thinking in mathematics respects these principles. The turf wars of mathematics education in the early twenty-first century are witnessing a negotiating team: critical points in a humanistic approach to critical thinking in mathematics.

From Critical Thinking to Critical Mathematics Education

Teachers of mathematics have been searching for ways to describe and enact critical thinking in their classrooms for a very long time. On the one hand, mathematics itself is often held up as the model of a discipline based on rational thought, clear, concise language, and attention to the assumptions and decision making techniques that are used to draw conclusions. In the nineteenth century, a view of the mind as a muscle that could be trained and strengthened in particular skills (this has come to be called "faculty psychology") led many people to justify a central place for mathematics in the school curriculum simply based on the belief that mathematics would train the mind in clear, logical thinking. To this day, employers often hire mathematics majors fresh out of college under the presumption that they have been trained to "think." On the other hand, teachers of mathematics have always been disappointed in the critical thinking that their students demonstrate. And there have been many research studies that point to a dismal chance of any sort of transfer of learning of critical thinking from the mathematics classroom into other realms of intellectual effort. Indeed, research has failed to document any consistent transfer of what might be called "critical thinking skills" from even one branch of mathematical inquiry to another.

Back in 1938, Harold Fawcett was asked to publish his work with geometry students as the yearbook of the National Council of Teachers of Mathematics. This book introduced the idea that students could learn mathematics *through* experiences of critical thinking. This was a big leap from older ideas that promoted specific ideas about how to "teach" the skills of critical thinking. Fawcett wrote that there had never been a time in the history of education when the development of critical and reflective thought was not thought of as an important outcome of school; but he thought that this particular outcome had assumed increasing importance in the 1930s, and, further, that this importance held strong implications for the nature of the school curriculum itself. Fawcett effectively demonstrated that it was pointless to try to teach critical thinking skills (e.g., comparing, contrasting, conjecturing, inducing, generalizing, specializing, classifying, categorizing, deducing, visualizing, sequencing, ordering, predicting, validating, proving, relating, analyzing, evaluating, and patterning; see O'Daffer & Thornquist 1993). Better would be to take advantage of the critical thinking skills that students bring with them to school mathematics, in order to learn the mathematics.

Years later, in 1989, the National Council echoed this earlier call for critical thinking in its *Curriculum and Evaluation Standards*. Calling for classrooms that place critical thinking at the heart of instruction, the new *Standards* stated clearly that a pervasive emphasis on reasoning would be an essential aspect of all mathematical activity. Presumably, more than a half century after Fawcett's lovely yearbook, thanks to years of research and accumulated teacher-lore, the Council had a clear set of ideas for how to accomplish this. Perhaps more interesting is the lack of any direct attention to "critical thinking" in the more recent *Principles and Standards*, published in 2000. Critical thinking is still present in the goals, but it has been subsumed by more holistic notions of what it means to teach, do, and understand mathematics. For example, in a discussion of *communication* in any K-12 mathematics classroom, we are urged to design experiences that enable students to:

- organize and consolidate their mathematical thinking through communication;
- communicate their mathematical thinking coherently and clearly to peers, teachers, and others;
- analyze and evaluate the mathematical thinking and strategies of others;
- use the language of mathematics to express mathematical ideas precisely.

How similar these ideas are to those promoted by Fawcett so many years earlier! We can see that little has changed in the mainstream ways that people tend to define critical thinking in the context of mathematics education. Yet careful attention to the details in the *Standards* does reveal increased sophistication in what the Council means by these goals. For example, there is increased interest in the idea that students must truly understand the strategies and mathematical thinking of others in the classroom community. Students are expected to search for the strengths and weaknesses of each and every strategy offered. It is no longer good enough to reach an answer to a problem that was posed. Now, students are cajoled into communicating their own ideas *well*, and to demand the same communication from others. A shift has occurred from listing skills to be learned toward attributes of classrooms that promote critical thinking as part of the experience of that classroom.

One way in which teachers can create such a classroom is by designing good ways for students to communicate with each other. More specifically, it is now recognized that reflection and communication are intertwined processes in mathematics learning. The Council recommends explicit attention and planning by teachers, so that communication for the purposes of reflection can become a natural part of mathematics learning. And the Council further urges that teachers focus on the building of a classroom community in which students feel free to express ideas; teachers are asked to look for evidence that the students understand that contributions to this community include the skills of listening, and developing an interest in the thoughts of others. It seems that the Council is asking teachers to spend more time on developing mathematics as an evolving literacy, rather than as a set of conventions and techniques to be mastered. Rather than rush to formal mathematical language, mathematics teachers should recognize mathematics as a group experience that requires reading, writing, listening, speaking, and the use of various modes of representation.

> Students who have opportunities, encouragement, and support for speaking, writing, reading, and listening in mathematics classes reap dual benefits: they communicate to learn mathematics, and they learn to communicate mathematically.
>
> (NCTM 2000: 60)

Nevertheless, from Fawcett in 1938 to the Council in 2000, we can identify a strand of undeveloped theory. Students are supposed to be communicating freely in some form of ideal democratic environment. They not only have freedom of speech, but indeed are guaranteed an audience from the other members of their community. In this perfect democracy, students know that when they talk, people listen to them. And they grow to understand that it is this sort of communication that leads mutually to richer understanding of the material and increased sophistication in talking about this mathematics. What is undeveloped, however, is a critical consideration of the context in which this perfect democracy is supposed to take place. The classroom is embedded in a society that determines a wide range of ways that students do not come to the same table with equal opportunities and resources at their disposal. Political theorists have noted that ideal speech communities require as much attention to these contextual realities and to responses to them as they require the generation of ways to organize democratic participation.

For the last century teachers of mathematics have been figuring out how to drop the teaching of critical thinking in favor of establishing environments that allow for the critical thinking that is possible through discussion and interaction. This has meant abandoning long cherished notions about what mathematics classrooms look like and what the products of such classrooms should be. Instead of timed tests with the number correct circled at the top, students now bring home portfolios of material amassed over time, or complex reports on open ended investigations. Yet in the new millennium, teachers of mathematics are now beginning to realize that this can only go so far. They are likely to *still* be disappointed in the ways some of their students participate or contribute. The response to this current malaise, growing out of the more nuanced political awareness, is a movement called *Critical Mathematics Education*.

Critical Mathematics Education demands a critical perspective on both mathematics and the teaching/learning of mathematics. In doing so, it takes one step further in questioning our assumptions about what critical thinking could mean and what democratic participation should mean. As Ole Skovsmose (1994) describes a critical mathematics classroom, the students (and teachers) are attributed a "critical competence." A century ago, we moved from teaching critical thinking skills to using the skills that students bring with them. We accepted that students, as human beings, *are* critical thinkers, and would display these skills if the classroom allowed such behavior. It seemed that we were not seeing critical thinking simply because we were preventing it from happening; through years of school, students were unwittingly "trained" *not* to think critically in order to succeed in school mathematics. So we found ways to lessen this "dumbing down of thinking through school experiences." Now we understand human beings more richly as exhibiting a *critical competence*, and because of this realization, we recognize that decisive and prescribing roles must be abandoned in favor of all participants having control of the educational process. In this process, instead of merely forming a classroom community for discussion, Skovsmose suggests that the students and teachers together must establish a "critical distance." What he means with this term is that seemingly objective and value-free principles for the structure of the curriculum are put into a new perspective, in which such principles are revealed as value-loaded, necessitating critical consideration of contents and other subject-matter aspects as part of the educational process itself.

Keitel *et al.* (1993) together offer a new way for teachers to think about the mathematics that is being taught. New ideas for lessons and units emerge when teachers describe mathematics as a technology with the potential to work for democratic goals, and when they make a distinction between different types of knowledge based on the object of the knowledge. The first level of mathematical work, they write, presumes a true–false ideology and corresponds to much of what we witness in current school curricula. The second level directs students and teachers to ask about right method: Are there other algorithms? Which are valued for our need? The third level emphasizes the appropriateness and reliability of the mathematics for its context. This level raises the particularly technological aspect of mathematics by investigating specifically the relationship between means and ends. The fourth level requires participants to interrogate the appropriateness of formalizing the problem for solution; a mathematical/technological approach is not always wise and participants would consider this issue as a form of reflective mathematics. On the fifth level, a critical mathematics education studies the implications of pursuing special formal means; it asks how particular algorithms affect our perceptions of (a part of) reality, and how we conceive mathematical tools when we use them universally. Thus the role of mathematics in society becomes a component of reflective mathematical knowledge. Finally, the sixth level examines reflective thinking itself as an evaluative process, comparing levels 1 and 2 as essential mathematical tools, levels 3 and 4 as the relationship between means and ends, and level 5 as the global impact of using formal techniques. On this

> **Box 4.3**
>
> **Critical Mathematics Education**
>
> a) Design a five-part unit that assumes students and teachers come with a "critical competence" and which uses this critical competence to develop a "critical distance."
>
> b) Design a mathematics investigation that focuses students' attention on levels three, four, and five, as described by Keitel *et al.*
>
> c) Identify three new ways of working in your classroom that you can use to help the whole group reflect on the issues of equity that are hidden behind some students performing more successfully than others in mathematics. How can this become part of your curriculum, so that tools of mathematics are developed in order to research these issues?
>
> d) Go back to one of the five-part units that you have recently designed, and consider how you will introduce level 6 into the archaeology portion of your unit.
>
> e) The most valuable part of this set of tasks is to share your designs with those of others who are reading this book with you. Compare the strategies you each devised in order to introduce critical mathematics education into your work with students.

final level, reflective evaluation as a process is noted as a tool itself and as such becomes an object of reflection. When teachers and students plan their classroom experiences by making sure that all of these levels are represented in the group's activities, it is more likely that students, and teachers, can be attributed the critical competence that we envision as a more general goal of mathematics education.

In formulating a democratic, critical mathematics education, it is also essential that teachers grapple with the serious multicultural indictments of mathematics as a tool of post-colonial and imperial authority. What we once accepted as pure, wholesome truth is now understood as culturally specific and tied to particular interests. Philip Davis and Reuben Hersh (1981) and David Berlinski (2000, 2008), for instance, have described some aspects of mathematics as a tool in accomplishing a fantasy of control over human experience. They use the examples of math–military connections, math–business connections, and others.

Critical mathematics educators ask why students, in general, do *not* see mathematics as helping them to interpret events in their lives, or gain control over human experience. They search for ways to help students appreciate the marvelous qualities of mathematics without adopting its historic roots in militarism and other fantasies of control over human experience. Arthur Powell and Marilyn Frankenstein (1997) have collected valuable essays in *ethnomathematics* and the ethnomathematical responses that educators can make to contemporary mathematics curricula. Ethnomathematics makes it clear that mathematics and mathematical reasoning are cultural constructions. This raises the challenge to embrace the global variety of cultures of mathematical activity and to confront the politics that would be unleashed by such attention in a typical North American school. That is, ethnomathematics demands most clearly that critical thinking in a mathematics classroom is a seriously political act.

One important direction for critical mathematics education, as discussed in Chapter 3, is in the examination of the authority to phrase the questions for discussion. Who sets the agenda in a critical thinking classroom? *Problem Posing* was offered as a variety of powerful ways to rethink mathematics investigations through, and in doing so this pedagogical strategy gives us a variety of ideas for enabling students both to "talk back" to mathematics and to use their problem solving and problem posing experiences to learn about themselves as problem solvers and posers. In the process, problem posing helps us to frame yet another dilemma for future research in mathematics education: is it always more democratic if students pose the problem? The kinds of questions that are possible, and the ways that we expect to phrase them, are to be examined by a critical thinker. Susan Gerofsky (2001) has recently noted that the questions themselves reveal more about our fantasies and desires than about the mathematics involved. Critical mathematics education has much to gain from her analysis of mathematics problems as examples of literary genre.

And finally, it becomes crucial to examine the discourses of mathematics and mathematics education in and out of school and popular culture (Appelbaum 1995). Critical thinking in mathematics education asks how and why the split between popular culture and school mathematics is evident in mathematical discourse, and why such a strange dichotomy must be resolved between mathematics as a "commodity" and as a "cultural resource." Mathematics is a commodity in our consumer culture because it has been turned into "stuff" that people collect (knowledge) in order to spend later (on the job market, to get into college, etc.). But it is also a cultural resource in that it is a world of metaphors and ways of making meaning through which people can interpret their world and describe it in new ways. Critical mathematics educators recognize the role of mathematics as a commodity in our society; but they search for ways to effectively emphasize the meaning-making aspects of mathematics as part of the variety of cultures. In doing so, they make it possible for mathematics to be a resource for political action. We will come back to this in Chapter 5.

The history of critical thinking in mathematics is a story of expanding contexts. Early reformers recognized that training in skills could not lead to the behaviors they associated with someone who is a critical thinker. Mathematics education has adopted the model of enculturation into a community of critical thinkers. By participating in a democratic community of inquiry, it is imagined, students are allowed to demonstrate the critical thinking skills they possess as human beings, and to refine and examine these skills in meaningful situations. Current efforts recognize the limitations of mathematical enculturation as inadequately addressing the politics of this enculturation. Critical mathematics educators use the term "critical competence" to subsume earlier notions of critical thinking skills and propensities. A politically concerned examination of the specific processes of participation and the role of mathematics in supporting a democratic society enhances the likelihood of critical thinking in mathematics. The serious work ahead of us involves supporting the use of critical competencies in the development of critical distance.

Response to Chapter 4:
It is Critical to Think

David Scott Allen

Many of the issues of this book are not unique to mathematics education. I find myself thinking that I could substitute Social Studies, English, or Science class for mathematics at a number of points. This is no less true for this section. For me in my everyday world, the key juxtaposition for this chapter is found in the stated truth that standardized tests have not caught up with reform, and the quote that critical and reflective thought had always been important, but now recently has also had effect on the nature of the curriculum. I guess I would want to split the difference and say that maybe critical thinking isn't the goal because students do think critically already. Maybe assuming that they do isn't the context because it would lead to too much off-topic conversation, but maybe the goal is to get them to apply their already critical minds to the objects of mathematics which the district values. Then this opens up the question of whether or not we should problematize the fact that this is the case.

In other words, it would seem that there is actually a set of tensions here: the standardized test criteria for success, the problem solving as goal approach, the district's expectations, and each student's own personal journey. It is truly a pluralistic mathematical society we live in when students can choose technique, thinking skills, passing tests and going to class as optional. This affects our culture. All things are optional unless I make them mandatory for myself. The only thing we can be sure of is change, which is exacerbated by the shifting sands of pedagogy, curriculum, and social attitudes and politics within districts and states. At the national level, No Child Left Behind (NCLB) stares me in the face on a daily basis, but I am left to give it many faces in return. Let us look at these eight points in summary:

Point #	I Used To . . .	Now I . . .
1	enable students to invent algorithms and present multiple perspectives	support students' use of other's ideas to discuss strengths and weaknesses of those and their own
2	structure activities to include the collection of critical thinking skills	provide situations where students use the skills to draw conclusions or accomplish a task
3	enable students to collect data and look for patterns	support students in recognizing how changes in one data set impact other data sets
4	enable students to present strategies for finding answers	support students in offering several possible answers based on calculations
5	offer options for learning a particular skill or topic	present the teaching of a particular skill or topic as a controversy among educators
6	enable students to gain an appreciation of the role of math in every day life	support students' study of how the data is represented, and consideration of why the author chose that method
7	allow students to make decisions	support students' study of how math is used to turn ourselves into objects of study
8	use many assessment strategies	help students see that what they have learned is transferable to new contexts

Are *you* using these eight points as you read about them? In other words, are you doing what it says to decide whether or not you agree with what it says? We have now reached a kind of formalization of what good teachers have done all along. They don't swallow the product or theory whole, but they blend much together to make the recipe complete.

A key for the latter part of this chapter is the fact that mathematics is a human endeavor. Mathematics is a humanly constructed technology of meaning. It is a way of thinking about our world. What can you do in your classroom to make it more conducive to learning? Open communication lines, encourage question asking, leave time for reflection, provide rich objects of mathematical discussion. Doing this seems difficult. So just pick one of these things and try to change it. One thing I have concluded after examining other ways of doing things for a few years is that it can't get worse! State Standardized Tests are powerful drivers to our whole way of thinking, but what if we shift our perspective from that of looking at our achievement through a microscope and dissecting every technique and manipulation, to that of looking at the whole examination table itself and noting what makes the examining possible. This is the context for learning. What kind of microscope is it? How are its examinees chosen? What are the skills required to look into the microscope? These truly are the tip of a much bigger and more wonderful iceberg.

Action Research 4:
Lesson: World Population and Wealth

Ada Rocchi

The following lesson is based on articles in *Rethinking Mathematics: Teaching social justice by the numbers*, by Eric Gutstein and Bob Peterson. Each of the articles in the book shows the importance of integrating social justice issues into the current mathematics curriculum being studied in schools today. Although the lessons addressed provide a general outline for the social justice issue, each one can be specifically adapted to address the many math topics identified in any math curriculum. Additionally, these concepts parallel the math standards set by both the NCTM and PA state standards. Examples of questions from the PSSA (Pennsylvania System of School Assessment) test that I include also support the correlation between this lesson and those topics which are assessed by this test.

Although any number of lessons could have been chosen, the one I was most interested in includes that of World Population and Wealth. As a math teacher of at-risk youth, I want my students to see the power in math. By correlating the importance of understanding the factors behind why the old adage of "the rich just keep getting richer" holds true, I hope lessons such as these will enable my students to break the cycle of poverty. Good decision-making as a productive member of society is a right of all, and education plays a crucial role in attaining this right. Students from affluent families generally learn the importance of good financial decisions through parents or classes they choose to partake in. My students do not have this advantage, as most have parents who do not understand these concepts themselves or continue to model an ineffectual financial choice.

As discussed in this book, mathematics and the understanding of its applications "make it possible for mathematics to be a resource for political action" (see page 145). By its sheer understanding, mathematics can be a catalyst for change for many, if it is learned and understood. This can then propel students to feel that their voice, participation, and leadership will be recognized as an integral part of our democratic society (see Chapter 7). Through their voice, they can then make informed changes needed to provide a more equitable balance throughout society. And in the short run, students can begin the journey to acting and thinking like a mathematician. By embracing the mathematics that relates to our world we eliminate the need for the problems with one correct answer and explore the beauty of the math all around us.

Currently I am teaching in an alternative education setting, servicing grades seven–twelve. The maximum number of students in the school is seventy-five, with five classes of fifteen students each. The classes are not divided by grade level or ability level, so instruction is to be on a more individualized basis, more of a tutoring than a traditional lecture classroom. This book allows me to generate group lessons, in order for students to benefit from the interaction of math instruction, while satisfying the specific curriculum goals of each subject taught. Those specific subjects include: General Math, Pre-Algebra, Algebra 1, Geometry, Algebra 2 and Business Math.

Lesson Plan Title: World Population and Wealth

(*Rethinking Mathematics*: 67)

Standards Addressed

NCTM Standards

- The Number and Operations Standard
  - Developing a deeper understanding of very large numbers and various representations of them.
- The Measurement Standard
  - Understanding measurable attributes of objects and the units, systems, and processes of measurement.

PA Standards

2.1 Numbers, Number Systems and Number Relationships
2.2 Computation and Estimation
2.5 Mathematical Problem Solving and Communications
2.6 Statistics and Data Analysis
2.8 Algebra and Functions
2.9 Geometry

Specific Objectives

- Division of Whole Numbers (General Math)
- Scientific Notation (General Math and Algebra 1)
- Exponents (Algebra 1 and 2)
- Properties of Triangles (Geometry)
- Logarithms (Algebra 2)
- Graphs and Data Analysis (Business Math)

General Goal(s)

Understand the differences in population and wealth between different areas of the world.

Required Materials

World maps, Different colored chips, wealth data table, calculators, *USA Today* snapshot, graph paper.

Anticipatory Set (Lead-in)

Go to http://bioweb.wku.edu/courses/Biol115/wyatt/Population/pop1.htm. As a class, put in some ages of students to see how the world population has increased since they were born.

Step-by-Step Procedures

1. As a class discuss:

   - How many people live in the U.S.?
   - How many people live in the world?
   - What do you think might be some of the consequences of a rapidly growing population?
   - Of the world's 6.5 billion people, more than 1.2 billion live on less than $1 a day.
   - About sixty percent of those who live on less than $1 a day, live in South Asia and Sub-Saharan Africa.

   (How can we represent large numbers in a more efficient way?)
   As a class, review scientific notation.

2. Give a copy of a world map and twenty-five chips to each student. Have them identify the continents.

   - Calculate how many people are represented by one chip if all twenty-five chips represented the world population?
   - Have students place chips on the continents based on where they think people live.
   - Give students twenty-five different colored chips. Have them place them on the continents based on the wealth produced.
   - Look at the World Population and Wealth Data Table and compare actual data to estimated data from students.
   - Use the table for allocating treats in Gutstein and Peterson's World Population and Wealth lesson, calculating the number of students and number of treats columns, based on number of students in the class.

Plan for Independent Practice

General Math

Total world population: 6,525,000,000.
   Total habitable land on Earth: 52,000,000 square miles.

- If the world population was evenly distributed over Earth's habitable land, how many people would live on each square mile of Earth?
- How does that compare with the number of people per square mile in your state? In the U.S.? In other countries in the world?

Research the personal and political backgrounds of the U.S. Representatives and Senators from a specific state.

- Summarize the characteristics of the general population of that state.
- Compare and contrast the ethnic, gender, age, and political party of the state representatives with that of the population.
- In your journal, make inferences about the representation of Congress for that specific population. Do you notice any patterns? If a population is under-represented, what social factors may be affected for that population?
- Formulate a plan that might help with social equality.

Algebra 1:

www.col-ed.org/cur/math/math51.txt

From the site mentioned above, find the world population from the years 1650–2000. Plot them on a coordinate graph and explain whether your graph is linear or exponential.

In your journal, based on the data studied so far, predict which continents might have a life expectancy of seventy-five years or older. Which might have a life expectancy of less than fifty years?

Research the actual results, and compare and contrast your list with the actual list.

Synthesize which factors might influence life expectancy.

Do you see any patterns formed by life expectancy and wealth of the continent or nations?

Geometry

Using the *USA Today* snapshot:

Explain where you would build a medical center in the triangular region formed by Niger, Afghanistan, and Angola that would be convenient for all three.

Explain why you would build the medical center at that location.

How would you find the centroid/balance point for the region formed by Niger, Malawi and Angola?

Choose two other continents, or nations:

Pick three locations similar in distance to the above mentioned example.

Compare and contrast the number of medical facilities within these areas.

Draw a conclusion based on the location you have chosen, as to why one nation may have more or less medical treatment centers available.

Algebra 2

Enter data from the above site, for world population figures from 1650–2000, into List Editor on the TI-84.

Determine the Linear Model using the regression capabilities of the calculator.

What is the correlation between years and population?

What is the rate of change in population per year?

Using the linear model, predict the population in 2010.

There are concerns that as the world population increases, shortages of food, water and quality of life will be experienced.

In your journal, prior to discussing as a group, debate which percentage of the population will be affected the most.

What patterns do you expect to see based on these concerns?

Describe and graph what will happen to the wealthiest and the poorest of a specific population as these shortages occur

Business Math

You are four years old. A rich aunt wants to provide for your future. She has offered to do one of two things: Option 1—she would give you $1000 a year until you are twenty-one; or Option 2—she would give you $1 this year, $2 next year, and so on, doubling the amount each year until you are twenty-one. Which would you choose? Why? Graph the two sets of results. Think about the connection to world population.

Find at least one article in the newspaper referencing a specific continent or country. Analyze the article based on the wealth and population of that area. In your journal, write a summary that explains what factors are being reported that might affect the wealth for the people of that country. Evaluate what conclusions could be made today based on changes in technology, jobs, population, wealth, and so on.

Closure (Reflect Anticipatory Set)

What trends in population do you notice?

What about the trends in the wealth for those countries?

Ten Chairs of Inequality Activity from Gustein and Peterson's World Population and Wealth lesson

How do you think the global wealth inequality compares to the unequal distribution of wealth in the United States?

Assessment Based on Objectives

Individual work and classroom activities mentioned above. (Specific assignments from individual texts and PSSA sample questions can be used.)

Literature Circle: Divide up into groups based on the choice of the following books: *If you Made a Million*, by David Schwartz; *A Place for Zero*, by Angeline Sporagna LoPresti; *Big Numbers and Pictures that Show Just How Big They Are!*, by Edward Packard.

Using small groups and literature circle roles, analyze how the the use of large numbers described in each book correlate with the topic being studied. Each group showcases their results to the class.

Journal Activity for the class: Draw conclusions based on activities completed individually and as a group, that would evaluate the factors that influence population growth or decline and how it affects the wealth of a nation.

Individual Projects: Choose one of the countries discussed in the World Population activity. Research the mathematical contributions, either a mathematician, game, mathematical topic, or whatever. Present a mini-lesson to the class based on your findings. Examples of things students could research might include:

1. A timeline of the use of calculating machines: the Roman abacus, the Japanese soroban, the Chinese suan pan, Korean number rods, the Peruvian quipu, Napier's rods, and the modern calculator.
2. The Lo-Shu magic square from China.
3. Ancient systems of numeration: Egyptian symbols, Babylonian symbols, and Mayan numerals.
4. Compare and contrast Egyptian multiplication and Russian peasant multiplication with our lattice multiplication.
5. Egyptian rope stretchers and the introduction of Pythagoras.
6. The golden ratio.
7. Chinese tangram puzzle.
8. Mancala game.

Some examples of assessment questions for this lesson that also would be aligned to the PSSA were retrieved from the following site: www.pde.state.pa.us/a_and_t/cwp/view.asp?a = 108&q = 103267

1. The number of bacteria in a culture doubles each hour. Which graph below **best** represents this situation?

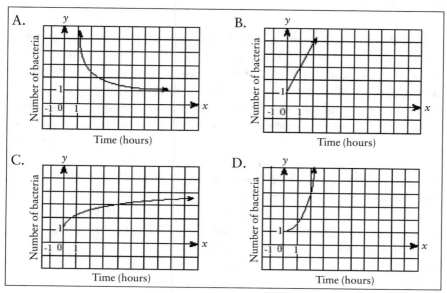

2. This table shows the 1997 and estimated 2002 populations of the four most populous countries in the world. Jessica and Katy are using the data for a presentation of their world geography report.

Country	1997 Population (in millions)	Estimated 2002 Population (in millions)
China	1221.6	1272.3
India	967.6	1042.3
United States	268.0	279.5
Indonesia	209.8	225.6

a. Jessica says that between 1997 and 2002 the population of China increased faster than that of the United States. Katy insists that the United States population increased more quickly. Explain how **each** of them can be correct. Justify your answer mathematically.

b. Suppose the table was extended for additional five year periods (2007, 2012, 2017, etc.). If the population of the United States continues to grow at the percentage rate that is predicted in this table, after how many five year periods will the population reach 300 million? Show or explain how you found your answer.

3. Dave's Electronics is having a sale on radios. Each day the price of every radio will be reduced five percent from the previous day's price. If the price of a radio before the sale was $50, which expression can be used to find its price on the nth day of the sale?

A. $50 - 0.05n$
B. $50 - 50n(0.95n)$
C. $50(0.05)^n$
D. $50(1 - 0.05)^n$

Based on the individualized nature of my classroom, with so many varying skill and grade levels of the students, I think this lesson would best fit at the beginning or middle of the unit. The ideal for me would be at the beginning, with all students completing the activity regarding world population and the distribution of wealth. From there, after a class discussion of large numbers and their representation using scientific notation, further study of the topic could then be investigated based on the math course they are currently enrolled in. As an ending, each student or group of students would then share their piece, explaining what they found regarding the future trends of population growth. This would hopefully then lead into a discussion of a smaller scale investigation of wealth distribution in the United States.

I could also incorporate the lesson in the middle, opening with showing the video, "World in the Balance – The Population Paradox" by NOVA. From there I could use a classroom discussion to look at large numbers, then introduce the activity and end with the individualized work to further understand how their specific math curriculum is applicable to this topic.

Although not conducive to my setting, I could use this activity at the end of a unit, but it would be difficult based on the transitional nature of my classroom. I would have to make sure all students were at a point in their text to contribute something to the lesson. I could definitely use the activity as an investigative search at the end of each of these topics on an individual basis, although I think some of the power of the lesson would be lost. Even though I would prefer to use the lesson as a class activity, I do think this lesson is crucial, and I would probably implement it on an individual basis for those students who may have come to my class later in the year, after the lesson was investigated.

This lesson satisfies the CIEAEM (2000) goals of using mathematical reasoning and tools in order for people to think critically as citizens. In this specific lesson, the idea of unequal wealth distribution and world population numbers can be explored as part of the lesson that would allow students to be informed citizens while using their higher order math skills. In addition, the lesson shows the connections between real-life problems, crucial themes, educational policies, and math practice in schools. This goal can then further include the obligation of globalization with mathematics education and societal concerns. Finally, this lesson would establish teaching and learning practices guided by social justice principles.

The two gatekeepers I've identified, who might resist a change in my classroom, are the director of the program, and the school districts we service. Currently, the program accepts students from grades seven to twelve, from a variety of school districts throughout Bucks and Montgomery counties. In each class, students would be working individually in the text that is required by the school district that referred them to our program. To satisfy the concerns of our director, I would invite him to visit my classroom to observe the lessons, as well as provide a copy of my lesson plans identifying how the lesson ties into each curriculum, along with the standards that would be met. I would highlight the part of the lesson that would be completed as a group, as well as the part that would remain individualized, providing students an opportunity to not only work cooperatively but also independently. Additionally, I would ask him to participate in a lesson, providing a social justice concern he may be interested in, or one that may have affected him or his family at one time. I would ask that he participate as much as possible in the research, with the students, of this topic. I would also have the students in the class keep in touch with their progress, maybe having the director reading some of the students' journals and responding to specific tasks.

As for the school districts that are used to students working independently for the time they are referred to us, I would have students keep a detailed portfolio of the topics they've addressed, including class work and assessments to show mastery of the lesson and topic. These journals and/or portfolios would be sent with the students when they return to their school. Also, I might invite various math teachers and/or department heads of the school districts we service to participate in a lesson, similar to that described above with our director. If time and distance are a concern, students could interact with their teachers via email, not only to keep in touch while placed with us, but also for those teachers to see the progress of their students. Another possibility might be a panel presentation of specific social justice units, similar to those we do for senior projects. We could invite public school math teachers to a presentation of the unit, several times a year. Additionally, prior to implementing the lessons, I would send copies of the lesson plans, with the goals/objectives and the standards that are addressed to the math chairs of the schools we service, asking for any concerns or suggestions they may have for the students in their district.

Ultimately, I would like to develop a year-long curriculum with a series of lessons that incorporate social justice issues that would allow my students to have more opportunities to work together in the classroom while seeing a real connection to the math they are studying. This specific lesson segues nicely into the unit on Understanding Large Numbers, from *Rethinking Mathematics*. Topics that would easily be covered by this unit include: Mean, Median, Mode and Range; Scientific Notation; Measurement and Data Analysis. Although there is some overlap, due to the transitive nature of my students, the review of certain skills is necessary.

Also I see the easy transition to the unit on Home Buying While Brown or Black and How Do You Spend Your Wages? Both of these lessons would meet my Life Skills Activity requirement. As a journal activity, I would use the cartoon on page 51 of *Rethinking Mathematics* to discuss budgets, wages, and savings. We would discuss the current trend of foreclosures and sub-prime lending practice and the effect these topics have on the population. Additionally, I would have them research the number of Rent-to-Own type stores that are located in certain areas and why this makes sense to those who own these franchises. We would look at the relationship between individual racism and structural racism. What math topics do students see as applicable while studying this phenomenon?

Another activity I would like my students to experience is the one on page 29 of *Rethinking Mathematics* that addresses global concerns. Individually I would like them to think about the list of global problems and what they feel is most important to address. In their journal I would have them write which concerns they would eradicate, if they could. A survey of people outside our school would then have students comparing and contrasting the varying opinions of the importance they see from the people they interviewed. Discussion of their findings and why, based on the representation of their polling group, would lead to a discussion of what factors influence a person's opinion of the most important social concern. All of these ideas hopefully would exhibit the power math has on the students' lives and the important decisions that will need to be made throughout each person's life.

"Math has the power to help us understand and potentially change the world" (Gutstein and Peterson 2005: 5). Not only does *Rethinking Mathematics* promote this ideal, it also provides teachers with lessons to connect math to real-life applications while integrating social issues into their curriculums. Overall, it provides students an opportunity to become informed citizens, with an active voice in their futures. Revolutionizing math curriculums in this way not only provides a framework for students to see the applicable uses of the math topics they are studying, but enables them to continue to use that knowledge to better their lives, long after they have left the classroom.

Initial resistance to change and thinking outside the box when it comes to math curriculums was a result of my initially narrow vision for how I could integrate these topics in order to cover all the current requirements. But with an open mind and time, the innovative lessons presented in this book not only creatively connect math to very important issues in students' lives, but also parallel math curriculums in a number of ways. Additionally, the social justice issues span a variety of curriculums, providing a framework for interconnectedness of the courses students are enrolled in.

Examples include "The War in Iraq: How Much Does it Cost?" which clearly connects math and social studies. "Environmental Hazards: Is Environmental Racism Real?" provides a connection between math and science. "Home Buying While Brown or Black" and "Sweatshop Accounting" shows math and economics being commingled. Also, "World Population and Wealth" easily includes lessons applicable to math, social studies, and economic curriculums. Providing a link between curriculums' strengthens students' understanding of the subject area as well as increasing their interest.

Although the lessons described generally give a broad overview, they can specifically incorporate a number of mathematical concepts. Unemployment rates provide a basis for the study of percentages, ratios, and proportions. The war in Iraq allows a perfect segue into a study of data analysis and scientific notation. Environmental Hazards incorporates measurement and graphing, while Home Buying While Brown or Black allows for the connection to varying graph depictions, which then can be broadened to a study of slope, rates of change, and best-fit equations. World Population and Wealth includes most of the previously mentioned topics as well as exponents, properties of triangles, and logarithms. Understanding Large Numbers not only connects budgets, that are real-world numbers in our society, but supports a lesson in mean, median, mode, and range. The list could go on and on.

As mentioned earlier, my initial response was one of disbelief that this integration of social issues could become a reality. But once I took the time to actually reflect on the parallels and opportunities presented, more and more correlations were revealed. The primary gatekeeper in making this a reality is time, the time needed to adapt these general lessons to each teacher's specific curriculum. As with most change, it will not happen overnight, but it needs to happen.

Our students, especially students of color, women, and students from working-class and low-income families, need to be provided with the power to change their lives. Education, and specifically math education, is the tool for achieving that power. *Rethinking Mathematics* provides a recipe for creating that tool. All we need to do is add the time. John F. Kennedy once said, "The great French Marshall Lyautey once asked his gardener to plant a tree. The gardener objected that the tree was slow growing and would not reach maturity for one hundred years. The Marshall replied, 'In that case, there is no time to lose; plant it this afternoon'" (see M. Moncur's website, www. quotationspage.com/quote/1928.html). We should not wait any longer to implement this change in our curriculums. We need to find the time; we need to give our students, all of our students, a chance to have an equal voice to truly be a democratic society.

As stated in Chapter 5, everyone is a mathematician and everything can be mathematized. This text supports *Rethinking Mathematics* in the following excerpt:

> Critically aware teachers of mathematics recognize the role of mathematics as a commodity in our society; but they search for ways to effectively emphasize the meaning-making aspects of mathematics as part of the variety of cultures. In doing so, they make it possible for mathematics to be a resource for political action (see page 157).

Allowing students to see the parallels between what they are studying and the impact it has on their lives allows them to take a more active part in their own education.

Ada Rocchi

I teach grades 7–12 at Community Service Foundation/BuxMont Academy, a private, non-profit, alternative education school. Students attending the school struggle with behavioral, emotional, and substance abuse issues. I have taught at various alternative education programs, in many states, and feel education is a key component for this population to achieve success in their lives. This chapter is the result of research for the course Math and the Curriculum, which I took at Arcadia University during the spring of 2007. The course renewed my hope for all students to experience the beauty of math and to have the opportunity to become their own mathematician.

MathWorlds 4:
Pitching Questions at Various Levels

For each of these topics, compose two questions that you think would be good ones for early elementary grades K–3, two good ones for upper elementary grades 4–6, two for middle grades 7–9, and two for high school grades 10–12—one each that you think would be fairly "easy," and one that you imagine would be a "challenge." Explain why you think these questions "fit" the grade levels you assign them to, and why you expect them to be easy or challenging.

1. Subtraction
2. Fractions
3. Comparing parabolas, circles, ellipses, and hyperbolas
4. Functional relationships
5. Prime numbers
6. Determining whether to use mean, median, or mode of a set of numbers
7. Infinity
8. Drawing inferences from graphs

five
Consuming Culture: Commodities and Cultural Resources

We live in a consumer culture, where everything we experience is influenced by the ways that we are led to want to buy things, the means that we have to buy things, and the kinds of desires we are told to want. Given the pervasive nature of consumer culture and its influence on our lives, it is important for us to reflect both on how this culture affects our view of mathematics and mathematics education, and how it affects our students' expectations for mathematics and mathematics education. In Chapter 4, I introduced the idea that mathematics is simultaneously a collection of "commodities" and "cultural resources." Mathematics is a bunch of commodities in our consumer culture because it has been turned into "stuff" that people collect (knowledge) in order to spend later (on the job market, to get into college, etc.). But it is also a cultural resource in that it is a world of metaphors and ways of making meaning through which people can interpret their world and describe it in new ways. Critically aware teachers of mathematics recognize the role of mathematics as a commodity in our society; but they search for ways to effectively emphasize the meaning-making aspects of mathematics as part of the variety of cultures. In doing so, they make it possible for mathematics to be a resource for political action. In this chapter, we explore what this might mean for our everyday practice.

First of all, consider how we "sell" mathematics to young people—we tell them that it is important to spend their time on it because it will pay off for them in the long run: in college admissions, on the job market, in their everyday life decisions, and so on. We blend right in with the language of a consumer culture when we do this, and when we do so, we do not just support the values that consumer culture encourages; we also turn mathematics into a commodity for the students, so that it is hard for them to imagine it in any other way. But there are important differences between a consumer of a commodity and a user of a cultural resource. The differences are actually more theoretical than physical, which makes them harder to "see," because a user and a consumer are often involved in exactly the same activities; but the ways in which we understand the activity are very significant in how the activity affects what we learn, what we perceive ourselves to be doing, and the potential importance of what we are doing—that is, the meaning of our involvement in the activity shifts in very profound ways when we move from a consumer of a commodity to a user of a cultural resource.

Buying and Selling Mathematics?

Commodities dominate all facets of our life, so it is not surprising that we so easily allow mathematics to be understood as simply another commodity. Mathematics and mathematics education swim in an amalgam of late capitalism characterized by the degree to which commodities create and maintain desires. Desires in turn provide the purposes of our lives, and the fulfillment of desires is often merely a temporary oasis in the perpetual pursuit of desire, a psychoanalytic point not to be ignored. Desire is never completely satisfied in consumer culture—indeed, it cannot be if we are to maintain a capitalist system where people need to continually buy things—so we are always in a permanent state of satisfaction denied. How does mathematics in school fit into all of this? It would be impossible to escape commodities, including the "commodification" of mathematics as a symbol and mathematical skills and concepts as items to be bought and sold, even if we wanted to. In the economic sphere, writes the cultural theorist John Fiske (1989), we more readily see how commodities ensure the generation and circulation of wealth, and how they can vary from the basic necessities of life to inessential luxuries. By extension, we can include non-material objects such as television programs, a woman's appearance, or a star's name as a commodity. But more important, perhaps for our purposes in this chapter, are the ways that commodities serve other material and cultural functions, so that the economic sphere is interwoven in complicated ways with the material and cultural. When we sell mathematics to students as something they should study, or more advanced mathematics as something they should buy into, or groups of students to potential teachers as maintaining their jobs, or when we buy and sell math textbooks, we are living the experiences of consumer culture. When we divide students into categories by gender or ability or types of academic programs, or by race or ethnicity, we are constructing market niches to which different mathematical commodities can be advertised and sold. We run the risk of turning mathematical ideas into special luxuries that one might opt out of buying, rather than embracing them as ways of being.

Fiske is well known for his example of torn jeans as an illustration of material and cultural functions of commodities. The material function is to meet the needs of warmth, decency, comfort, and so on. The cultural function is concerned with meaning and values: all commodities can be used by the consumer to construct meanings of self, of social identity, and social relations. Describing a pair of jeans or a TV program as a commodity emphasizes its role in the circulation of wealth and tends to play down its separate but related roles in the circulation of meaning. Jeans originally were worn by working men; they were associated with rural labor. But in the 1960s they became popular as a statement of freedom when worn by young people; soon they were understood as a counter-cultural statement about agreement with the political and cultural freedoms of the sixties. Later, it became important to be seen in appropriately worn and torn jeans, so one could easily know who was cool, as these were more authentic. Yet in the 1970s, pre-washed jeans were sold as new commodities, so one really needed to be able to see the difference between real, worn jeans, and pre-washed jeans in order to judge a person's coolness. In the 1980s, dress jeans were sold, along with designer jeans, transforming what was once a counter-cultural message into a new form of commodity. When we treat mathematical skills and concepts as commodities, the same processes unfold. Mathematical skills are associated with certain branches of labor and economic identities, so that mathematics, through at least algebra and beyond if possible, is essential if one is to end up in a white-collar profession. Yet students use elaborate mathematics outside of school when playing role-playing games, or when establishing a character in a video game, or when planning a big party. Knowing the alternative language of outside-of-school culture means that one is part of the group, just as much as the specialized language discussed in Chapter 3 demonstrated inclusion in the classroom community. The construction of mathematical power to be bought or won through schooling emphasizes its role in the circulation of wealth and tends to play down its separate but related role in the circulation of meaning. It becomes difficult even to imagine mathematics as a cultural resource, outside of buying and selling situations of everyday life. Students "learn" through school that their mathematical ways of thinking and making meaning need to be shut down in order to do the work of school mathematics, which in turn often becomes a lifeless and meaningless experience separated from real life. In the research literature, the closest examples we have for real-life mathematical thinking are in the work of Jean Lave, Etienne Wenger, and their colleagues on the mathematics of grocery shoppers or dieters (Lave 1988; Lave, Wenger, *et al.* 1991; Wenger, McDermott & Snyder 2002). Their important insight is that the mathematics is embedded in particular communities of practice through which identity and meaning is mutually forged with skills, concepts, and ways of working. In popular culture, the clearest illustration is the use of the variable X by Malcolm X as a statement about identity and racism. Even here, though, the resource is easily a commodity. We now have X-caps, X-shirt, X-jewelry, X-brand potato chips, and so on. (Also, Generation X, but I am not sure this fits here—is this use of the X yet another layer of cultural resources being appropriated?) There exists a small market for recreational mathematics books of puzzles and math-games, but these more abstract ways of making mathematics part of everyday life are marginalized for virtually all of our students.

The difference of emphasis—money or meanings—carries with it a corresponding difference in the notion of "power." Fiske suggests that the commodity-consumer approach places power with the producers of the commodity (factory owners, professional mathematicians, mathematics teachers), because they make the profit out of its manufacture and sale. In one reading the consumer gets to enjoy the benefits of the product (jeans, mathematics). A more critical reading describes the consumer as exploited insofar as the price he or she pays is inflated beyond the cost of the materials to include as much profit as the producer is able to make. The exploitations can even take on a second dimension, as when the consumer is also a member of the workforce and so finds his or her labor exploited to contribute to the same profit.

Whether conservative or progressive, such a reading, when it begins to consider meaning, does so, according to Fiske, through the means of production. In the case of jeans, someone might argue that jeans are so deeply imbued with the ideology of white capitalism that no one wearing them can avoid participating in it and therefore extending it. By wearing jeans we adopt the position of subjects within that ideology, become complicit with it, and therefore give it material expression. We "live" capitalism through its commodities, and, by living it, we validate and invigorate it. Even in the case of Malcolm's X, the meaning might be lost in the process of commodification; it is less clear to us, either because we are still living the politics, or because the reading really is ambiguous. When we sell school mathematics, and when students buy it, mathematics as a commodity has the same sort of ideological function, returning consumer culture to the forefront and dissolving the potential transformation of culture that is the result of something as powerful as Malcolm's statement about his name being stolen and lost, and thus open to any optional slave name. What I find exciting in thinking about mathematics as a cultural resource, though, is the ways that the mathematical ideas totally transform our understanding of the world and our lives in that world when we use mathematical ideas in this way. The concept of a variable as well as our perspective on the legacy of slavery will never be the same once Malcolm takes X as his official name. Similar new conceptions of ourselves and our world are possible when we make active use of key mathematical ideas such as "function," "ratio," "system of equations," "algorithm," "problem types," "connections between numbers and geometry," and so on. Yet school mathematics does not currently emphasize its power to transform our interpretation of everyday life or the potential things that we might do in our lives. Instead, mathematics is persistently advertised and sold as the key to success, on the job market and in social status, even as it serves to demarcate those who have not progressed to the more advanced levels as "lesser" in some way (creating in turn biases and prejudices).

Now consider the role of mathematics in the creation and use of commodities, especially the pervasive technological commodities that mathematize so much of our lives even as they make the mathematics invisible to us—bar codes and GSP systems, PDAs, and computerized cash registers, each of which make mathematics a hidden way of living even as they do the mathematics behind the scenes. Here the cultural function is seemingly invisible, and therefore, I would argue, is even more important to study as part of the mathematics curriculum. Take the concept of time and how it is measured. The insightful German mathematics educator Christine Keitel (1989) explored the history of our measure of time and space with youth who began to apply their knowledge to a critique of the ways that we have begun in recent centuries to be ruled by the assumption that time exists in such supposedly precise increments. Such a study can introduce students to interesting mathematical concepts and skills, for example when students build their own clocks. They might fill a bucket that has a small hole in it with water, placing another bucket underneath; they mark the height of the water at hourly increments and then can use the bucket as a clock by noting where the water level is when they check the bucket. They might attach a pole to the outside of the school building, and mark on the wall where the shadow is at its shortest, where it is at hourly intervals, and so on, in order to create a sundial. They might build a clock by studying pendula and their periods, then using a pendulum of a particularly appropriate length to run a simple clock. Lew Romagnano (1994) did this very pendulum exploration with his class; they investigated how changing the weight at the end of the string affected the way the pendulum swings, and the length of the string, developing a model that led to a functional relationship via charts and graphs. In Keitel's class, the students took this one step further and explored how the increasing ability to divide time into smaller increments led historically to employers' abilities to monitor and supervise workers, to governments' abilities to control the ways that people traveled with public transportation, and the need to standardize the workday, family life, and public encounters to fit the increasingly standardized ways of keeping track of time. Her class also examined the historic role of standard algorithms in arithmetic in determining how bookkeeping is maintained for businesses, and thus how people interact through the buying and selling of basic needs. Her students were fascinated by the ways that they could see the effects of such mathematical practices on their own lives, such as in the ways that their "workday" at school is structured and predetermined by time and market forces, and how their experiences of the curriculum are measured in small increments in order to fit standardized periods of time and accountability. The nature of the educational encounter in Keitel's class is significantly different from the encounter in Romagnano's class, even though many of the same activities occur in both. It is not enough to introduce open ended exploration, or to encourage students to simply apply what they are learning to everyday life. There needs to be a clear purpose of meaning making beyond the mathematics in order to achieve a vibrant sense of mathematics as a cultural resource that mirrors the resources in popular culture experiences.

Intercultural Mathematics

We find similar arguments concerning ideology and power voiced by some ethnomathematics proponents, who lament the imperialism of Amero-Euro-centric mathematics on, for example, Native-Americans who perceive the world in an entirely different space-time frame (non-Cartesian notions of time and space), or Africans whose mathematics is literally "lived" in their own bodies (body gestures as counting and calculating). Munir Fasheh (1990) is the most convincing of this sort of reader of mathematics and culture. He dramatically communicates the politics of "Orientalism," as implicated in the practice of teaching Palestinians "Western" mathematics and its associated conceptions of identity, rationality, and technology, by juxtaposing it with the clearly complex and mathematical work of his mother when she is weaving traditional textiles.

Ron Eglash (1997: 79) defines ethnomathematics as "the study of mathematical concepts in cohesive social groups, with an emphasis on small-scale or indigenous cultures." I find it useful as a teacher of mathematics to place what my students are learning about and not learning about, what has been identified as part of the curriculum and what has been officially "left out," to organize my thoughts in terms of Eglash's five subfields of ethnomathematics.

First of all, I need to realize that there are many non-Western forms of mathematics; unless I or my students introduce such material ourselves, my students are unlikely to ever be aware of the legacies of entire civilizations such as the Chinese, Hindu, Muslim, and Mayan empires, which created other ways of thinking mathematically. Some of these ideas can be directly translated into Western mathematical ideas, but others are really different and might lead to different insights about the world. Second, mathematical anthropology uses mathematical modeling and a mathematical way of thinking to understand the implicit mathematics in cultures, such as kinship classification systems. Third, we can study the work of mathematicians with the tools of sociology and anthropology, understanding mathematical work as a social enterprise. How does mathematical knowledge come to be known as such, for example? How are mathematicians implicated in the general public's ideas of value, society, and so on? In "vernacular mathematics" Eglash notes the realms of folk mathematics, informal mathematics, and non-standard mathematics that are found in everyday life experience and until recently have been ignored in school mathematics classes. Today, such vernacular mathematics is seen in at least two ways. One way is as a rich collection of techniques that can be mined by students as they develop skills of calculating, organizing, and interpreting information; the second way is as a body of activities that can be analyzed in school for its mathematical meaning. The final subfield of ethnomathematics is "indigenous mathematics," where we can explore how local cultures have their own forms of mathematizing the world, for example, in African cultures where counting and calculating might be done by pointing to various locations on one's body.

The dangers of ethnomathematics are when we may be tempted, as in early anthropological work, to either denigrate local mathematical cultures as somehow "less advanced" than "Western" mathematics, or to romanticize it, as somehow more genuine than the abstractions or uselessness of much of contemporary school mathematics. And the less obvious pitfalls of ethnomathematics are found in our hopes, for example, to provide multicultural relevance by a brief connection to students' cultural heritage, or by knowing so little about the non-Western mathematics that our presentation of it is so primitive as to caricature it as primitive and elementary; by only referencing "ancient empires" like Egypt, China, and Mayan cultures, thereby reinforcing the myth that small-scale indigenous societies had no sophisticated math; or by applying only vague notions of self-esteem that link everyday mathematics and popular culture mathematics to ways of by-passing cultural subgroups' rejection of mathematics as "white and middle class" in turn only perpetuating students' perceptions of school as not believing in their abilities to actually learn genuine academic content (Eglash 2001). These last concerns point to the problems that arise when we label mathematics as belonging to one culture, or when we assume that (Western) mathematics is universal to all cultures.

I have always agreed with a parallel view of culture suggested by the anthropologist

James Clifford (1988). "The 'exotic' is uncannily close." For those of us in the United States, there are no longer any distant places where the presence of "modern" products, media, and power cannot be felt. "Culture is no longer a stable, exotic otherness," writes Clifford;

> self–other relationships are matters of power and rhetoric rather than of essence. A whole structure of expectations about authenticity in culture and in art is thrown in doubt. . . . identities no longer presuppose continuous cultures or traditions. Everywhere individuals and groups improvise local performances from (re)collected pasts, drawing on foreign media, symbols, and languages.
>
> (14–15)

It is easier to register the loss of traditional orders of difference than to perceive the emergence of new ones, as Clifford notes. I suggest we adapt his "ambiguous" yet attractive "Caribbeanization of Culture," a creolized "interculture" modeled after the "neologistic" cultural politics of Aimé Césaire. (Césaire, a French-speaking poet,

---

Box 5.2

Intercultural Unit Planning

For each of Ron Eglash's (2001) Four Principles below, design a unit that emphasizes the principle.

1) *Deep design themes.* When examined in their social context, indigenous mathematical practices are not trivial or haphazard; they reflect deep design themes providing a cohesive structure to many of the important knowledge systems for that society. Examples: fractals in African cultures, Cartesian organization in Native American cultures.
2) *Anti-primitivist and anti-racist representation.* By showing sophisticated mathematical practices, not just trivial examples (e.g. "African houses are shaped like a cylinder"), ethnomathematics directly challenges the cultural stereotypes and genetic myths most damaging to both minority and majority ethnic groups.
3) *Translation, not just modeling.* Often indigenous designs are merely analyzed from a Western view. Ethnomathematics, in contrast, uses relations between the indigenous conceptual framework and the mathematics embedded in related indigenous designs. In effect, it uses modeling as a tool to provide a "translation" from indigenous knowledge systems to Western mathematics.
4) *Dynamic rather than static views of culture.* While evidence for independent indigenous mathematics is crucial in opposing primitivism, it is also important to avoid the stereotype of indigenous peoples as historically isolated, alive only in a static past of museum displays. For this reason ethnomathematics includes the vernacular practices of their contemporary descendents.

politician, and playwright from Martinique, celebrated the identity of the colonized and displaced African through an intercultural mixture of language and symbols.)

Such a way of understanding culture destabilizes the identification of "intention" in power relationships, because any commodity can become a resource for political action. Fiske believes the producers and distributors of jeans do not "intend" to promote capitalist ideology with their product. They are not deliberate propagandists. X crafts people further problematize (raise critical questions about) the relationships expressed, because the politics of the X is what is so powerful in the marketing of the products, thus commodifying political action as well, even as consumer culture is politicized. I am not sure, however, that we can problematize mathematics educators in the same way. Many clearly argue the crucial importance of school mathematics for a capitalist economy; others, such as Bob Moses in his Algebra Project (2001), intend to proselytize a particular form of the "mastery of reason" within a local "interculture." The realization for us is that the "intention" need not be a focus of our concern right now. The "system" or "the way things are" is what we label common sense notions of "good" goals, which in a strong sense regulate the field of possibilities that can be imagined for practice; interrogating the naturalization of common sense is more crucial than placing blame. Ethnomathematics changes our sense of what is possible in mathematics class, as does an awareness of how we and our students experience mathematics outside of school.

"Reason," something at the core of what we believe we are working with in mathematics education, is an excellent example of the common sense naturalization ideology. As the British educational theorist Valerie Walkerdine (1989: 3) writes,

> the government of reason works by assuming both that reasoning is natural and that some sections of the population are themselves profoundly unnatural. . . . Those others, such as the working class, women, the mad and colonized peoples, were understood as being less civilized and more animal, especially as when, in some accounts, reason was understood as the highest pinnacle of an evolutionary sequence of development.

She traces femininity and reasoning from early accounts of women's sexuality that claimed the womb as the site of madness and unreason—the very fact of the possession of a womb made women unsuitable to reasoning. Today we have arguments about a lack of male hormones. The idea of a natural femininity antithetical to reasoning is repeated again and again in data on gender differentiation in performance: girls' hard work and rule following are compared with boys' "natural" talent or curiosity.

"Reason" was linked to "power" more than a decade ago by the National Council of Teachers of Mathematics through the cultivation of mathematical and logical thinking. It was through this connection that the above sort of construction of difference becomes an outcome of practice as a "technology of power." What I mean by this is that our ways of thinking about difference become implicit techniques for maintaining unequal power relationships in society. As teachers of mathematics, we have a serious obligation to understand how our work contributes to the relations of power in the larger society, and to think through what we might do in response to these relations of power! Developmental approaches to mathematics education provide further examples of power implicated through practice as opposed to "power" as an object produced in students. We are confronted continually with children and "difficult-to-teach" older students who "do not fit" and who are often understood as somehow "both over-mature and underdeveloped." A four-year-old middle-class white girl does not understand that a window cleaner cleaning the windows of her home has to be paid; she becomes in Walkerdine's analysis too likely to be an example of the "power of the puzzling mind" in contrast to those children who do not yet "puzzle." Several working-class children in research are said to "not yet" puzzle over money; this is acknowledged by researchers but left unaccounted for in their description of the "generic four-year-old." "We must assume that it cannot possibly be the case that *they* could be more developed or that this evidence questions the whole notion of a universal sequence of development itself," was Walkerdine's sarcastic response (1989: 4).

From her own personal experience and from her research with working-class children, Walkerdine knows all too well that these children do not puzzle over money because it is not a puzzle for them: money and large numbers are part of their everyday lives of struggle. She challenged a number of other research-based theories as well. "Exchange value" is commonly thought to be an "abstract concept" beyond the grasp of young children; supposedly these counter-pose abstract to concrete, but fail to engage with the idea of "lived" practices and relations. Yet researchers can give numerous examples of the skills of "third-world" children far ahead of their "first-world" peers, regularly acting on abstraction and sophisticated reasoning. Finally, Walkerdine has argued extensively "that mathematics provides a clear fantasy of omnipotent control over a calculable universe, which the mathematician Brian Rotman called 'Reason's Dream', a dream that things once proved stay proved forever, outside the confines of time and space" (1989: 6). Ranging from the ways that we classify young children, to the ways we judge adult students, to the expectations we hold for mathematics, power is at the core. What are we ready to do in order to make this clear to our students? On the one hand, studying explicitly the centrality of power may provide the kind of motivation that selling math as a commodity strives to achieve; on the other hand, we and our students are one small part of the creation of new mathematical knowledge, knowledge that may in some ways be different and therefore be part of different kinds of relations of power.

Analyzing mathematics education, then, in terms of popular culture and commodification, raises the problematic nature of "power" and "intention" in a new light. It is not the point to find a scapegoat to blame for the propagation of "reason," or to whine about how school mathematics contributes to societal inequities; but it is imperative to understand the origins of reason as a fantasy, and to collaborate with our students on a different kind of mathematical experience that forges new relations of power. This intellectual twist avoids the conflation of empowerment with traditional (often gendered) models of "hypostasized" power to be "held," "hoarded," or "spent"; empowerment instead has more to do with imagined possibility (both "good" and "bad") than potential application or release. In other words, if we continue to emphasize the study of commodities, we can become mired in pessimism, because the economic system (which determines mass production and mass consumption—of jeans and mathematics) reproduces itself ideologically in its commodities, whether indirectly or directly. Nothing will change. A commodity is ideology made material. And ideology works in culture as economics does in its own sphere: it naturalizes the "system" so that it appears to be the only one possible. An emphasis instead on cultural resources promotes readings that allow for agency on the part of subjects. When we think about mathematical as cultural resources, we and our students become political actors working to change our own lives and the lives of others for the better.

Finding Mathematics: Commodities and Object Relations

Where, then, can we find mathematics as a cultural resource? Fiske looks for a "refusal of commodification . . . an assertion of one's right to make one's own culture out of the resources provided by the commodity system" (1989: 15). This is Clifford's open and uncertain "local futures." Wearing torn jeans. Metaphorically "tearing" or disfiguring the image of Judy Garland in order to create a heroine for the gay community in masquerade. Similarly, we can look to a group of Minnesota college students who appropriated probability to form an alternative culture of fantasy games that later became known as *Dungeons and Dragons*, and spawning an entire industry of role-playing and virtual reality. We can consider the note-passing of eighth grade girls in a mathematics classroom. They rate each potential class "hunk" in a variety of categories, on a scale from 1 to 10, while also ranking the categories themselves; later, they determine their ideal boyfriend while simultaneously disappointing their teacher with their mathematics performance on a test of basic statistics. Or the rigidly geometric and repeated patterns of skinheads and punks in the 1980s in alternative dance clubs. The symmetry of the "peace sign."

Box 5.3

Is Mathematics Inherently Democratic?

Arturo Sangalli quotes Canadian mathematician André Joyal in his article in *New Scientist* (1992):

> the value of teaching students how to carry out mathematical proofs, demonstrating a truth by a logical argument rather than imposing it is at the heart of democracy . . . Whoever claims that a certain proposition is true has an obligation to prove it, if he or she expects others to be convinced. Mathematics is democratic in the sense that it convinces without resorting to force or to an authority argument.

Has this been your experience in school mathematics? What do you think of this quote? Can mathematics education be a model of democracy? What specific ways of working in your classroom would contribute to mathematics as democratic?

At these junctures, it is clear how mathematics can be part of what Fiske calls "excorporation," the process by which the subordinates make their own culture out of the resources and commodities provided by the dominant system—Clifford's "compost for new orders of difference" (1988: 15). In an industrial society, this is central to popular culture, since virtually all resources available are those provided by the subordinating system. Because there is no "authentic" folk culture to provide an alternative (as in Munir Fashe's mother's weaving), popular culture is the "art of making do with what is available." This means that the study of popular culture requires not only the study of the commodities out of which it is made, but also the ways that people use them.

Let us consider this duality of any object constructed in or through consumer culture. It was in John Fiske that I first came across this idea: that consumer culture does not create a dichotomy, but rather a duality, of commodities and cultural resources. Dichotomies are either-ors; one might want to ask, "Is mathematics a commodity or a cultural resource?" It turns out it is always both; how we look at it and what we believe we are doing with it makes the difference. Any "thing" in consumer culture—from a tube of toothpaste, to a fifth grade student, to the concept of balance, from test prep materials to yogurt containers to the image of the yogurt container, to the lifestyle promised by the image of the yogurt container juxtaposed to the test-prep materials—any such "thing" is, at once, a commodity to be sold and consumed, *and* a resource through, with, and mediated by which cultural workers (that is, people) can create meaning and construct new relationships. For example, Malcolm's inventive use of "X," of the notion of variable (as a powerful application of the concept of variable in order to clearly communicate the multiplicity of surnames that might be assigned to him, given the particularities of histories and politics of race) can be compared to the essentially "unknown" value of his family name post-slavery. This was "variables" as cultural resource, even as X-brand potato chips, X-hats and shirts, and other X-commodities were and are marketed in ways that turn Malcolm as a resource into a vital commodity: EXXON—one of the first corporations to recognize the market potential of the X; Generation X, a predictable outcome of consumer culture's need to place people into categories of market demographics; X-Games and X-Boxes; and our most recent symbol of the duality of commodity and resource, which makes clear its unwillingness to distinguish between the resource (Xtreme sports, as Xpression of cultural desires) and commodity (the selling and marketing of the games, players, accessories, etc.). The important question for curriculum is, given the cultural proclivity to commodify, what sorts of educational encounters are more likely to embrace and foster the playful use and aesthetics of cultural resources over the processes of commodification?

Here, I want to discuss ideas for connecting cultural resources with psychoanalysis and object relations. Having wrestled with consumer culture's historic moment, in which any cultural resource—a book, a person, a human gesture, or an ephemeral fragrance— is at once an object for marketing and consumption, I was delighted to find a psychoanalysis that is informed so profoundly by the idea that any conceivable "thing" is a potential object that a person might relate to in some way. The brilliant theoretical gesture of psychoanalysis, I believe, is to foreground that rather obvious "fact." There is never a moment in which any human being is not already steeped in a history of relationships—to other objects in and of one's environment. In fact, a person could be understood as an expression of the ongoing creations of relations with objects of self. The question for curriculum becomes, "Can there be relationships other than those of reification (of turning something into an object, and identifying that object through its personal assignment of its particular 'thingness'), and of production (of assimilating an object as a potential meaning-maker, capable of mediating and hence constructing new possible relations)?" Object relations as a theory sets me up to ask, "What sorts of educational encounters allow for an object to be 'taken' in relation as a cultural resource, and which types of encounters tend to make this less likely?" Quite simply, we want our students to experience a triangle or function or pattern as more than merely something to collect as future lucre. Relations with these mathematical objects must transform the ways in which they read their worlds of meaning.

Being Limited by "Vision"

It is within this framework of consumer culture and object relations that I re-start my thoughts on becoming a mathematics teacher. Given the hyper-reality of so much curriculum work (of writing and talking and teaching and learning *about* teaching and learning itself), if one spends any time interpreting one's efforts in the discourses of consumer culture and psychoanalysis, one runs up against a peculiar ideology of vision, in which "coming to know" is hegemonically equated with perception. Imagine first, serious efforts to understand teacher education in terms of consumer culture. In mathematics teaching, we see this very clearly: we are always trying to help students to "see" something, to be able to "see" it a certain way, to be able to "show" what they know. Everything is at once a commodity and a cultural resource: subject matters are marketed to students and family members, while practitioners of the disciplines use the content of their subjects to reinterpret and comprehend their life world; text materials, too are bought and sold, and "sold" to the students as "useful reading," even as writers either use textbooks to work politically for a new conception of their discipline or to make money selling what's wanted by the market; people (teachers, students, potential students) are both resources and commodities to be bought and sold; systems of evaluation, forms of institutional structure, special programs, and so on, are subject to the whims of salesmanship and the fads that subcultures use to establish identity even as they open up fields of possibility for new ways of understanding education and schooling. In particular, all are bought and sold in the rituals and practices of teacher education. Likewise, these very same participants are expressions of their object relations. The experiences of teacher education "events" are psychoanalytic encounters through which people are denied and enabled new objects and new relations. Choices and actions of a student, a teacher, a future teacher, and so on, are both expressions of these relations to the vast complex of objects in their environment (and by objects we mean *everything*, from things like string and blocks to other people, from the idea of a hexagon or a recursive function to the notion of justice) and opportunities to create relationships with objects in the environment.

But suppose we are actively working to use these discourses as resources in our educational work. At some point we can no longer deny the authority of vision in our metaphors: the way we make things into objects that can be seen, or could be seen "in our mind's eye," the way we turn ourselves and others into unitary, individual standpoints in a Cartesian or Euclidean "space" with all other aspects of our environment, of that space, as objects floating around in our imagined space. Some of these objects hover in particular locations, attached to us by webs of relation. Others whiz by so fast that we can't grab hold of them; still others we can't "see"—they are behind us, are too far away, they're just not in our field of vision. In general, all of our language for speaking about learning can be reduced to an assumption that "coming to know" is pretty much a process equivalent to perceiving. Sure, there have been philosophical variations on this perception, some of which emphasize touch over sight, or smell and taste, or hearing and listening. Each is a different perceptual metaphor and offers fresh perspectives on perception itself. But they all come down to perception as *the* modality and perception for us is dominated by ocular perception. The real sign of the thorough hegemony of perception is our inability to come up with any other ways to conceive of coming to know. I often think of education as a social institution that is a symbol and tool of the ideology of perception. Maybe this is what the French philosopher Michel Foucault had in mind in recognizing the ways in which social institutions are technologies of the panopticon: tools of the "gaze" where we always feel like we are being observed and never can believe we are safely alone. Perhaps a new theorization, a new epistemology, will render education a mere ruin of the old regime of truth.

The Author as Teacher-Educator

Now, I would like to work through some issues that emerge when we look at the discourses of consumer culture and object relations within and against the environmental "screening" of an ideology of perception. The institutional positioning of such work defines the experiences in rather curious ways. To begin with, I am never hired in order to bridge these terrains or to redefine them in light of each other. Instead, my efforts are construed as part of a certification process, in which all concerned imagine that we are achieving precise expectations in the refinement of technical skills. Because of this, *all* of our work can be reduced to a clear moment of commodification, in which discourses and practices are sold and bartered as what Donna Haraway (1992) has called "technologies of vision." Rather than a search for a new practice of epistemology parallel to metaphors or perception, the events of curriculum theory *as* teacher education are strongly characterized as an expert improvement on the old technology—as new way of "seeing" or "feeling" the educational encounter. Students are commonly encouraged to learn about "this new way of thinking about learning," or say they do not feel right teaching in a certain way. For a number of years, I seized the opportunities of the duality this offers: I would use reified terms of education, such as "lesson plan," "assessment," "evaluation," and "concepts," even specific packages of these products, such as "multiple intelligences," or "developmental stages of psycho-social learning," as recognized commodities, only to play with them with my students in ways that changed their meanings and purposes toward political action that assumed cultural resources. I got this idea from Cornell West's appropriation of "self esteem" in *Race Matters* (1994), in which he takes a pop cultural commodity and turns into a cultural resource for important social justice efforts. The dilemma of such efforts, of course, is the very duality it perpetuates: even as I might be using ideas and theories as cultural resources for new conceptions of schooling, teaching, and learning, others are again appropriating these terms as commodities to be bought, sold, and applied as "capital" in the consumer culture of educational practice and the teacher job market forum.

Second, the technology of vision, as a key commodity and cultural resource, is constructed as the result of enculturation. I believe Winnicott, the psychoanalyst, declared like others that the most important preparation for child psychiatry was personal individual experience in psychoanalysis. Through such "work," an adult was prepared to interact with children psychoanalytically. By adopting the discursive tools of psychoanalysis, labeled, for example, splitting, counter transference, fetish, one is slowly enabled to "perceive" a child through this expert technology of vision. Similarly, we want to declare important concepts of curriculum theory to be key discursive tools of teacher education, and in turn, for teaching in general. *Currere*—curriculum as the experiencing of the running of the track rather than the track itself—is one fundamental term of this discourse. Others might be autoethnography, queering, and youth culture; or, problem solving and problem posing. Commodification and production of meaning through cultural resources are processes of education, just as countertransference and splitting (see Chapter 2) are processes of individual development. Yet if objects are *related to*, if these processes of relation constitute the ongoing development of an individual—teacher, child, administrator, or whatever—what does this suggest about the complexities of teacher education about education? If people can be understood as expressions of multiple and conflicting object relations, what does this suggest about the ways we can enter into encounters and work together in educational contexts?

In a course that I teach now, there are several core experiential components. On-campus discussions of assigned readings, workshops on alternative assessment, theater games focusing on student-selected "moments" of curriculum encounters, and a weekly "field experience" are some examples of the kinds of organized structures employed. Just as any potential psychiatrist should have experiences in analysis, according to Winnicott, we should say that any potential teacher should have experiences as students with the sorts of curricular encounters we expect to be crucial elements of the "teaching/learning" they will be engaged in as teachers some day. The experience of analysis is not considered equivalent to an ordinary conversation, or other possible forms of communications between one person and another. It is a particular *kind* of experience. Likewise, we would imagine future teachers to have had specific sorts of experiences *as students* that they will later construct with their own students. Oddly, the students I work with have not been fortunate enough to come to teaching with such experiences. Even more strange is that almost no prospective teachers have had such experiences. They have never studied as an inquirer into the subject matter they are interested in. They have rarely if ever planned a performance or exhibit to communicate to an audience outside of their classroom some interesting or provocative aspect of their inquiries. They have rarely if ever written reflections about what they have read. Rarer still is the prospective teacher who has regularly, as a student, discussed with others with whom they have worked the group processes that they are experiencing and developing.

Yet these students are assigned the job of interacting with small groups of children once or twice a week for a semester. Will they "buy into" inquiry structures for learning? Will they read *with* their students? Will they pose provocative mathematical questions with their students, instead of answering formulaic questions and drilling on procedures? What are they packaging and selling the children they work with? Inquiry, inquiring *as* an inquirer, writing, reading, mathematizing: as concepts and as teaching gestures, they are objects to be related to.

The discourse of "reform" generates several positionings for the teacher educator. One of these is the required embrace of a Bourdeuian "*habitus*" *as* teacher education curriculum. *Habitus* is a term that refers to the routines that define taste and expectations in one's life. For example, children in families that routinely use libraries, museums, and comic books as places to visit and enjoy where ideas are found and then used as resources for problem posing, are more likely to use libraries, museums, and comic books for these purposes in their lives. There's an implicit "deficit model" inherent in the use of *habitus* as a discourse. Thus, mathematics reform efforts consistently assume that teacher practices shift before teacher ideology. Teachers will start to use journal writing and impact theater techniques, clinical interviews and problem posing approaches before they understand such curricular practices in light of reform ideology. In our course for pre-service teachers, they are required to lead a small group of eight–ten students through a mathematics/science inquiry experience that lasts a full semester. The assumption is that they can adopt the skills and practices of inquiry facilitation and participation, skills and conceptual understandings that they "lack," through the practice of such experiences. Of course, we do not posit such a deficit: it is imagined that students always come to an educational encounter with prior understanding of the concepts and skills involved. We would never choose to take a "basic skills before application" approach to learning and teaching. And why not assume this? Don't people become writers through apprenticeship, by acting like a writer? Mathematicians by acting like a mathematician? Magicians and clowns by acting like magicians and clowns?

The deficit of apprenticeship is, I think, only constructed in practice when curriculum is "designed" independently of the participants. In such a situation, the encounter assumes no prior accumulation of experiences rich in *habitus* unrelated to the apprenticeship. I find object relations apt for thinking through this dilemma: the curriculum question might be, "What sorts of educational encounters can help the curriculum workers—often 'teachers' in such contexts—gain a richer comprehension of the object relations at stake for the student?" Notice I didn't ask, "object relations that exist or that can be 'made'"; the psychoanalytic challenge is to fathom the relations at stake in the encounter. Rather than training pre-service teachers in how to use a new technology of vision, the job becomes one of "holding" students, of providing a safe place where they can repair some of the damaged relationships with objects that are primarily influencing their ongoing creation of relations, their potentially new relations, and their avoidance of certain forms of relation. "You see," technologies require training. To see through a telescope or microscope one has to be taught how to see and what to see—historically early users saw *nothing* and could make no meaning. Nowadays such nostalgic technologies are taken as "transparent," whereas new technologies require generations to reach the stage of nostalgia. The psychoanalytic stance is, for the teacher, a new technology of vision as well, if perceived as such. This is the danger. We only see "splitting," "countertransference," and so on if we are trained or apprenticed to do so.

But: being taught how to use a telescope or microscope, and being apprenticed to psychoanalysis as a way of knowing, in order to use it as a model of the educational encounter, are very different from each other, even as they share an ideology of perception. In the first case, the student/apprentice and the teacher/trainer are in parallel, looking at a common object of study: the technology and what it can help you "see." We discussed the rich potential of such a standpoint as eloquently articulated by David Hawkins in Chapter 2; he explains how a teacher and a student need a common object—an "It"—through which they can build a relationship. As a scientist, Hawkins applies the modernist conceptions of science practice as a tool toward new object relations: the teacher as object for the student, and the student as object for the teacher, as well as the teacher as object for the teacher, and the student as object for the student, are related to in fundamentally important ways (referred to as an "I" and "Thou" in a nod to Martin Buber) *if* a particular *other* object of study is taken by both as an "It." We can see the modernist, scientific origins of psychoanalysis in this respect, in that the analyst and the client together "look at" the client's objects and relations. But here the purposes are reversed in a sense: instead of using objects to build relationships between people, psychoanalysis, growing out of a particular relation, makes it possible to relate to the objects in new ways. We also discussed several sources for further consideration of this theoretical shift, including Julian Weisglas and Brent Davis, who have posited *listening* as an alternative to *seeing* as the primary technique of perceiving. They note significant differences between visual and auditory relations. If the crucial relation that the teacher has with the student as an object of self is one of observing the student, they are distant from an external stimulus. If that relation is one of listening and hearing, then the analogy extends to the inner actions of the ear in making it possible to listen in ways that are not comparable to the eye. Genuine listening requires dialogic participation, physically, metaphorically, and intellectually. Weisglas and Davis are following in the path of numerous epistemologists. Sartre, for example, was fond of the hand and touch as the metaphor of coming to know, instead of the eye or the ear. This, too, changes the relation of distance between knower and known; touch also claims a change in the knower in response to the known, and a fundamentally different encounter.

Smell and taste are yet further metaphors we could explore. By literally *consuming* the known, the knower is fed and transformed internally, as the "known" is now ecologically part of the knower. In this act of consumption—inhaling and eating one's objects of study—the knower now "knows" in a new sense. What would it mean for prospective teachers to taste or consume "inquiry" or "alternative assessment," as ideas, as objects of relation? What would the educational encounter be? What might it mean for people to breath in and physically metabolize the fragrance, stench, aromas, and odors of inquiry or assessment, literally, to be "in-spired"?

The fact is, even as these new epistemological metaphors get me excited, I remain fascinated by the ways they continue to entrap us in the ideology of perception. I think this may be why folks as divergent as Evelyn Fox Keller and the Michel Foucault have searched for contrasting *images* of knowing and coming to know. We're locked into perception, but maybe we can use perception to envision relations other than perceptual images! So Fox Keller and Foucault teach us about ancient Greek uses of sexual interaction as models of knowing and coming to know. Any conceivable structure of sexual encounter is a potential model. Typically, in this approach, there is an accent on the relationship between the knower and the known: any member of a sexual encounter can become, metaphorically, the knower; in the process of focusing on a knower the others or other are then the known. The relationship between the learner and the known may often have as an attribute a relation of dominance or subjection. This certainly has rich metaphorical potential. Except: if we go through the exercise of categorizing the variety of potential sexual encounters, we find the resulting taxonomy to be overpopulated by "relations of consumption." A large percentage of the options position the knower as reifying the known, as an object of pleasure, a commodity to be taken and used for the knower's purposes. There is little space for "perversity"—something Deborah Britzman once described as pleasure without utility.

From Complicity to Heterarchy

My usual theoretical tact when feeling frustrated or hegemonically maeopic is a "Deleuzian nomadic epistemology." Rather than assume that my theoretical axes are the defining axes of a Cartesian reality, I search for a new, independent trajectory that is coexistent with, yet not complicit in, the original hegemonic categories. Deleuze and Guattari (1991) called this "nomadic epistemology" when using the example of the dichotomy of the homeless versus the sheltered. Nomads are neither homeless nor sheltered, yet they are both homeless *and* sheltered. In that example, there are two categories. The *new* term, nomadic, overlaps with the original categories yet does not fit in or out of them: it is coexistent and independent. Can we perform a similar theoretical maneuver with our *four* categories? Consumer Culture, Object Relations, Ideology of Perception, *Habitus*.

What we are looking for are categories that not only coexist with the problematic concepts and practices, but which also subsume them in a way that displaces—literally robs them of their place. One trajectory that I associate with critical pedagogy offers three "nomadic" opportunities: Youth leadership, Voice, and Participation in democratic institutions. Imagine teacher education and curriculum theory blurring together in the actions of youth leadership, voice, and democratic participation. Instead of quibbling over the idea of technology of vision, or suffocating in the deluge of commodification, "teacher-education-curriculum-theory-practice" would use some technologies of vision, loop in and out of commodification and cultural resource work, understanding but not being confined by object relations, toward student leadership within democratic forms of institutional and community organization. Neither "selling out" nor "buying in," such a political turn can be both post-modern and the "pause that refreshes." Is there anything more going on in this work than finding ways to mask critical pedagogy as a technology of vision? How would such a project manifest itself, given the lack of interest that so many people have in youth leadership, voice, or democratic participation? The questions remain open to me.

Here are two other nomadic terms with potential: disparity and desire. It seems to me that much of the malaise of consumer culture and object relations discourse, and much of the dissatisfaction with technologies of vision, grow out of the ways in which they posit *differences*, that is, ways in which they set up "norms" by clarifying the ways in which things are "not the same." Disparity emphasizes the "unequal in difference"; as Sharon Todd (1997) has written, a notion of disparity in the material conditions that structure differences *differently* enables us to avoid the collapse of diversity into a rendering that is individualized and psychologized. It is in the struggles against the disparities of injustice that *desires* are produced, mobilized, and frustrated in the pedagogical encounter with difference. Thus, *desire* itself can be our new, independent term, referring to that which ceaselessly circulates through the *unsaid*, manifesting itself in expectations, hopes, visions, and fears (even as it intersects with the symbolic and spoken discourses uttered by teachers and students). Deleuze (2002) wrote that "desire never needs interpreting; it is it which experiments." Todd (1997: 239) writes, "desires are not only 'handled' or 'dealt with,' but . . . [are] also produced and constituted."

I am suggesting that consumer culture, object relations, and the ideology of perception are embraced but also left to do what they wish while other important conceptual work is accomplished by recognition of difference, disparity, and desire. These three "D"s are outcomes of events instead of causal origins of entrenched problems. In casting them as outcomes, their identification offers possibilities for coalitions and multiple levels of possibility that cut across the lines of difference and power. Rather than standpoints in a Cartesian universe, disparities are *enunciations*.

In a teacher education experience, we would set up contexts in which the prospective teachers are not taught techniques, but are given outcomes to strive for: work with these children, or with this group of teachers and students, in order to facilitate Youth leadership, Student and teacher voice, and Democratic participation. Each field placement could be examined as a case study in action research, Tools will be requested: "How can I understand what is going on?" "Who can help me solve my problem?" "I don't even know where to begin, let alone how to formulate a question."

The response will be a set of discourses: first the duality of commodity/cultural resources; how and what is commodified, advertised, marketed, bought, and sold in this case study? How is something treated as a commodity by some in the group, yet as a cultural resource by others? What does this disjunction mean for what proceeds in this group? How can any commodity be turned into a resource? How can any resources be treated as a commodity? Another discourse will represent the object relations that are comprehended in the ongoing construction of community in the group: Can individual relationships be established in order to

> **Box 5.4**
>
> **Nomadic Research**
>
> Your assignment is to work with a group of students to support the development of leadership, voice, and participation in democratic institutions, and to come back with your questions. Go to it!
>
> When you come together to discuss your questions, analyze the ways that disparities and desires have created expectations, norms, assumptions, and differences which ended up affecting the outcome of your work with the students.

facilitate the understanding of potential relations? What forms of encounter can allow for psychoanalytic work to be accomplished by members of the group? How do individual group members use each other as objects of relation? And the third primary discourse examines the ideology of perception as enacted by the group: what do people in the group employ as metaphors for knowing and coming to know? Next we are asked to interrogate difference, disparity, and desire: How can our discourses of consumer culture, object relations and the ideology of perception elaborate our descriptions of difference, disparity, and desire? How does each *discourse* create differences, disparities, and desires?

Response to Chapter 5:
Emphasize the Meaning-making of Mathematics

David Scott Allen

There are cultural commonalities which the students bring to us on the first day of every school year. We look out over our freshest cast of characters, and try to discern their potential for being rejected by the group. There are always the students who like to put others down, and then there is the group-think that happens in the dynamic of every new collection of twelve year olds I see. They just have a way of deciding who is in and who is out. If given the opportunity, they will go to great lengths to tell you why these other people are in or out. You will see the in ones being gathered around by the lesser lights, so they can hang on their every word in order to attempt to be more in. You will also see the out ones being progressively isolated. I believe that everyone has a gift to offer society at large as well as to the society in microcosm. It is up to the teacher to make room for the out kids to be heard, so that the in kids can see what they are missing.

The ins also have a way of dominating the landscape of how we understand the math we are working on in class. They will speak up in favor of a particular algorithm or method, and then it will become the thing to do. The out kids might even take up the ideas, but to no avail because they do not think that way. They are wired differently from the popular. Ultimately, all of us are wired differently from one another. One of the first ways this begins to take shape is when students begin to blurt out phrases like, "Who doesn't know that?!" or, "You don't know that?!" They are establishing themselves as the Joneses to be kept up with. This can get especially nasty in the female sub-community, because girls mature before the boys catch up. It is very easy even at an early date in September to have the students mentally opt out of math, because they see that they cannot keep up with the ins.

So it is difficult to sell a non-hierarchical structure to our students. That is, one in which each voice is not only valid, but equally valid. Mathematical ability, at least to the extent that the students have experienced mathematical ability pre-middle school, is every bit as much a commodity as jeans styles are for serving as leverage for the ins against the outs. Yes, students do pay attention enough to know what is important so they are not left out. Many go through school paying attention solely to such things.

Their goal is simply not to be left behind socially, and if that means learning math or wearing jeans, or playing role-playing games, then so be it. The mathematics of these role-playing games is not taught in schools, at least not the way that it is used in the game. Intuition is not a part of the scene in formal mathematical schooling. And girls understand that. It is absorbed apprentice-style by those who dare to attempt to participate. So students as kids make choices to learn bodies of mathematical metaphor much like kids as students learn the jargon of school. If they have to jump through teacher-district-state-federal hoops to achieve their goal, then that is fine. But some still struggle because their way of understanding the mathematics is not the way it is taught.

I admit that I do sell school mathematics to my students. But I try to sell it as a parody of the standardized testing industry that so pervades all of what we do and how we do it. The standards are important because we must look carefully at what we are teaching our children and decide purposefully and specifically what the focal objects of mathematics will be. But I tell them that in my opinion, these tests are easy hoops through which to jump. If this is all the adult world expects of us, then we will meet it easily. And with all the time we have left over, we will study truly interesting and challenging mathematical concepts.

Tapping into our students' sensitivity to irony, parody, and satire is an important skill to master for a teacher who really is old enough to be his students' father. They see me as part of the caricature of mathematics. Sometimes I play into the nerd image. I show them that everyone is a nerd to the things that interest them. Mathematics just happens to be something that isn't considered cool by American culture, and we limit our use of the label nerd to people who do unpopular things. For example, I could use the label football nerd, but since football is popular, it would be a cultural oxymoron. If you show signs of acting too mathletic instead of athletic, you are labeled as a nerd with a negative connotation. That is, unless you are on the football team!

I want my students to stand tall with their analytical skills, and the Math Team does this well. I am the Math Team coach for my building, and I try to incorporate these types of challenging math problems into my classroom whenever I can so that the students can see outside the box. We tend to teach not so much to the test, but to the box. Students don't realize all of the interesting ideas in life which are, in fact, mathematics. Ideas which are not on the lists given to me by the district or state as requirements, but are nevertheless on my list of things to which students should be exposed at some point, lest they think of mathematics as a series of pointless equation solving techniques to be used at the whim of the quiz-wielding teacher. There are many worthwhile mathematical objects which do not show up on tests. Some teachers can ignore these things, but educators can't afford to. And yes, perhaps Math Team is yet another way of selling math to my students (it is optional and after school). But it does get them thinking more creatively than they would need in order to solve the cookie-cutter exercises we have all seen in textbooks. My Math Team is inclusive—unless you choose not to participate. Sometimes it takes a little marketing to get a shy middle school kid to come to a meeting, but when they arrive they do not receive a commodity, but real mathematics and a membership in a community of unrequired fun. It gives problem solvers a place to be cool. Two dozen students come once a week after school for two hours, to do interesting but difficult math-type problems. They earn varsity letters just like a sport, and have a place in the hierarchy of activities. In addition, many of them participate in sports and theater, both of which are a big deal at my school.

I said earlier that mathematics is not taught in a vacuum. There are social concerns to be deliberated over. But perhaps Western or at least American mathematics is taught in the vacuum which is being created by standardized testing. Western civilization has sterilized and sanitized mathematics and mathematics education so much, that the students are turned off by the culture it represents. They look at us and say, if that is what mathematics is all about, then I don't want to have anything to do with it. There is a definite cultural difference in the way our students think, and so we need to shift our approach to keep up.

Perhaps a metaphor for this situation would be that you can lead a horse to water but you can't make him drink because he may see other uses for the water. He might want to sprinkle you with it (social recreation). He might want to take a bath (practical) or get a suntan by pond-side (water as atmosphere or prop), but he may choose not to drink no matter how thirsty he is. It's like this with my students. I can provide laptops, notes, textbooks, paper, but they will adapt these materials to their own needs, wants, desires, and imaginations. A worksheet can become a paper airplane as often as it generates thoughtful consideration. How many students at the college level choose to sell their books after taking required courses? In my experience, they sell the books from classes that had no impact on their careers or their lives: that is, neither their minds nor their hearts. This usually includes mathematics. Perhaps you will face the same choice after you read this book for a class. Authors do not write with such reselling in mind, yet understand its reality. I wonder, however, if textbook writers write with changing the student's world of meaning in mind. I want my students and you readers to experience mathematics and this book as a way to change your perception of the world.

I am attracted to school for many reasons. I love to learn, and that drives my education every day. I've never really thought about why I love to learn, other than that the world seems so interesting. I have developed a mathematical mind more from looking carefully at and into objects of my own appropriation than I have devoured math classes. I used to play chess by passing notes to a friend in my math classes until we finally got caught my junior year of high school! And I didn't take math my senior year. The math was secondary. I was good at it, and I enjoyed the structure. But I don't really remember learning any of it. We didn't have courses labeled Algebra and Geometry, but rather non-descript titles such as Intermediate Mathematics. Branches of mathematics should be written in lower case and have fuzzy boundaries. I don't view them as necessarily crystallized, static structures but as tools for achieving goals. One of the goals I set for myself (yes, for myself, as opposed to that which a teacher would present) is that of aesthetics. Mathematics should be beautiful and fun and interesting.

So I take graduate courses because I love to learn. You could argue that I pay the big bucks for college credits because it moves me up the pay scale (a.k.a. the food chain). But while I may be forced into it for this reason, I actually do the work because I love it. I choose the courses I take carefully, and attempt to do what I can to enhance my work in the classroom. If my students don't benefit, my personal benefit is no benefit at all. Sometimes I appropriate assignments to my own ends, and hope they meet the teacher's requirements. Perhaps my own students really do get out of class what they want from it. Is this a good thing? How much should I be at odds with their own perception of what they are doing versus what I say we should be doing in class?

This happens with student teachers and student interns in medicine. We achieve Level I Secondary certification, but do we have any rubric criteria for competence? Such it was in the ancient apprentice social structure. Peter has succeeded in suggesting to us that our curriculum is within us. We must look to ourselves to contextualize our practice. Only we can *currere* (run the track) ourselves by determining its course as we go. And yes, children tend to learn what we teach them. The question is: do we learn what children are trying to teach us? We tend to imitate that which we know. If we don't get exposed to an object in the context of respect and responsibility, it will never be owned. Math has a tendency to be kept at arms length by most of the population, even as they go about their mechanism- and computer-driven lives. Teachers of the future will teach in the ways to which they are exposed. They can't imagine it being taught any other way, so they can't teach it that way. They've never seen or heard anyone being taught well that way, so they can't imagine it working. Skills must be applied. No different for a math teacher than their math students. No different for a math student-teacher than a math teacher-student.

Students don't need another way of thinking about the material. They need a *reason* to think about the material. They need to be immersed in the culture of our classroom environment where this is just what we do—asking what if and going to find out the answer. You don't need a teacher present to accomplish that, you need motivation and translation to the world at large. Yes, if teacher education-curriculum-theory-practice were merged there would be better teachers and better students, but you can't begin with the whole bicycle except when riding it. In order to build one you must also understand the parts and their distinctives, just as understanding all the points of view in an argument helps you to decide what your own opinion is. Knowing yourself and your strengths and weaknesses allows you to read this book and decide where to start with all of these new ideas crowding your head and blowing your mind. Multi-tasking means doing more than one task without blurring the lines between them. There are in fact many differences among the above topics which need to be recognized for clarity's sake, but in the end they all merge into one praxis in the classroom.

My next step is to assign a current math events project to my students. Go find math in the news and report on what you find. I might pick a certain day of the week for them to present their findings, and have a weekly or monthly contest for the best presentation. Students need to see math all around them, but they won't try unless we encourage their efforts.

Action Research 5: A School-wide Survey

Colleen Murphy

I wanted to promote mathematics education in my building for both students and teachers. This idea came from class discussions and my own desire to learn more about what happens to students mathematically before and after they are in class with me. I believe all teachers have a common goal—to facilitate students' learning and help them to become successful contributing adults in society. Current legislation also mandates that we, as teachers, ensure that all students demonstrate a level of proficiency on a state-wide test. We try to do this using many resources including the materials in our given curriculum. However, many assumptions are made each year about what students should or should not know and what material has been covered in previous years.

Each of us learned about the importance of prior knowledge when teaching a lesson. Students have a better understanding of new skills and concepts presented, demonstrated, and experienced in a lesson if they can draw upon what they know already. Many of us read in our Teacher's Manuals the students' expected level of prior knowledge. For example, students may come to a lesson with beginning, developing, or secure knowledge about a certain concept or skill. But is this really true? Do most students have this expected amount of prior knowledge? If so—wonderful—forge ahead! If not—why not? How are we expected to cover material at our level if we are reteaching concepts developed in earlier years? What if the students have the building blocks but are unaware of their own knowledge? How can we bring this to the forefront without reteaching each year?

Mathematics is flexible and occurs every day in every place; therefore, students and teachers are constantly experiencing mathematics. How can we recognize these experiences and give them meaning so that, when presented with variations of the situation again, students (or teachers) draw upon prior knowledge to demonstrate success? How will the student relate to mathematics? What approaches does each bring to school? I believe students and teachers will experience their own perceived successes, whether it is higher test scores, greater understanding and application of concepts, more meaningful experiences, and so on when both groups recognize that they are working, doing, teaching, and learning together. I created a survey to gather information from both teachers and students to find out their current feelings about our mathematics curriculum and how we each believe we could create a more mathematical environment in our school building. Will a change allow for more student discovery or ownership of mathematics?

I have summarized some of the most significant information gathered from a sample of twenty-three teachers representing grades Kindergarten through sixth grade.

The majority (> 50%) of teachers categorized students' prior knowledge of these areas as:

Number Sense—Strong
Number Operations (Computation)—Strong
Geometry—Average
Measurement—Average
Algebraic Concepts—Above Average
Reasoning and Problem Solving—Above Average
Data Analysis—Average

These same teachers wished Measurement, Computation, and Reasoning and Problem

Solving had a greater emphasis in the earlier grades. When asked how we, as teachers, could better meet student needs the responses were as follows:

- Increase math time to allow for more flex grouping.
- The games should be stressed more with more opportunities to play various levels of games. We should also have more time to let the kids really have a chance to use different manipulatives.
- Advanced students should be allowed to skip some of the easier pages to move into more challenging explorations.
- Students should take a pretest at the beginning of the year. The curriculum could be adjusted or adapted to meet student needs based on this assessment.
- I believe more time should be built into our lesson plans to allow for game playing. It is the basis of student learning.

Teachers felt that students could benefit from additional materials for the following reasons:

- Multi-level instruction, pre-teaching of concepts, drill, and practice (there is not enough built in to the program when a concept is first introduced): 1) to present confusing material in a more "kid friendly" way. 2) to provide additional practice on difficult concepts.
- Some students can grasp the concept in one lesson. However, the majority of the students need additional practice in order to learn a concept and especially to store it for long term retention. It's also important to teach concepts in different ways to address different learning styles. Additional practice is very helpful.
- Our program presents one or two ways of learning concepts. Some students benefit from or require different modalities of learning. Additional resources allow me to supplement for those students.
- There just isn't enough practice with the newer, harder concepts.

The student results were as follows:

What are your favorite parts of math class?
Skills: Partial Products, Lattice, Multiplication, Graphs, Fractions, Division, Geometry, Money
Activities: Playing Games, Math Vocabulary, Math Boxes
Methods: Working with Partners, Learning Many Strategies for Solving Problems, Charting to Find Answers

What are your least favorite parts of math class?
Skills: Division, Fractions, Subtraction, Percents
Activities: Games, Tests, Math Boxes
Methods: Writing the Answers, Graphing Results, Using Different Ways to Solve Problems

What are your strongest areas in math?

> Skills: Area, Perimeter, Fractions, Addition, Telling Time, Multiplication, Subtraction, Pictographs, Geometry

What are your weakest areas in math?

> Skills: Fractions, Division, Graphing, Measuring, Multiplication, Finding Perimeter and Area, Converting Fractions to Percents without a Calculator, Decimals, Subtraction with Regrouping, Formulas
>
> Methods: Explaining Answers, Not Checking My Work and Making Silly Mistakes

How can teachers help you to be a better mathematician?

> Strategies: Math Flex Groups, Individual Time with Student, Different Approaches to Solving Problems, Study Skills for Math, More Patience when I don't Understand, Modeling
>
> Tools: Extra Worksheets, Flash Cards, More Problems on Concepts

How can parents help you to be a better mathematician?

> Strategies: Provide Examples, Do Homework with Me, Study Skills, Use Free Time to Create Math Examples, Introduce New Concepts before they are taught in School, Show Us Shortcuts
>
> Tools: Flash Cards, Practice Tests, Worksheets

What can you do to be a better mathematician?

> Strategies: Try Different Ways to Figure Out a Problem, Check Work, Pay Attention, Take Time, Try Best Every Day, Practice, Do My Own Homework
>
> Tools: Flash Cards
>
> Methods: Come to School Every Day

The results of the surveys illustrate the benefit of the Five-Part Inquiry Unit. Many students expressed frustration when they had to relearn information that they already knew regardless of the fact that most of the class needed a review. I see a strong need for both pre and post testing concepts at the beginning and end of the school year as well as designated checkpoints throughout the school year. The information gathered from the pre and post tests would provide information about students' prior knowledge and experiences with different math concepts. Within my building we have three–five classrooms at each grade level. Each grade level could designate fifteen minutes a day toward mathematics investigations. Each classroom would host a different activity (standard) for a designated period of time. Students could be divided according to the pre or post test results and go to the various classrooms to experience mathematical learning on their level for a short period of the mathematics class. The main instruction would be delivered in heterogeneous classrooms where students of all ability levels would be able to learn from each other. Then, during the fifteen minute flexible group time, teachers could focus on working with students to discover their own methods of solving problems. This could eliminate much of the general reteaching that occurs in mathematics class. Students would have the opportunity to create problems that would be meaningful and beneficial to their needs in an investigative setting.

I also believe teachers and students should develop a consistent mathematics vocabulary to be used across grade levels. A consistent vocabulary will help the students recall prior learning to allow them to progress and further develop conceptual understandings of concepts as they increase their mathematics abilities. This language could be associated with a kinesthetic activity to provide a recollection for students as they prepare for learning each and every day. Copies of these activities would be shared with parents throughout the school year to reinforce learning at home as well as in school.

Colleen Murphy

I am a first grade teacher in the Abington School District in suburban Philadelphia, Pennsylvania. I look for collaborative work efforts in my professional life and academic life, and enjoy spending time with others.

MathWorlds 5:
Turning "Puzzles" into "Problems" or "Exercises"

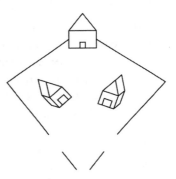

1. **Park Paths.** It is said in a tale that three neighbors who shared a small park, as shown, had a "falling out." The owners of the large house, complaining that the neighbors' children were bothersome, built an enclosed pathway from their door to the gate at the bottom center. Then the owner on the right built a path to the gate on the left, and the owner on the left built a private path to the gate on the right. None of the paths crossed.
   a. Can you draw the three paths?
   b. Change attributes of the given situation or the question asked to develop a list of new questions that could be explored here.
   c. Which of the questions in (b.) appear to *you* to be the most interesting, and why?
   d. Suppose our criterion for "interesting question" was that we believed that knowing how to answer it would help us to know how to answer a number of other questions we have asked in the past. Which of your questions in (b.) best meets this criterion?

2. **Hanky Knot.** Grab one end of a handkerchief in each hand and, without letting go, tie a knot.
   a. How!?

b. Change attributes of the given situation or the question asked to develop a list of new questions that could be explored here.

c. Which of the questions in (b.) appear to *you* to be the most interesting, and why?

d. Suppose our criterion for "interesting question" was that we believed that knowing how to answer it would help us to know how to answer a number of other questions we have asked in the past. Which of your questions in (b.) best meets this criterion?

3. **Whythoff.** This game is played by dividing a handful of buttons between two saucers. Two players take turns taking buttons according to *these* rules:
Each player must claim *either*

   i. *any* number of buttons from *one* saucer, or
   ii. an *equal* number of buttons from *both* saucers.

   The winner is the player who clears the saucers on her or his turn.

   a. First try playing from *these* situations to make sure you'll win:

   b. Now consider a "trickier" version: three saucers, rules are the same, except each player must claim: i. *any* number of buttons from *one* saucer; or ii. an *equal* number from two *or* three saucers.

   c. Describe the strategy involved in winning this game.

   d. Change attributes of the given situation or the question asked to develop a list of new questions that could be explored here.

   e. Which of the questions in (d.) appear to *you* to be the most interesting, and why?

   f. Suppose our criterion for "interesting question" was that we believed that knowing how to answer it would help us to know how to answer a number of other questions we have asked in the past. Which of your questions in (d.) best meets this criterion?

4. **Tossing Dice.** Someone is told to throw three dice and add the faces. The participant is then instructed to pick up any one die and add the number on its bottom to the total. *That* die is rolled again, and the number on top is added to the total. The magician turns around and mentions s/he has no way of knowing which of the three dice was thrown for a second time. Then, the correct grand total is announced.

   a. Can you figure out how the trick works?

   b. Change attributes of the given situation or the question asked to develop a list of new questions that could be explored here.

   c. Which of the questions in (b.) appear to *you* to be the most interesting, and why?

   d. Suppose our criterion for "interesting question" was that we believed that knowing how to answer it would help us to know how to answer a number of other questions we have asked in the past. Which of your questions in (b.) best meets this criterion?

5. **Cutting Cakes.** Pictures a and b show how each of two shapes can be divided into four parts, all exactly alike.

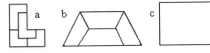

a. Your task here is to divide the blank square c into *five* parts, all alike in size and shape.

b. Change attributes of the given situation or the question asked to develop a list of new questions that could be explored here.

c. Which of the questions in (b.) appear to *you* to be the most interesting, and why?

d. Suppose our criterion for "interesting question" was that we believed that knowing how to answer it would help us to know how to answer a number of other questions we have asked in the past. Which of your questions in (b.) best meets this criterion?

6. **Chain of Fools.** You've taken up jewelry-making and you want to join the four pieces of silver chain shown to form a circular bracelet. Since it takes a lot of work to cut a link and weld it together again, you naturally want to cut as few links as possible.

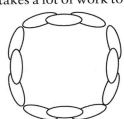

a. What's the minimum number of links you must cut to do the job?

b. Change attributes of the given situation or the question asked to develop a list of new questions that could be explored here.

c. Which of the questions in (b.) appear to *you* to be the most interesting, and why?

d. Suppose our criterion for "interesting question" was that we believed that knowing how to answer it would help us to know how to answer a number of other questions we have asked in the past. Which of your questions in (b.) best meets this criterion?

7. **The Big Race.** Musky and Quashanda race each other for one hundred yards. Musky wins by ten yards. They decide to race again, but this time to even things up, Musky starts ten yards behind the starting line.

a. Assuming both run with the same constant speed as before, who wins?

b. Change attributes of the given situation or the question asked to develop a list of new questions that could be explored here.

c. Which of the questions in (b.) appear to *you* to be the most interesting, and why?

d. Suppose our criterion for "interesting question" was that we believed that knowing how to answer it would help us to know how to answer a number of other questions we have asked in the past. Which of your questions in (b.) best meets this criterion?

six

Metaphors for the Classroom Space

What does it feel like to be in a mathematics classroom? This is a very different question from "What is the best way to get students to sit still and learn how to find common denominators?" or "What is the best sequence of tasks for students to do if I want them to be able to use determinants to predict the distribution of a population in ten years?" When the teacher thinks mostly about what to tell students, either what to tell them to do, or what to tell them that they should know, the experience of the classroom is one of following directions, listening and memorizing; students in such a classroom are expected to figure out what the teacher wants and then to provide it. Some students get very good at this; others do not, either because they have not yet figured out how to play this game, or because they have not been convinced that it is worth playing it. In general, though, no matter how creative we are as teachers, the students perceive what is happening in the same way: a lot of waiting, a lot of listening and trying to understand what the single authority wants, and a lot of trying to please the authority. I am not presenting this picture as something to condemn. There are times when this is precisely the kind of experience that is called for. But I do want to ask whether or not we could make the experience different sometimes, or most of the time. One reason for arguing this is simply to offer a diversity of experiences, just for the sake of trying something new; another reason is to explore possibilities for learning and teaching not experienced in the typical classroom.

This chapter explores different ways of experiencing mathematics class through the application of metaphors for the space of learning. In order to do this we need to look first at the curriculum concept of "space" and what it means. Then we will apply different metaphors for the different possibilities of "space" in order to come up with ideas for planning lessons and units. At the end of the chapter we will also look at another curriculum idea, "place" and how it is different from space. Place, like space, can help us to be more creative with our teaching and thus help to make the experiences of our classrooms more engaging and rewarding.

Spaces of Learning and Teaching

When I write "space of learning and teaching," I am thinking of how we describe what the experience is like for everyone involved. We need pictures of what is possible, what it feels like to be there, and what goes on—how the classroom works. Unfortunately, when we turn to curriculum guides, state frameworks, and standards documents, we mostly get lists of content, with little advice on how to make the content "come alive." We tend, most of us, to think of the school, or the classroom, as a space in which things happen. People do things; students learn stuff. We fill the space. What I am finding in my own practice is that it is not so much the subject matter content or the particular things that are done in the space, although these things are clearly important. What is really crucial is how people "relate to the space of learning." This is where we get into curriculum theorizing rather than instructional psychology: we start talking less about the things we touch or the exact things we say, and shift our attention to a "metaphorical space," the space "of learning."

Educators have written a great deal on "spaces." One way that they approach this is to describe in an anthropological way what the people are doing as "performing." This does not mean that people are putting on a play or a show, but that we can understand what they are doing as acting in certain ways that present messages to others and which communicate more than the simple action; in other words, the actions and words convey collections of cultural meanings and political effects. How others respond to what someone does and says would let an observer know a lot about what the shared meanings are in a particular setting. Of course, it is not always the case that each person in a situation shares the same meanings and understands the actions and words as communicating the same things; this is part of the performance that is unfolding. Alan Block (1999) writes that some performances define a space made "real" by the performance. For example, saying a prayer at the edge of a stream turns the bank of that stream into a Temple. Tying a person to a tree turns a wooded lot into a prison. In the same way, "doing school" in some classrooms makes those classrooms prisons or torture chambers, or parties or important board meetings. What we want to do is think of still other possibilities: and we could play around with a lot of images. What would be going on in a classroom that is experienced like a videogame arcade? A theme park? A soothing walk through a nature preserve? A business meeting? An art studio? An encounter group?

Jan Nespor (1997) writes about how bodily movement organizes space: we can distinguish between a body generating space, and the body as an object in a space defined by administrative or technocratic practices. Many teachers that I work with in study groups and courses are quick to speak in this way, too. Like Nespor, they might describe much of contemporary school experience as like sending youth up to the top of a skyscraper to learn about a city, whereas an important part of the process of living in a city is *being there* in the city, controlling or experiencing your body in that city. The point for mathematics educators would be to note the overuse of abstract, distant "places" of mathematics defined by texts and representations. A visceral experience of mathematics would be an antidote to this. Perhaps this is why we are attracted to manipulative materials or computer software that allows us to manipulate mathematical objects. What has emerged in my own work, however, is a realization that teachers, along with Nespor, can add a different dimension to all of this: it's not just getting out of our seats or using manipulatives. To take this one step further, with Alan Block, somebody could go to the top of a skyscraper or live within the city, and either experience might be educative, miseducative, or non-educative. What is important for education is the "space of relation" to that event in either case.

The axiom for all of this could be expressed in a quote from Henri Lefebvre (1991: 299): "The concept of space is not in space."

How annoying! Why did I introduce the idea of space in this chapter if I want to say that the important idea is not actually in the space of learning itself? Because this is important to think about: we need to think about the space of learning and teaching, and we also need to understand that what is particularly important *about* this space that we are creating with our students is not actually in the space itself but in how each of us is relating to that space. How we relate to the space determines a great deal of how we are experiencing that space. Even though there are some recently fashionable theorists of "space" these days (Edward Soja, A. Vidler, Doreen Massey, David Harvey, Patricia Yaeger), I still find Lefebvre's writing insightful and powerful for our work with teachers and youth in schools. He describes the history of social space as a kind of "pernicious ideology" that, being an ideology, warps our ability to think differently.

> The most pernicious of metaphors is the analogy between mental space and a blank sheet of paper upon which psychological or sociological determinants supposedly "write" or inscribe their variables.
>
> (Lefebvre 1991: 297–98)

Lefebvre writes about how technicizing, psychologizing, or phenomenologically oriented approaches displace the analysis of social space by immediately replacing it with a geometric (perceived as neutral, empty, blank) mental space. In this way, objective space and the subjective image of space—the mental and the social—are "simply identified." The ultimate effect of these sorts of descriptions is either that everything becomes indistinguishable or else that rifts occur between the conceived, the perceived, and the directly lived—between representations of space and representational spaces. I take this directly to the heart of educational practice: the rifts or collapses of distinction between the lesson plan and the lesson, the curriculum outline and the lived curriculum. We start taking one *as* the other. For example, in our work, teachers are forced to contend with pressures to collapse their mathematics curriculum into the State Curriculum Framework's Standards. What is your lesson?—Standard 10.2.3. The point, though, is not to bash the Standards, although some people want to. In general, I *love* the Standards. They promote meaningful mathematics education. The purpose of this chapter is to help us enact the Standards in ways that create "space" differently.

This is how we do it: We think of an experience of a space. Then we think: What are the critical features of this space? What are the significant emotions associated with this space? How can the design of the instructional event make it likely that the nature of the experience will be very much like the metaphorical space we are using as a model?

The trick is: We don't want to pretend we're somewhere else. We just want the experience to be metaphorically analogous to the space.

Theme Park

What if we wanted the experience of our classroom to *feel like* a theme park? Once you get in, spending a great deal of something of value to enter, there are a lot of choices of where to go. Rides do things to you. They *scare* you, but you know you'll never get hurt. There are lots of surprises. If this was our generating image for the classroom space, then we should make it cost a fair amount in some way to get into the classroom; students should feel like they invested a great deal to see what is in the room. And what they invest must be something that they could have spent elsewhere. Time is one thing they can invest in our classroom. Effort is another. They must have given up something else in order to be in the room in the first place. Then there should be a wide range of choices of things that they can experience once they are there. Mostly they would give themselves over to the experiences they choose. Perhaps there would be centers based on instructions to follow; interesting surprises would arise from going along with what the worksheets say at each center. Students would talk amongst themselves about which centers were better than others—which were more surprising, more thrilling, more interesting, more entertaining—and they would be given time to select centers to return to before the classroom theme park was closed for the day.

We might instead choose a specific theme park experience as our metaphor. For example, have you ever seen the Third International Mathematics and Science Study video of American, Japanese, and German teachers teaching mathematics? There is one lesson, in which a Japanese teacher leads students through an application of ideas about areas of triangles to obtain areas of other less obvious shapes. I have always felt that this lesson is like a roller coaster, with higher and higher peaks. The lesson starts out slowly, with a short review of previously learned material, just as a roller coaster starts off slowly on a level track close to the ground. Soon, the teacher gives the students a challenge, which I find analogous to the first peak in a roller coaster. Now, on the ride, you can scare yourself while knowing in the background that you are not going to fall out and get hurt. In this Japanese classroom, hint cards and talking with others provide a safety net. Students are not going to fail, and they know it, because they know the assistance is available should they need it; however, they can make the experience as thrilling and scary as they want. The question the teacher poses is not like anything they have ever seen before, and I believe it feels a lot like balancing at the top of that roller coaster peak. As they figure out solutions and share them with each other, they are eventually led to a new and higher peak, a new question that challenges them again; and again they know that the hint cards or the option of talking to someone else is available. One interesting feature of this Japanese lesson, by the way, is how the teacher asks two students who have taken different approaches to the problem to share their solution with the class. The teacher's job in this room is to pose the challenge, and then to identify two students who take such different approaches, but who together can help him to move the class along the track toward the next challenge. In this classroom, the teacher summarizes what the students' different approaches have in common, and uses that as the basis for the next question that everyone is to work on. The time during which the teacher is talking and summarizing is again, for me, a period of flat track upon which the students can get their bearings before the next thrill or peak of the roller coaster experience. How might this classroom be more like a theme park instead of just like a roller coaster? For a theme park we need choices, to mirror the different rides that are available. Right now it is like a single roller coaster.

Birthing Center

Suppose I used the metaphor of a "birthing center" to think about my classroom. Then, as the teacher, I am a midwife. I can't do it for you. It's going to be horribly painful. Really frightening. But it will be exhilarating, it will change you forever, your life will never be the same again. And, you have this new responsibility, to take care of what you have created and brought forth for the rest of your life. As the teacher, I can provide a comforting and supporting environment, and I would want to plan ahead to have whatever tools and comfortable materials and working arrangements might be necessary. But I myself am not giving birth to new ideas; the student is the one who is the parent of the new ideas. The midwife's role is to witness and to ease the painful, horrifying, joyous event. My experience working with many new teachers searching at first for concrete lesson plans, working with them on issues of classroom space, has indeed been the creative and threatening experience of birth. And indeed the births of new ideas and new forms of relationship have transformed each of us after the life-changing encounter.

Is there a step-by-step way of making this be the primary way that students experience mathematics class? A good first thing to do is simply to tell students that this is what they are doing:

You are mathematicians. What this means is you are creating new ideas that will take on lives of their own. Your job is to give birth to these ideas, nurture them until they are mature and fully grown, and then see how they apply themselves on their own in the world of mathematics.

When I work with my class in this way, I start off with such an introductory statement. Then we talk together about where they can go to get started on creating new ideas. I give them possible topics and questions to work on as potentially interesting starters. As they work, I encourage them to make lists of their own questions and their initial ideas about what they are working on. I conference with individuals and small groups as they begin to develop an interest in a certain topic or mathematical object. At the end of the first day, I ask students to write down on index cards three things that they are willing to talk with others about based on their work that day. The next day, I assign them to groups based on common interests and similar questions from the cards. I tell them this is only to facilitate their work; they are not permanently assigned to these groups, and they may change what they are interested in after today's work. On this second day, our goal is to help each other to identify something that we would like to work on, either together or on our own. At the end of the second day, I ask the students to again list two or three things, this time that they want to work with on their own for homework between now and the next time we meet. When we get together for the third meeting, I assign students to groups based on this second round of what they wrote down. I ask each group to report by the end of the day to the whole class any questions or possible explorations that they think would be interesting or worthwhile, along with their justification for why it would be interesting or otherwise worthwhile. The next homework assignment is to select a project to work on and begin to work on it for next time. As a midwife, my job is to support and ease each student's entry into the experience. What do they need to be able to do this? Would another person in the room be able to help them think through their ideas? What would be a good way to organize work groups? What materials or books might I find that could help them in these initial stages when I know much more about what they are working on than they do? Just as expecting parents share advice and strategies with each other, we hold whole-class and combined small-group meetings to compare ways of working, things we have learned, and plans for future inquiry. But my job soon moves into the next phase, where I need to help them through the gestation process of exploration on their own. They will be nervous; they will have lots of questions about what they are doing and how to do it, just like pregnant parents have about the preparation for childbirth. I encourage them to record their experiences in their mathematics journals and notebooks.

Now, the birth of a new idea is as unpredictable and as predictable as the due date of a new baby. It will take time, at least a couple of weeks of work. But the exact date can never be known. When the idea is born, it is surprising, shocking, joyful, maybe scary. I need to remind students of this constantly as they are working, and ask them for constant updates, just like a midwife meets regularly with expectant parents to check on the status of their baby. At some point, when the new idea is born, I have to remind the students that this new idea means a lot of responsibility for the parents: Are they nurturing and taking care of the idea? Are they finding ways for the young idea to connect with other mathematical ideas, and to forge relationships with these other ideas? (Think of playgroups and daycare for young children.) How is the idea being used by others? Responsibly? Have they raised the idea to maturity by searching for how it fits into the world of mathematics, and teaching the idea how to get along with everything they know about the world of mathematics? And the next question becomes: are you ready for the experiences of birthing and taking care of another new child, that is, are you ready to start a new mathematical exploration?

Counseling Session

In most of my classes, I find that everyone would benefit from learning how to get the most out of group work. I take "counseling sessions" as my model for such interaction. When people join a counseling group, they need to learn how to contribute, participate, and learn from the experience. This takes a lot of work and effort. I expect the same or analogous challenges for my students in mathematics classes. How does a group work together in counseling? One primary goal is for participants to listen carefully enough, and to support one another's developing ideas enough, so that they can truly understand what others are experiencing and understanding. Each participant must learn to suspend what they are thinking about themselves in order to be open to others' ideas. We need to practice this a lot before we can do it. Suppose two students in a group have very different ways of calculating an answer to a mathematics question. At first, each one will tend to want the other to see it in the way that they themselves did the problem. In a counseling session, however, we need to think from the other's perspective. So, in our classroom, working in small groups, each of us must interview the others long enough to be able to think in the way that they are thinking, to be able to calculate the answer using their methods as well as our own.

But there is more to the group experience than sharing strategies and being able to reproduce each other's work. In a counseling group, the group collects each member's perspective and analyzes the collection for insights, patterns, and significant themes. This, too, must be part of our small group meetings; indeed this is a crucial element of the work that is done. The counseling metaphor helps me recognize how serious this part is. Suppose one pair of students solve a problem using a system of linear equations and the elimination of a variable through substitution; another pair solves the same system of equations by subtracting one equation from the other; and yet another student graphs the equations to estimate the point of intersection. In a typical classroom, each method would be accepted as leading to the same answer. In a counseling-like classroom, students are required to probe how each perspective on the original question interprets the question, models the situation, and provides a response. Furthermore, each method must be compared with the others within the details. Is subtracting one equation from the other related *in meaning* in any way to substituting for one of the variables? How? How do they accomplish similar results, even as they are different strategies? What part of solving for one variable is directly analogous to graphing of the two equations? . . . to looking at the point of intersection of two graphs? How do visual solutions change the way we understand the problem as compared with symbolic solutions? Why do some of us avoid one or another approach when it would be more effective? Do some of us select what appears to be a more complex method even when a simpler one would do? Why? The group needs to create together similar situations in which they can project a good reason for choosing one approach over the other. For example, if one of the equations makes it very easy to express one variable in terms of the other, then substitution would make sense; if one of the variables has the same coefficients and exponents in each of the equations, then subtracting one equation from the other might make the most sense. Yet, if either of these situations is not evident, we might find it easier to graph the two equations. Or, perhaps we think it would be more useful or fun to practice graphing, or to practice solving by substitution; should this affect our strategy choice?

The results of such counseling-like group work, then, need to include the "answer" as only the very beginning of the group's efforts. What is "checked" as evidence of successful completion of the task is whether or not each member can successfully use more than one approach, whether the group has come to a consensus or amicable disagreement on the reasons why they would select one or another strategy on their own, and whether or not the group has helped each member to understand something about themselves as a problem solver. The teacher's role is that of the group facilitator. Instead of judging the quality of work as the single focus, the teacher must make an important focus the monitoring of group processes. Is every member contributing? Is every member being listened to? If not, the group needs to have this pointed out to them. Is the group avoiding one of the goals of group work (e.g., they did not identify a reason for using each method they discussed in some appropriate context)? If they are avoiding something they need to have this identified for them, and to be asked to reflect on whether there is an underlying reason for this avoidance. The teacher also gets the fun job of pointing out the successful ways that the group is accomplishing its work, by noting particularly good questions that one member asked of another to clarify something, or by observing ways that members use body language and eye contact to push the group to consider important ideas, and so on. Eventually, as the groups get more proficient at interacting and accomplishing the group goals, the teacher can help groups to perform these monitoring tasks themselves.

Listening, mentioned in other chapters, is an important aspect of counseling groups, and deserves attention here as well. Other chapters stressed the importance of the teacher's listening, and this is obviously important in the analogy of the counseling group as well if the teacher is to develop good facilitation skills. But in this kind of a classroom, the students also need to work on listening. Here are four ideas for helping students to analyze how to improve their group listening skills.

1. *Talking Stick*: to get better at focusing on one person without interruption until they have completed their entire thought, and to practice allowing others to say something that is on your own mind without feeling bad about the lack of credit for the idea. A special stick or other object is passed from one member to the next around the circle; only the person holding the stick may talk, and the order of talking may not change as the stick is passed around the circle. After working in this way for at least fifteen minutes, small groups or the whole class should debrief what they noticed about how the stick changed the kind of conversation they had. They should discuss ideas for how to make the talking stick conversation more successful, and

try these ideas again at a later time. Even if members do not like this way of talking, they should work using the stick for several practice sessions, trying each time to make it as successful as possible. They should always identify what they learned about how to make a conversation more successful from the experience.

2. *Five-Chips*: to get better at choosing when to talk and when to listen. Start by passing out five chips to each member. Each time a person speaks, he or she places one of the chips in a pot in the center of the table. When a person runs out of chips, he or she must wait until everyone else has used all five chips before the chips may be redistributed and everyone may speak again. Like the first exercise, students must be led through a debriefing, and identify ways to make this as successful as possible, using it multiple times before they can dispense with it as listening practice.

3. *Only Questions*: to get better at identifying what is important in what others are saying. Start with one person sharing something in the form of a question. The rule here is that each person may only ask a question of the previous speaker. One may never answer a question or divert a comment to someone who did not just finish speaking. This one takes a lot of practice but it is always interesting and provides much to talk about. As with the other exercises, it should be practiced multiple times with debriefings in-between where participants identify things to try in subsequent practice days.

4. *Connections*: to reflect on what you want others to know about you and what you want to talk about with others. Members describe what they are thinking about *not* to have a conversation right now but to share things they would welcome a conversation about at *another time*. Start with a required thirty seconds or full minute of silence. Be strict about making sure that nobody has the conversations during connections but saves the conversation for later and that people always pause for a minimum of five seconds after someone has been talking before they themselves start speaking, to make sure that the previous person has completed their thoughts. This, too, can be frustrating, but deserves much practice because it teaches how silence can be an important part of communication, and usually generates much more lively conversations later, after folks have thought about ideas, than the different, more immediate spontaneous conversation that we are used to.

Box 6.2

Group Listening

Get together with a small group. Alternate working on a challenging mathematics problem with a discussion of your group problem solving processes. Use each of the four exercises, *Talking Stick, Five-Chips, Only Questions,* and *Connections* at least three times each. There are two ways to do this: a) rotate through each exercise 1, 2, 3, 4 three times, with debriefing in-between each exercise; or b) do the same exercise three times in a row, 1, 1, 1, 2, 2, 2, 3, 3, 3, 4, 4, 4, with debriefings in-between each one. If you are in a class on teaching mathematics, it would be great if different groups tried each way and compared them.

With each exercise, the teacher must patiently help students. Comments and questions should be on-topic, having to do with mathematics, group processes, thinking skills, communication strategies, and so on, and not just about anything at all. One approach that works well for me is to alternate assigning the groups to work on particular mathematics problems in their conversation, or to speak directly about group processes and/or mathematical problem posing and solving.

Art Studio

I like to balance group learning experiences with ways that the group can support an individual's private work. One good way to do this, I have found, is to use the metaphor of an art studio classroom. Think about an art studio class in which each student is working on her or his own project during class time. An important feature of such a class is when once per session or once per week a student shares their work-in-progress with the rest of the class, describing the decisions they had to make in their work so far, and asking for advice on particular problematic aspects of their work. For example, if a student is painting a portrait, he might describe how he went about choosing the color palette, the orientation to the subject, and so on, but ask for help in getting the shadow just right under the chin; students would ask questions related to their own concerns, offer comments on what they find to be very promising in the portrait in progress, and suggest ideas for the chin's shadow. The teacher might note interesting aspects of the work from which others might learn. This student would take the chin advice, and also be made aware of the ways that his work has influenced the work of others in the class. We can offer mathematics students similar experiences if they are working on their own mathematics problems or investigations. They can share their work in progress. Perhaps one student per day, or two every Thursday, or five every Friday, or whatever works best for our class. During each studio day, students would spend time in class working on their investigations. At a designated time, perhaps the beginning or end, one student would present her work so far, noting the decisions she has made in her work, her choices of investigation strategies, and her results so far; she would then ask the class for advice on particularly challenging aspects of her work, or for help in choosing among different strategies for solving a problem. As in the studio class, the teacher and students would note aspects of this student's work that others might learn from, and they would point out aspects of her work that are particularly insightful or surprising. They would make suggestions based on their own experiences for what might work well in her investigation. And they would get a chance to ask questions about what is of interest to them in her work so far.

Arcade

Here is an example of a way that I once used the idea of a videogame arcade to generate my students' experience. We needed to begin a unit on ratio and proportion. I wanted this opening class to be full of excitement and I wanted to move the students around a lot using different approaches to and models of ratio. I set up seven centers around the room analogous to different video games in an arcade: the ball bouncing investigation I have mentioned before, where students are encouraged to notice that balls tend to bounce back to a height proportional to where they are dropped from, if they are simply released rather than thrown; making a percent calculator from a rubber band that is marked at one-inch intervals when stretched next to a ruler—it can now be stretched to any length that would be the 100 percent distance and other objects can be measured using the ten inch-marks as a percentage of the whole; scale models of simple building block structures made with cubes to be double or triple in size—their volumes and surface areas can then be compared to the change in the length of a given side; snap cube patterns based on ratios that can then be used to represent rhythms and clapped or otherwise performed for others; reading of the book, *Zoom*, by Istvan Banyai (1998); and the challenge to design an eye chart for determining whether someone needs corrective lenses or not. I thought about a videogame arcade. Visitors go to a certain game and play it, but before they get really good at it their time is up and they need to decide whether to play it again or to go to a different game. I told my students to go to one of the centers that looked interesting to them. They had ten minutes before they needed to stop and decide to stay or go to another center. If there were more than five people at a center, then there was no room and they had to wait for them to be finished before they

Box 6.3

Illuminations Shed Light on Metaphors

The NCTM's *Illuminations* website http://illuminations.nctm.org/ is designed to:

- provide Standards-based resources that improve the teaching and learning of mathematics for all students;
- provide materials that illuminate the vision for school mathematics set forth in *Principles and Standards for School Mathematics*.

There are tons of super activities and resources at this site. For our purposes, we can explore the metaphors for the classroom space that are implicated constructed in these materials.

(a) Choose several activities and resources from different sections of the *Illuminations* site, and describe the metaphors for the classroom space that are possible to enact using the materials provided.
(b) Check out the **Video Reflections** found in the *Standards* section of the *Illuminations* site (http://illuminations.nctm.org/Standards.aspx). First, use the discussion questions to think about issues that come up in your viewing of the videos. Afterwards, discuss the metaphors for the classroom space that you think could be used to redesign the activities you witness in these videos.

could play at that center (this was supposed to be analogous to waiting until someone else is done with their quarters at an arcade before you get your turn—you have to play something else in the meantime). After several rounds, I asked students to pick a center to focus on for a while so they could explain it like an expert to the rest of the class (analogous to someone deciding to play one videogame a lot and becoming very good at it, and then being able to tell others the fine points of how to play well).

Dinner Party

I once taught a college course on mathematics education using the metaphor of a dinner party. Here is an excerpt from my teaching journal that semester:

> When everyone arrived, I had laid out a couple of options for them to try. But I did so in a way that was both inviting and not inviting, for I had no desire to require them to taste my treats. I just hoped they would find them intriguing. The group straggled in after their long day at work, finding friends and new people they had not yet met. Soon people started to talk, about the traffic, about their ongoing assignments, about the upcoming comprehensive examination. "Would you like to try this new way of playing mancala?" I asked Viola. I asked this in the same way that I might offer *tartines chevres* to guests at a dinner party in my home. "Come, Nancy," called Viola, as she turned her head. "Let's try it." "Shareefa might want to try this one too, have you two met yet?" I asked. "No, somehow we haven't had a class together yet," noted Shareefa, as she joined the group; "Let's find one more person, and we can play in pairs." I walked away, content that they had met through the mancala game. Who else might find a friend here? Perhaps I could suggest *Traverse* to both Sal and Fatima . . . Later, we would sit together for the main meal: a *Jasper Woodbury* episode on laser-disk.

Salon

I also enjoy using the metaphor of a "Salon," where people are invited to gather and discuss topics that they themselves suggest as interesting to the group. Here is an excerpt from one student's journal from that semester, reflecting on the Salon experience.

> When Peter told us we would be discussing this book, I figured I'd scan it over the weekend and come in with a few questions that would provoke him. I know it's easy to do: just start in on the fuzzy math versus meaningful math debate again, and the class will fly out from under him. Then we can see how he amusingly returns us to the discussion he wanted us to have. So I picked up the book Sunday night and—yikes! This stuff was incomprehensible. I called MaryEllen. What is this stuff *about*? No, I don't get it either. Yeah, one page of each chapter is enough for me. I don't know. Well, see ye' on Tuesday.
>
> Tuesday night we sat in a circle around the table—somebody had brought chips and guacamole; somebody else had brought ginger ale. The discussion

was funny: I keep expecting the lecture, but instead here we were chatting about a book we had all (?) "read," and the questions were the ones the *students* posed. Peter would take his turn declaring an opinion, but he never tried to teach us anything or win an argument with us. He seemed to relish having a good time listening to each of us. One thing he seemed to be working on was on getting each of us to feel like what we had to say was at the center of the discussion we were having—I think everyone felt like they had made a good point. I know *I* wanted to come back next time, either for the conversation or just to be there with the others. It was like going to my great aunt's house when she had people from around town over for coffee and cake. They would talk about *important things* and leave knowing Aunt Edith wanted them to come back.

Using Metaphors

There are at least two types of ways to use metaphors for the classroom space: "How might I plan an experience, an event, using a particular metaphor?" and using a certain metaphor, I can ask, "How is a classroom experience like this? How could it be more like this? What if I thought about the same experience using this *other* metaphor instead?" I usually do not tell my students what metaphor I am using. It is mostly just a tool for my own planning. But it is sometimes interesting to ask students what they thought a class or unit was like. Do they have a metaphor for the experience? Often it is very different from my own, prompting me to think about whether I like their metaphor better, or whether I should change the way I use a particular metaphor.

Box 6.4

One Possible Two-week Schedule:

	Monday	Tuesday	Wednesday	Thursday	Friday
Week 1	Birthing Center— day 1: Starter activities	Birthing Center— day 2: Groups based on initial interests	Birthing Center— day 3: Groups based on common questions Listening skills practice in prep for tomorrow	Roller Coaster Challenges Counseling Groups to implement ideas from listening practice	Art Studio
Week 2	Dinner Party	Art Studio Counseling Groups Birthing Center— Checkups	Birthing Center— Checkups Listening skills practice looking back on yesterday's Counseling Groups	Birthing Center— Checkups Salon	Arcade

---

There are some difficulties in using metaphors for the classroom space common to everyone working in education. These difficulties arise from the common sense approach to curriculum, which emphasizes a very *concrete* relationship to the teaching/learning encounter. But there are also ways in which a teacher's understanding of the subject matter of mathematics, and his or her conception of—or beliefs about—education, make it easier or more challenging to use metaphors of the educational space in their practices. Let's first talk about teachers and their use of metaphors, and then come back to the more entrenched, common sense discourse of education and my argument that a fundamental shift is necessary.

Teachers overall are surprised by my interest in metaphors. Their initial reaction is bewilderment that someone might think this way. This is not part of typical courses about teaching mathematics. Yet this initial reaction is something worth moving beyond, because every group of teachers that I have worked with has found my metaphorical interpretations of their work fascinating, and each teacher was willing to apply metaphors of their own in retrospect after it was suggested. In a project where I talked with some teachers once each month, each of these teachers quickly began to use metaphors for the educational space in their planning and in their ongoing reflective assessment. It became routine for them to praise this "new way of working," which, given our frank relationships (in which, I believe, they would have been comfortable telling me otherwise), seems to support the strength of metaphorical approaches in providing a new dimension to the work of experienced teachers. Several of these teachers have moved into increased leadership roles in mathematics at the school or district level, and one has been nominated for Teacher of the Year based on a number of projects she initiated using the metaphorical approach.

Some teachers struggle more with the ideas of metaphors for the classroom space. Many of these teachers are very frustrated by the ideas of metaphors and metaphorical interpretations of their work. Such a teacher tends initially to think of the metaphors in a very concrete way. For example, if they imagine themselves as a tour guide leading a group of students on a mathematical journey, they orchestrate activities in which they and their students actually *pretend* they are going on a trip. We take turns facilitating class activities in one of my university courses. On a certain night, for example, a group of students provided us with passports and tickets for an adventurous journey through "Fractionland," and then led us to several "sites" or centers around the room. They then gave us unstructured, free time to return to our favorite site for a longer, more personal visit. On another evening, a hostess met us at the door, and seated us at tables in a French Café. The menu included a variety of mathematical appetizers, main courses, and desserts that we could choose from. The servers brought appropriate materials depending on what we ordered. These were "fun" activities, but they were not examples of what I have been describing in this chapter. Instead of the class activity being metaphorically analogous to a tour or a meal at a café, we were actually pretending to be on a tour or at a café. In some ways, this could be *good*: using such activities in a classroom would be fun, at least when they are novel. But there is an issue that needs to be addressed: If we are pretending and performing roles in a pretend space, then we are just doing the same thing that always happens in school. We are playing the parts of people being there. If we use such a form of organization too often, then this is no different from a "regular" classroom, in which teachers and students play the roles of teacher and students in school. The *experience* is artificial and not authentic in either case. This is important to think about because there are times when students might design a role-play as part of class, or times when people might pretend to go on a trip through "Fractionland," but the context is different enough to make this an authentic experience, because the metaphor for the space is one in which groups of students, or individual students, compose such fantasies as part of, say, a writers' workshop metaphor for the space of mathematics learning. But this, again, is a significantly different experience than "going on a trip through Fractionland" as planned by your teacher as the way you learn about fractions, being a student in this class.

Let me give you another example from the same graduate course: A group gave us a complex and ambiguous logical reasoning puzzle, and asked us to spend ten minutes reaching our own conclusion. Then we formed groups based on our individual conclusions. We were asked to present a convincing interpretation of the problem to the facilitating group twenty minutes later. After our public presentations were made, the facilitating group met by themselves and discussed the presentations. Then they returned to our class and explained why they found each "convincing argument" confusing, how information from one presentation seemed to contradict another, and so on. They asked us what we wanted to do: should we try again to present a convincing argument, or should we just declare that there was no way we could reach a consensus? We tried once again, and again waited for their return. Again they found inconsistencies and conflicts. Discussion following this included: all people in the room felt like a "mistrial" had been declared, as if the facilitating group were a "hung jury," even though we were not told to pretend we were in a court room. Most people could list a number of mathematical and logical reasoning strategies that they learned or understood for the first time that night.

One pedagogical question is, "How do we get better at understanding the difference between pretending and metaphorically reflecting on our practice?" I find it works well to go with the groups facilitating activities for each other, even if a lot of the events are counterexamples rather than models of what I am talking about. We discuss each event afterward, and as I noted, even the ones that are not models of meta-phorical reflection are fun and valu-able experiences. It's O.K. to go to the Math Café, also, because it helps us "see" the difference, it help us under-stand that this is *not* the same as a metaphoric use of a café as a tool for planning a problem solving context in a mathematics classroom. And, as I interject my own Ratio Arcade, or Salon, evenings, over time students in these courses can begin to critique *my* events and improve on *them* as well.

> **Box 6.5**
>
> **Design an Event**
>
> Design an event for a group based on a metaphor for the classroom space. After the event is over, ask the partici-pants what the experience felt like. What metaphor was it like? In what ways did the experience feel this way? Compare their responses with your own plans and design for the event.

Another pedagogical question is, "Can we expect new teachers to think this way, or does this kind of abstraction demand a certain amount of maturity, experience working with young people, or depth of reflection that is developmental in its evolving sophistication over time?" I am exploring this in my work with emerging professionals in mathematics and science methods courses, but especially in my work with student-teacher/cooperating teacher pairs. In this field based context, the cooperating teacher agrees to explore new ways of thinking about their teaching, and shares that experience with the student-teacher intern. Just serving as a model of a teacher, as a person always challenging himself or herself to rethink their practice, has a significant impact on the intern's understanding of the intellectual life of a teacher. But I also find that metaphoric interrogation of the educational space is a particularly accessible and powerful intervention for emerging professionals.

If we look at a typical classroom, it turns out that it is hard to recognize metaphors that are appropriate to describing the experiences. The only images that come to mind are different forms of *training*. School itself is a metaphor that permeates our culture and constructs ways of interpreting life experiences as "just like school." The paucity of variation in school experience is the source of this! And the message for thinking about schooling as an "authentic experience" is pretty clear: classrooms are our cultural metaphor for a lack of engagement, a loss of a sense of self, of delayed gratification, of a regulation of our presentation of emotion. This is the point of every episode of *The Simpsons* ever aired on television.

At the same time, new teachers are asking for *techniques*: "How do I plan?" "How do I know if I am successful?" "What should I look for in students' performance?" Given the narrow range of metaphoric descriptions for the schooling new teachers themselves have experienced, as students and as interns, where can they go for help? So we can provide options: A Madeline Hunter Theory-into-Practice lesson plan model is one option. 4MAT is another. We can ask, what if you took into account multiple intelligences?

And we can ask, what is the experience like? What if it were like something else, such as a birthing center, a *rave*, a birthday party? And suddenly people start thinking of all sorts of new ways for organizing activities in their classroom! By considering metaphoric spaces that feature choices, open ended situations, and multiple possibilities, we find ways to facilitate more open ended problem posing structures, more student projects, and just plainly, more engaging, possibly more "authentic" pedagogic encounters.

Box 6.6

Observing Metaphors

Observe a mathematics lesson that you are not teaching yourself. Think of two metaphors to describe the nature of the experience during the lesson. For each metaphor, note how the features of the experience are analogous to aspects of the metaphorical description.

a) How could the lesson be modified to make the experience even more analogous to each metaphor that you are using?

b) What do you learn about this particular lesson when you compare the two different metaphors that you have applied to describe it?

Relate to the Space, Stop Creating Spaces

There is yet another, subtler pedagogical question. This question grows out of the work of educators who want to empower students and teachers to create new spaces, as in new ways of being and learning together. "Empowering" or "liberating" pedagogies see the creation of new spaces as a tool for empowerment or voice. The fact that these discourses are repeatedly criticized for their failure to make a difference, for their failure to lead to significant changes in institutionalized schooling—or many students' school experiences—suggests, to me, that we may need to re-think the ways in which these discourses use "space" to "make space." How can we avoid the traps and pitfalls of the common sense expectation that a teacher, or a teacher and her (or his) students together, or students, the teacher, and community members, *create a space* for learning? I am finding it very helpful to analyze curriculum discourse with teachers, and to deconstruct this discourse in order to understand the relationships enabled or constrained by the perception of space within the discourse. Just to take one example, Hilary Povey, a British mathematics educator, wrote a marvelous chapter on gender and mathematics, entitled, "'That Spark from Heaven', or 'Of the Earth'" (1998); here she identifies three characteristics of a mathematics classroom predicated on an inclusive epistemology:

1) the learners make the mathematics
2) mathematics involves thinking about problems
3) difference and individuality are respected.

(139)

From her title we can see she's constructing mathematics and the doing of mathematics as "of a particular place"—either sparks come from heaven, or hard, gritty, earthly efforts lead to accomplishments in mathematics. By "the learners make the mathematics," she is envisioning a place where students work together to produce as well as criticize meanings; she sees the mathematics being co-constructed by a community of validators. In the second characteristic, "mathematics involves thinking about problems," Povey conveys the understanding that a problem centered curriculum involves a need to take risks, which is a precondition for imagining a different and more just world; posing and reposing problems is a *spatial event* because it helps uncover, or reveal, hidden linguistic assumptions. She sees these first two characteristics as demanding that we *make room* for students to move and breath, as opposed to experiencing the current and increasing demands of performativity and patterns of surveillance. Finally, respect for difference and the individual challenges discourses of ability. Here I believe she is applying theories of difference and identity politics growing out of the work of homi bhaba, bell hooks, and educators working in cultural studies. These people conceptualize identity in terms of the "location of culture." In this formulation, we apply Cartesian visions of space to culture, where there are an indeterminate number of axes or dimensions of cultures and the complex, linear combination of all of these is imagined to express the dynamic motion of identity within this space. It all fits with Lefebvre's deconstruction of the ideology of space mentioned earlier. The ideology incorporates people as "centralities," as sources of energy crucial in the production of space.

"Bodies—deployments of energy—produce space and produce themselves, along with their motions, according to the laws of space" (Povey 1998: 171).

Now remember: I'm suggesting that we challenge that ideology, because these "liberatory" pedagogies, by fitting into the ideology, by reproducing the hegemony, do not give us a way out—I'm suggesting that we study the different ways in which educational discourses may be reproducing that hegemony, that maybe this is why we're disappointed with the impacts of these discourses. In the case of Hilary Povey, I suggest that the ways in which she uses the location of culture reproduces an insipid preservation of a Cartesian space of education. In this way, she is limited in her options for creating spaces in which students work together to make the mathematics, understand mathematics as thinking about problems, and respect difference. A space does not need to be created: a teacher and students *are* in a place together. Now, how might they relate to the space? What images come to mind that involve groups of people and a facilitator thinking about common questions and together seeking responses to those questions? A courtroom? A boardroom? A town council meeting? A divorce mediation? A community activist group? An amateur theater group? A religious group? A citizen's initiative task force? Any of these might provide a way to think of the space.

Here are some common ways in which space is perceived in educational discourse. A teacher might act as if the opportunity to work in certain ways is a gift to their students, as is the knowledge that they are learning. In this sense it often becomes a property of people, and can get confounded with issues of rights and responsibilities. It might be an intellectual space given by a principal to a group of faculty that feels excluded from the institution; or it might be a world of joy and ceremony bequeathed to others, as mathematics might be thought of by a teacher; or indeed it might be a generous chance to learn something in ways that are finally possible rather than exclusionary. Janet Miller helps us understand such a way of thinking about spaces: she heroically or stoically or romantically bestows a space upon teachers that is theirs to cherish (Miller 1980, 1990); teachers, for example, are given a space within which they can find their own voice. Here both the giving and the use of that gift express both the production of space and the laws of that space, the field of possibilities that emerges in the production of the space. What is possibly problematic about a gift is that you can never be sure how it will be taken and used. Suppose I give my daughter paints and brushes for her birthday because I envision the wonderful paintings she will produce. I say, "Paint!" What is going on in her mind? Does she want to paint? Does she want to paint in the ways that I am thinking of? Will she even use the paint? Will she use the brushes as arrows with a hand-made bow? Spaces of learning come with all sorts of baggage when they are conceived of as gifts, and this baggage does not always have the impact on the students that we anticipate. How can I bestow something on someone that is not necessarily interested or wanting or knowing how to use my gift, or has quite enough of it, thank you?

Perhaps then we could declare space an edifice: something that people build or craft, like a cathedral or a bridge. A mathematics resource center, a state-of-the-art smart mathematics classroom, or a feminist mathematics classroom may become for us a spiritual place, a connecting arch from one world to another, or a platform for political posturing. Now we can begin to talk about repairing the space, or desecrating it, or destroying it. Some spaces become closed or under construction. The students and teacher are constructing a new way of working together, or a new space where certain kinds of learning can take place. Feminist pedagogies, critical pedagogies, post-hetero-sexist and post-colonial pedagogies have been partly responsible for this "spatial turn." Each of these constructs space as something to be created. Often, these pedagogies collect in cracks—fissures of theory and practice—and then crystallize, splitting open a space of possibility (de Beaugrande, bhabha, Britzman, Grumet, hooks, Martusewicz, McLaren,

Box 6.7

Metaphor	What it looks like	Teacher's thinking	Students' experience (cast in a negative light)
Gift	Teacher does a lot of telling and showing; students do a lot of listening and practicing.	I have so much of value and enjoyment to share with my students!	Who cares? Why should I care?
Edifice (or the place where an edifice can be)	Lots of discussion about process for what will be happening next.	We can make new ways of working together, an oasis from the burdens and pressures of everyday life.	When are we going to get to actually learning something, instead of just talking about it?
Frontier	Teacher and students engage in collaborative and challenging work.	We are exploring the thrills of new knowledge together, as in Donald Duck's land of great adventure!	This is great, but I wish we could sometimes just relax and do things like most of my other classes.
Cartesian system of location	Students have individualized schedules and goals, moving in and out of small groups and teacher conferences.	People are labeled according to dimensions of their experience, such as race, class, gender, sexual orientation, preference, or distaste for math, etc.	Everything is usually at about the right level for me, but my teacher doesn't really know who I am, quickly jumping to fast conclusions about what I am capable of doing.
Distractor from the space-off	Classroom time is heavily taken up with organizational planning for what happens outside of class time, such as planning and report-ing on investigations taken on outside of school time.	My students are actors in the real world, not mindless pets that need to be trained. We can make a difference in the world, change the world, using mathematics!	What we do here is really important! But I am worried that my project might fall flat; I might make a fool of myself, or not have enough knowledge to make a good impres-sion on my audience.

Miller, Willinsky). Creating a space of possibility is also a feature of some phenomeno-logical and existential pedagogy. Here space is something preserved or inhabited, shared and experienced, possibly lost (Jardine, van Manen, Olson, Pinar, Smith). My main concern with space as an edifice is that we are creating spaces in spaces that already exist. I am not surprised that many feminist and phenomenological educators are exhausted from all of their efforts to build new spaces out of what was already is there, or to find cracks and fissures in the current spaces where their work can take place.

When we treat space as a "frontier" we are imagining education as a journey into uncharted territory where we know little about what we will find. Indeed, much of teaching feels like this; many new teachers find it very disconcerting that their course-work did not seem to prepare them for what they have to confront, because teacher training simply cannot know in advance what challenges and strange adventures the new world will offer. Teaching and learning mathematics would thus be a world of great adventure, where we explore this new territory. Here space is something found, not given or created. Like explorers of the past, we would be compelled to document what we find for those who will follow in our footsteps, to mark paths that take us to useful resources, local guides, and directions back home. One concern with space as a frontier is that the metaphor brings with it a host of ways of thinking that are part of the colonialist enter-prise of exploration and colonization. Is it imperialist of us to explore a new world and take it as our own? What right do we have to rampage through new territory as if only we have the authority and privilege to do so? This is stretching the metaphor a bit: I am not sure how imperialist or colonialist it really is when I fall into teaching in this way. But we do need to be on our guard for the discourses we select to interpret our work. If we are applying a language steeped in the traditions of colonialization and global imperialism, even if it is to talk about math classes and not to talk about real world geographic regions, we are perpetuating in our own way these discourses that carry with them forms of domi-nation and oppression and inequality that may have serious underlying effects.

I have already discussed the Cartesian metaphor of social space and Lefebvre's critique of it. Basically, we find ourselves identifying people in terms of dimensions that describe who they are, and then we locate them in a kind of social space determined by the dimensions. We might imagine plotting people as points in a social space where there are many axes labeled race, class, gender, ethnicity, sexual orientation, and so on in an extension of 3-D graphs. We then can imagine specific events in the classroom as trans-forming the social locations in this Cartesian space. If you have ever studied linear algebra, you could imagine that each classroom event would be like a linear transforma-tion that moves the people who are plotted as points in this social space to new loca-tions. My main critique of *this* conception of space is that we never know if we have covered all of the important dimensions. Social theorists who work with categories of race, class, gender, sexual orientation, and so on always need that "and so on" because they know there are always more categories that could be added. This kind of a theory makes me a little uncomfortable because it always needs tinkering with and new things to be added. Twenty years ago we added gender; ten years ago we added sexual prefer-ence; who knows what new dimensions will need to be added to the list in the next five years, once we realize that all of our ways of understanding social space would change if only we had thought of it?

Another way of thinking about space comes from film theory. In film the camera creates a sense of a space for the viewers of the film. Space is the *distracter* from the space-off—the space not seen through the lens of the space seen, or heard/not heard, and so on (Appelbaum 1999). The space is just a representation of reality, a fictional interpretation, and we could imagine that our own interpretations of what is happening in our classroom are just as "fictional" or up to alternative representations. In this way we begin to understand that there is no "true" story about what is happening in our classrooms, only many possible stories to be told about them, each of which might be like a film. Thus each story has its own space-off. As Teresa de Lauretis has written (1987: 25), if a view is nowhere to be seen, not given in a single text, not recognizable as a representation, it is not that we haven't succeeded in producing it. It is possible that what we *have* produced is not recognizable (to us precisely) as a representation that could be perceived. De Lauretis borrowed the term space-off from film theory to express the space of feminist theories of gender: the space not visible in the frame but inferable from what the frame makes visible. She noted that classic commercial film erases the space-off within a narration. But avant-garde cinema has shown the space-off to exist concurrently and alongside the represented space; this is done by remarking its absence in the frame or in the succession of frames, showing it to include not only the camera (the point of articulation and perspective from which the image is constructed) but also the spectator (the point at which the image is received, reconstructed, and re-produced in/as subjectivity). The (at least) two spaces are not in opposition or placed in a sort of competitive relation themselves; they coexist concurrently and in contradiction, simultaneously in harmony, counterpoint, and cacophony. To inhabit both spaces is to live a contradictory tension. This is, indeed, the condition of education: the critiquing quality of its theory, and the positivity of its practice. A negotiation of terms for a Math Center or a computer cart is really a study of the politics of school funding and organization; an entry into the teaching of mathematics is a statement of bureaucratic policy; a fantasy about gender and the ideal mathematics classroom is actually a careful commentary on class and race. The point here is that mathematics teaching is often about everything but the teaching of mathematics!

Box 6.8

Spaces and Space-offs

a) Observe a mathematics lesson or unit and take notes in four different ways: what do you find important when you think of the space of teaching and learning as a gift in this classroom? As an edifice? As a frontier? As Cartesian social space? Describe the space-off.

b) Plan a unit three different ways, using at least three different conceptions of the space of teaching and learning. Describe how your plans change when you make yourself think about the space differently.

c) Plan a unit so that the students are directed to study the space-off of teaching and learning as much as the space.

In any classroom, in any school, the people there are evoked by the space of the encounter, and if we are not pleased, it is because we hope for a different way of relating to the space. It is about time we stopped making recommendations for how to construct new spaces, and about time we started relating to what is there in a new way. This is yet another kind of space-off: it was there outside our frame the whole time. And by "not being there" in the creation of spaces, but instead being somewhere else, in the elsewhere, relating differently to the space, we and curriculum are performed in another way.

(*"Simply by not being in place"* . . . *I just used the spatial ideology to suggest a "way out" so there's probably a problem there that you can find! I have to use our spatial, ideological language to try to convey my ideas, though. So we'll just see "where this goes from here" to use yet another hegemonic statement!*)

I like to use a quote from Sun Ra, an avant-garde jazz musician famous for saying, "space is the place":

> I created a vacuum on this planet. Deliberately. I could be president of a college. I could be in college, I could be in Congress. I got the kind of mind to do it. But I stood back and therefore, where I should be, I am not!
>
> Now, they have a vacuum. Nature hates a vacuum, so other things are rushing in there that are not good. "Fools rush in where angels fear to tread . . . !" So what they gotta do? They have to listen to what I have to say. I can tell them what to substitute for the vacuum I created. I'm a scientist. I conquered a planet without a gun. Simply by not being in place. You know . . . for want of a nail, a shoe was lost, for want of a shoe a horse was lost, for want of a horse a battle was lost. Well, now, that can apply to me and my strategy to defeat a planet that's doing wrong.
>
> (Lock, undated)

Places, not Spaces

If we assume that the space of learning takes place in a classroom to which the students and teacher are assigned (of course, you could imagine all sorts of other possibilities for where and when people might learn, but this book is mainly considering school mathematics and a typical school situation), then we have two approaches to the design of this space: we can either say that, given the particular room, desks, chairs, tables, layout, supplies, and so on, only certain things are possible; or, we can say, given some things that I want my class to do, I must find ways with my students to use this room, the desks, chairs, tables, potential layouts, supplies, and so on, so that we can do what we need to do. The stuff listed (room, desks, chairs, tables, potential layouts, supplies, curriculum frameworks, building rules and regulations, and so on) constitute a *space* that may be taken as a gift, edifice, frontier, Cartesian dimensional social space, or a distractor from the space-off. What we do with these things in the space, the community we create, the aesthetic environment we enact, and the traditions and rituals we maintain, together breath life into a *place*.

David Callejo Pérez, Stephen Fain, and Judith Slater (2004: 1) describe a "place" as "the embodiment of a purposefully created space that is a creation and enactment of the cultural and social conditions of participants." *Education* results in a place, they write, when there is a purposeful creation of spaces that comprise learning environments and the aesthetic dimensions of the created space called school. When we are writing our lesson and unit plans, our fundamental question, then, is, "How do I, or should I, or could I, make it possible for me and my students to create a space that is a creation and enactment of the mathematical culture and mathematical society of the people in our class?" Callejo Pérez, Fain, and Slater recommend that the space becomes a *place* "when life is entered by those who understand that place is defined by boundaries and under-standing, or as dispositions of potential ready to be occupied and possibly used situa-tionally by the participants and authorities that run them." At first this sounds like a bunch of jargon. But if we spend some time thinking about the subtle differences between a space and the place that we make in that space, it can help us to plan mathematical places where students live rather than merely pass through on their way to someplace else. If we can work with the forces that shape the space of our mathematics classroom, we can increase the ability of that space to represent the needs and desires of each person living in that place.

Rob Helfenbein (2004), a teacher in North Carolina, presented a research paper at the annual meeting of the American Association for the Advancement of Curriculum Studies that made a very important point. Spaces are usually owned by adults in our society. Places are owned by the people who make them. So, as a space, a mathematics classroom may be a gift that students may not appreciate or use in ways that make the adults angry or disappointed; it may be an impressive edifice, again unappreciated (as when adolescents travel to Europe and beg not to be taken to yet another boring Cathe-dral or Castle) or used in ways that are not sanctioned by the adults who own it (as when a group of teenagers meet in a bathroom to buy and sell drugs); it may be a frontier, in which youth are trained in the dominant colonialist and imperialist practices of our time; it may be a Cartesian dimensional social space, described and studied by the adults, while the students are objects of analysis (examples of a category) rather than living, breathing people; or it may be the distractor from the space-off, in which the power is in the hands of whatever is analogous to the director, camera person, or film editor. In any of these ways, a mathematics classroom is owned and studied by the adults, and the students are left to buy into what the adults want, or left to use the space to make their own place. In most classrooms, unfortunately, the students make their own place. What we need to find are ways to support the creation by students of a *mathematical place* in the spaces that we have access to. "Space," says Helfenbein, "is the localized community filled with meaning by those who spend time there" (2006: 92).

Three ways to design lessons and units so that we foster mathematical places include: (a) use the space as a philosophical tool; (b) use the space as an aesthetic environment; and (c) examine the politics of the space (Lefebvre 1991). This is how I use the space as a philosophical tool: I work with my students in using the space as a way to begin a conversation, as individuals, as groups, and as a whole class community. Our goal is to establish and maintain the localized community. In a community, individuals interact with others in order to accomplish what they need and want to do, and the same sorts of purposeful exchanges must be part of the place we are creating in the classroom space.

Students need reasons to talk with one another, reasons for one group to communicate with another group of students, and reasons for whole-class discussions. And these reasons must grow out of their work in mathematics. Because we need to plan how next week's math museum will be organized, because we need to be ready with questions for the visitor who is coming to help us with our investigations, because we care about how good our part of the published book of math articles will be, because we each have been investigating different techniques and now we all need to find a solution to this question. Because we need to know the solution: we are going to use the solution to make a decision. When public spaces erode, individuals feel alienated. Some social critics say this is what is happening on a grand scale in our society, since we have lost those public places like squares, cafés, pubs, and other venues where people might have met and talked and worked through social, political, and cultural issues. I believe this is certainly what has happened in most mathematics classrooms. Individual students are isolated from one another and do not feel any attachments either to the group or to the mathematics. Even when they are in the classroom space, they are not mentally in that place; they are dreaming of other things, planning what they will do in other places. The teacher needs to empower individuals in the space to rally others to organize and meet, to create a place.

A popular urban renewal strategy is to support neighborhoods in creating places. On one block they might turn a small piece of concrete island in the middle of the road into a tiny patch of grass with two benches. Another neighborhood might use a vacant lot as a community garden. Yet another might install bike racks for use by anyone on the block. The city pays the costs of these simple renewal projects because they create places where people can meet and feel like they *belong*. Can we do the same thing in our classrooms? We need to claim small spaces where special places can be made. A place where two people can talk away from the rest of the class about a mathematics plan. Shelves where groups can store work-in-progress. This means our unit plan must include times where pairs of students meet together before speaking with others in the class—with a reason for keeping it private before publicly sharing, or long-term projects that demand storage over time.

When we use the space as an aesthetic environment, we use it to redefine what counts as knowledge and teaching, and to restructure our classroom or school. Think of planning as landscaping or architecture. When a park is landscaped, an aesthetic environment invites people to enter, play, connect, and explore. A building both constrains and enables activities to take place inside it. A classroom, like a park or a building, is not an empty shell waiting to be filled; it is a co-creator of the aesthetic environment, stimulating thoughts and ways of being and interacting, and shaping moral possibilities. Most of this chapter has focused on how I use metaphors for the classroom space in order to turn the space into a certain kind of place. Each metaphor we consider to inform our planning can help us to explore the variety of landscapes of learning that are possible; the metaphor acts as a co-creator of the aesthetic environment.

When we examine the politics of our classroom space, we foreground seriously the ways that "the space – physical, temporal, emotional, psychological space – in which one finds oneself is always a significant element in the equation linking learners and teachers" (Pérez *et al.* 2004: 4). The forces that shape the public space of the mathematics classroom ultimately engage individuals in creating their understanding of place itself. That is, how students experience this space has a big impact on what they expect of other spaces, what they imagine as possible places, and how they relate to other spaces of learning. Politics of the space asks us to think about where the power lies in the classroom. Who asks the questions, who moves the furniture, who can be alone, or who is allowed to lead the discussion? In my lesson plans, I can experiment with restructuring the space to accommodate ways that different members of the space have a say and make the decisions in the creation of place. For example, how can I find a way for each student at some point to be the person directing the activity? They must have a need to direct the activity in order to accomplish a goal of their own; otherwise, they are just leading activities because the teacher is requiring it of them.

Box 6.9

Brainstorming Grids

a) Make a grid with columns and rows. Label at least three columns with different forms of space as philosophical tool that produces conversations; label at least three rows with different forms of politics of the space, in which the power and authority is in the hands of different members of the community in different ways. Now fill in the boxes of your grid with metaphors for spaces—places where the different combinations of conversations and politics are elements of the space.

b) Choose a mathematics lesson topic and outline the major features of nine different possible lessons on this same topic.

c) Now you need to do the important work of a teacher: consider what circumstances would lead to the selection of each of the nine plans over the others.

Achieving Expectations through Metaphors

The peculiar thing about this seemingly *abstract* approach to a *practical issue*, the pursuit of Standards-based pedagogy, is that this apparently philosophical and fantastic way of thinking turns out to be remarkably *useful*, in the most practical ways. The biggest barrier to a Standards-based curriculum is that nobody knows what it looks like. We teachers are supposed to move towards something without knowing where we are going and what mode of transportation we should take to get there! Now, we could deconstruct policy because it too is locked into spatial ideology: moving toward something is falling prey to all of the pitfalls that Lefebvre and others help us understand. But why bother right now? The Standards sound pretty good to most of us. What's problematic is the policy discourse steeped in spatial hegemony. We can "achieve Standards" by *not* trying to move to a different place! And that's the point of this chapter. What we need to do is do what we are already doing, but relate to it in a different way. What the teachers I am working with are doing is finding pictures of what this might mean. And, in the process, they are achieving—indeed, surpassing—the expectations. It turns out that it's not about worksheets being evil: it's what people do with the worksheets. It's not that drill and practice are *bad* pedagogy: it is how students and teachers perceive the purpose of the drill and practice. It's not that we should dwell on strategies of problem posing or problem solving, and diminish the concern for correct answers: what's important is that a correct answer *matters in a particular context* if we want students to be concerned with the "answer." Increasing student talk is a good idea; but *only* if what they say *matters* to others, only if people want to hear what they have to say. *Why* would someone need to hear what I say? *When* would someone need to hear what I say? Well, let's think of group situations that share this feature: What are some other features of these situations? How can we promote people relating to the classroom context in this way, and in the process make the experience into this form of relation? In the same way that praying by a stream makes a religious space with all the trappings, in the same way that initial diagnosis of misconceptions, and subsequent assignment to remediation groups becomes a triage unit at a hospital emergency room, we search for ways to help people listen to each other. More importantly, we look for places in which people have something to say to each other in the first place.

Through metaphors, we can achieve the Standards, and a whole lot more. As I noted in the introduction to this chapter, there is a place for every kind of metaphor for the classroom space. Even for the most traditional and possibly "boring" metaphor of practice. Suppose we use the metaphor of coaching a team. Because the major theme of the experience is playing in the game that will be coming up again, and because we have repeated tests of our skills in every game, it is easy for me as the coach to say that I have noted a particular skill that I want everyone to practice for the next hour. Similarly, in my mathematics class, I can stop and say I want everyone to practice recalling multiplication facts, or drill on factoring polynomials, because I see these skills as something that will help my students to accomplish the important goals. But in order for the training of skills to take on this meaning, there needs to be that greater sense of some purpose that we are working toward, a project that is using these skills, or an investigation where I need to efficiently apply this skill.

Response to Chapter 6:
Take Ownership of Your Classroom Space

David Scott Allen

I have tried my hand at metaphors in the classroom on occasion. I haven't gotten to the point that I can say my students walk into a metaphor each class, but I can say that I have given lots of thought and experimentation to the idea of classroom space. It all started a few years ago when the idea was put into my head that sitting still and being quiet are not the goals of education. I agreed, but it did give me pause to think of what the educational goals actually are. I researched my district's stated goals, which were mostly the byproducts of NCLB. But in the fine print there were more important, if less evaluated goals related to attitudes and motivations, citizenry and character. No test short of life itself will be given for these notions.

My students are indeed very good at playing the game of pleasing the teacher. And some get exceedingly good at how not to please the teacher and still pass. I consider myself average at applied educational psychology, but yet somewhere along the way I got brainwashed by its premise: prepare a carefully sequenced set of well-articulated lessons, and the learning will happen. Well, that's fine until you try the same method for the fourth time in one week, and the students turn you off. I have not arrived at many conclusions in terms of what to do, but I have concluded that the way I approach the classroom atmosphere must be very flexible and have variety.

Some of this chapter you have to experience before you can understand, but by attempting to change the experiences of the classroom, you will understand it better. You will probably need to reach outside of your comfort zone of classroom spaces and your repertoire of acceptable lesson structures before you will hit your own stride in this area. I haven't yet done this thoroughly myself, so it is an ongoing journey. My journey has been enhanced because of the fact that one of my classes virtually stopped listening to me.

I noticed that my eighth graders were less than attentive near the end of the first quarter of a school year. They're at that age when they think they know everything, and this is exacerbated by my district's insistence on teaching two years of pre-algebra to the average level kid. My students' behavior was not unusual. I was determined to change something because it couldn't have gotten much worse. Yet there was learning going on. It just puzzled me how it could be happening in that chaotic environment. So, I decide to have the students teach the next unit. I divided them into six groups of four, and had each group pick one of the eight sections of the next chapter in the book. Even telling the story, it sounds mundane. They were to present the lesson and handouts, assign home-work, and be graded by their peers for their efforts.

I sat back and watched the class, not just the presenters. It was an amazing sight. There were more people engaged in learning during these lessons than when I had been teaching the months before. But what I noticed was that the atmosphere was more chaotic than I might have otherwise allowed. There were four presenters in each group, and at any given time the following was happening: one presenter organizing a group of students at the board to see who could solve a problem first, with (student purchased!) candy given for correct answers; one presenter looking around the room for behavioral issues, and who acted as the disciplinarian for such issues; one presenter circulating in the room seeing if the other (relatively) quiet students needed any help with a handout; one presenter giving a lecture to a small group of students about the finer points of something they missed; one student asking to go to the bathroom and another student changing seats to see the board; one student sharpening a pencil, two students making fun of another (friend) student and/or sabotaging the class, yet eventually turning to their handouts to see if they knew what they were doing. Seated students were doing problems or asking to be the next at the board, while others raise their hands for help. You get the idea. In a class of twenty-four, there were almost as many things happening. Some of the more subversive and antisocial activities notwithstanding, the students were engaged and wrestling with the material in the way they needed. I got to see ownership of the space in action. However, I still needed results. I had done a pre-assessment on this unit to check baseline knowledge. Then the post test came, with students doing fairly well. I then had a meeting/review with the students to discuss what we need to do to improve the situation, and they helped each other on the remaining difficult areas. Students wrote their own test for the chapter, and they did very well as a group on this challenging test.

There just happens to be many ways to interact with mathematical material. This is why there are so many variations on the way students operate. Math teachers need to think less about the objects of math and more on the context. What we need to think about is the pun of the day. Which analogy best describes what just happened, or what do we anticipate will happen? My school has eight periods a day (including study hall) plus lunch. Will my students have a clue what we did in my class when they are in the middle of verb tenses an hour from now? Will they recall the moment they "got" the concept of proportion when they are on the bus in four hours? Space helps fix these things in their minds, just like bad puns.

Thinking of myself as an architect of the classroom experience is actually a misnomer when you consider that teachers are typically thought of as the builders of the space rather than the architects. A true architect sets forth the plans and the builders actually build the building. In the case of the classroom, the building must be formed by those who live in it. We describe it with our plans (give the blueprint, so to speak), but the students build their own understanding using their own mortar and bricks. And while I don't think that class should be all telling and listening and following directions, listening for understanding and point of view are important to expand options in problem solving. And it will indeed take practice. This is where values become important. Where is listening in the curriculum? I'm not certain that it is taught anywhere in middle school. Yet it is a vital life skill.

I'm considering using the talking stick method for my eighth graders. They need to practice listening to each other one at a time. A stress ball would work better for my students, as it can be tossed to the next person and squeezed under the pressure of having to speak. I also like the connections and postponing conversations until a later time. This is something I do with myself all the time. I file ideas away until they actually happen or I get another insight. The very objects we purvey are metaphors, which we want them to remember in order to refer to them at a future date.

Student-teachering has to be the best metaphor of all. Students can see how the student-teacher is learning from the teacher. This is not always good, just always happening. I am in a project in my school for technology application in the classroom, and I am planning a unit for execution of what I have learned. I realized that I might have a student-teacher in the spring and what would I do? I decided that would be the perfect time to have a student-teacher, to have someone to share the new units and ideas with, and to have them help me improve and vice versa.

My favorite metaphor is the art studio. I literally do this with my students as part of a broader unit on geometry. As a culminating project, I have them create mathematical art, and the student-voted best projects are painted on heavy board and put up in the halls as a literal gallery of past accomplishments. As this has gone on over the years, however, the students of the past have upped the ante for those of the present. If I show them students' past work, they are now obligated not to duplicate it, but surpass it. They must do something different. Next year I am going to add to this by teaching regular geometric concepts using a recently purchased 3-D drawing book. I will teach them to draw 3-D objects, complete with artistic extras, using a kind of isometric grid. Then we will pull out the geometric concepts such as parallel, ray, perpendicular, and so on. I look forward to it myself.

All of the Cartesian talk can be very overwhelming, but all of us need to choose our emphases and do them often and well. Then you can expand into the other areas that are less often used and so less comfortable. You must be at ease in your own classroom. You must have a backup plan if your space takes on a life of its own. I do have another teacher right next door, and so I must be aware of my volume level. We may be having an arcade experience, while he is taking a test. But we have come to an agreement on what to do: whatever is best for the students is good, and if there's a problem, we talk about it. We especially need to make sure that one of us doesn't plan something loud on a day the other has a test. We just get it out in the open and adjust honestly to the situation. We are truly providing what the home should provide, but sometimes doesn't: a space to be safe. Students don't always have that these days. Teachers can create this in their own classrooms, and give students an opportunity to shine.

In thinking about place, I am reminded of do-it-yourself stores. Understand, I can't drive a nail straight, and I don't do cars either. But my two year old loves to go to our local building supply store. So I go. But secretly I love going, even if I can't use half of what's there. When I do go I appropriate their commodities for my own uses: mathematics and art. My daughter loves to draw, paint, and create. My son loves the tools projects. I go to see how I can use anything in a mathematics classroom. What could this *be*; what could it be *used for*? I am always looking for cheap ways to do something artistic with my students, and building supplies give me lots of ideas. I am taking their objects and using them for my own purpose. So it becomes my place, even though my wife can attest, it is not my space.

I believe that most students really don't learn certain mathematical material until they must use it as a means to an end in another situation. In other words, my eighth grade pre-algebra students are all but repeating what they had in seventh grade, and they catch this immediately. But if they were to do algebra at this point, they would be forced to think through pre-algebra enough to use it as a means to an end to solve the algebra problems which face them. Doing this may create a doubly artificial environment, but it is also reminiscent of the idea that students don't learn without a context. Students are so desperate for meaning and usefulness and seeing what it is good for, that they intentionally trick themselves into doing it because at least they can see what's going on. Why not add another class and have them all take algebra their eighth grade year? Or at least teach the second year of pre-algebra through the use of algebra. We never give them a glimpse of what is next, so they just see the exercises. This is boring at best.

Now go back and read the very last section of this chapter, "Achieving Expectations Through Metaphors." We need to keep this in mind as we keep the best of the old ways while integrating the new. And one last thought. If you think all this metaphor stuff is crazy, try listening to yourself after you have read this chapter of the book. How do you describe it to other people? Do I hear you saying, "Reading this book was like . . ."?

Action Research 6:
Linking Mathematics to Social Issues

Kristen Iaccio

Linking math to students' lives helps gain their attention. Students appear more interested when the teacher relates current social issues, such as income distribution, to mathematics. I will be discussing how such an activity can replace current topics in my eleventh grade Algebra II and tenth grade Geometry curriculum. The class will discuss social topics, explore possible solutions, and learn new material that matches the standards. Throughout the lesson, the teacher will act as a facilitator and allow the class to discover how mathematics is done.

First and foremost, before designing the unit, I needed to know what curriculum would be replaced and with what topics. I came across an interesting topic about income distribution while reading *Rethinking Mathematics: Teaching Social Justice by the Numbers* by Eric Gutstein and Bob Peterson (2005). I have found that many of my students relate more to topics concerning money, so I thought I could incorporate the new topic into my current curriculum.

In the unit, I plan on using current statistics from the Census Bureau of the 2005 income distribution in the United States. This one unit can replace my lessons on mean, median, mode, range, scatter plots, line of best fit, histograms, box-and-whisker plots, and using geometry to find the area under the curve.

It is important to not only know the curriculum that the unit will replace, but the placement of the activities that will be incorporated throughout the unit. As a teacher it is my job to determine where an activity should be placed in order for the class to receive enough time to explore, discuss, learn, and extend the application. If the activity were placed in the beginning, I found the class would gain most benefit. Below is Diagram AR6.1 and a short explanation of how the lesson is going to occur.

Diagram AR6.1

Beginning		Middle	End
Class discussion about income distribution in the U.S.	Activity—create scatter plot, histogram, box-and-whisker plot.	Show graphs, chart, and discuss mathematics terms.	Extension Application—Area under the curve.

When mathematics is taught in the beginning and then the activity is given to the students, it leaves out the exploration and learning processes (see Diagram AR6.2). Students are forced to memorize the steps of how to graph or find the mean rather than actually learning why it is the way it is in math. If the activity is placed in the middle, it may become difficult to get students interested and motivated to do the activity (see Diagram AR6.3). It is also difficult to place an extended application to tie in with the rest of the lesson when the activity is not placed at the beginning. Below are these two other possible ways to teach the lesson covering the same material, but I prefer the previous way.

Diagram AR6.2

Beginning	Middle	End
Teach the material—mean, median, mode, range, scatter plot, histogram and box and whisker plot.	Discuss why we are learning about this or what standards we meet.	Activity and Assessment

Diagram AR6.3

Beginning	Middle		End
Discuss why we are learning about this or what standards we meet.	Class discussion about income distribution in the U.S.	Activity	Teach the material—mean, median, mode, range, scatter plot, histogram and box and whisker plot.

Looking back at Diagram AR6.1, we can see that there are two parts in the beginning of the lesson, one in the middle and one at the end of the lesson. Throughout the lesson students will use data from the U.S. Census Bureau to create a scatter plot, histogram, and box-and-whisker plot. Using the scatter plot, students will find a line of best fit. Using the line of best fit, students will write an equation of the line and describe its domain and range. Using the data, students will find the mean, median, mode, and range of the 2005 income per household in the United States. After all the objectives are met in the beginning, the class will be shown graphs and charts and discuss how the new terms they learned can be matched with what they have created. At the end of the lesson, students will be pushed to explore a new concept that they will see in the next year.

The following is a more descriptive plan for the lesson.

Beginning

Part 1: While working in small groups of three or four, students will review the income distribution for 2005 put out by the U.S. Census Bureau. As a group, they will discuss and estimate the total income of 25%, 50%, 75% and 100% of households. A small class discussion will occur after they have reviewed the data and thought about what is meant by income distribution. This is the part where the class should be motivated to continue and discover.

Part 2: The groups will split up the data and create a scatter plot and will find a line of best fit. Since the activity is being used as a discovery lesson, the class may be unaware of the terms scatter plot and line of best fit, so they may need different directions. For example, plot the mean income to the number of households and draw a line through the center of the data. (The NCTM Illuminations has a great Line of Best Fit activity that can be used as a support tool for students.) Then, pick two points on the line and write an equation that fits the line. The groups will then be asked to find all the x-values and y-values that work for the function, hence finding the domain and range.

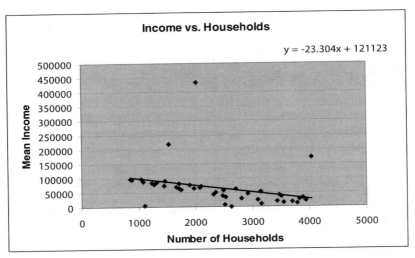

After the scatter plot has been developed, the groups will create a histogram and box-and-whisker plot. The groups will be instructed to divide the households' earnings into five equal parts, for example, $25k or less, $25k–50k, $50k–75k, $75k–100k and $100k plus. They will use the data to create a histogram of the income to the number of households. (The NCTM Illuminations Histogram Tool can be used here as a support tool for students.)

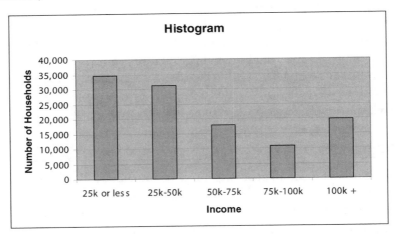

After the histogram is constructed, the class will discuss the differences between mean, median, mode, and range of the data. They will be instructed to find the average income, the middle income, the income that repeats the most and the difference between the highest and lowest incomes. The groups will then match the words mean, median, mode, and range with their findings and discuss which word means what.

Now, it is necessary that the class understand the terms mean, median, mode, and range before creating the box-and-whisker plot. The teacher should encourage the students to visit NCTM's Illuminations Histogram Tool for a better understanding of how to construct the diagram. The groups will be instructed to find the median income. They will then find the medians of the upper and lower half and plot the three numbers on the graph with labels.

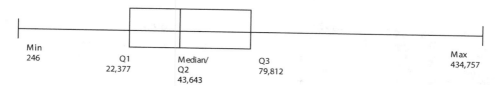

Min
246

Q1
22,377

Median/
Q2
43,643

Q3
79,812

Max
434,757

Lastly, students will plot the percentage of households in increments of 25% to the percentage of total income.

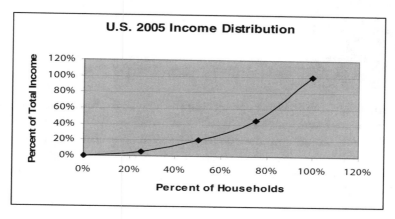

Middle

Once the class has completed the graphs and all the statistical analysis of the data, we will analyze each group and discuss their findings. As the facilitator, it is my duty to make sure the class is using the correct mathematical terms when talking about their graphs and results. Once it is understood that the class can have a clear discussion about their discoveries using the correct mathematical terms, the teacher will show other graphs of the data. The following graphs will be shown.

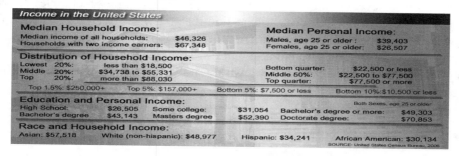

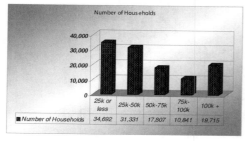

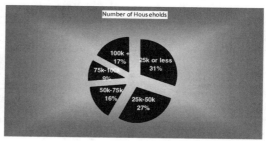

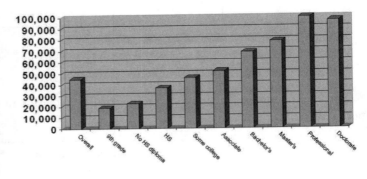

We will then view on an overhead explaining why it is important to learn this material. For instance, we will discuss what Pennsylvania standards and NCTM standards were met. We analyzed data taken from the U.S. and determined an equation of best fit. We compared the data using values of mean, median, mode, and range. We also analyzed scatter plots, box-and-whisker plots, and histograms. All of the previous statements meet the PA standards. According to the NCTM standards, most of the same standards were met. Throughout this lesson we are also "connecting mathematics education closer to other sciences, to the social reality and to the social mathematical practice" (CIEAEM 2000).

End

Now it is time to move towards the end of the lesson where the class will complete an extended application using the last graph they created. Using the income distribution graph, students will find the area under the curve geometrically. This is a great activity to end the lesson using the same data and graphs, but begin the next lesson or unit. It helps to keep the cycle going. Together as a class, we will draw rectangles below and through the curve. We will discuss how to find the area of the rectangle. Hopefully, the class will come to the conclusion that in order to find the area of the rectangle that intersects the curve they will need to cut it in half or they can use lower and upper limits and average the two. The figure below demonstrates how the graph will be cut up to find the area under the curve.

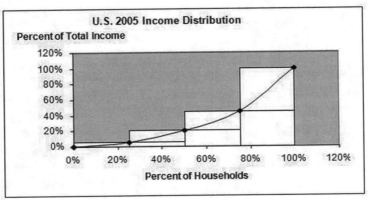

After I designed the unit, I used it with my sophomore Algebra II class. They worked in small groups creating graphs by hand or using Excel. They discussed their data and reported to the class. Each group was responsible for understanding each graph created. When it was time to present, each member needed to be prepared to answer any of the questions I or their classmates asked. For example, I asked "How did you make a line of best fit?," to a group who went above and beyond when they realized there were outliers that should not be included in their equation. The members described picking two points in the cluster of data, finding the slope and using the equation $y - y_1 = m(x - x_1)$ to write the equation of the line. I asked another member about the points not in the cluster of data and she responded that the points lay outside the data, so they should not be included when writing the equation. They showed me another graph where they used the outliers and realized the equation did not fit most of the data, so they excluded that from their line.

I was impressed that not only could they write the equation of the line, but they also learned how to exclude data that did not truly fit the line of best fit. They researched the material in their textbooks and were able to use the term "outlier" to the class. Not only did the members of the group learn a new term, but the class did as well.

I asked another group to describe how they got the percentage of total income when creating the final graph, which compared the percentage of households to the percentage of total income. They described ordering the data and splitting it up into 25% increments. They then found the percentage of each 25% increment and added up, for example, 25% + 25% + 25% to get what 75% of the households made up. In the same group, I asked what the point (75, 42) represented and they responded that 75% of the households only make up 42% of the income in the U.S. They began discussing how unfair that is and I informed them that education is the key to success, so stay in school if they want to get into that top 25% quartile.

My students were able to respond to most questions asked and the group was evaluated based upon their groups' participation throughout the week, presentation of graphs and materials, and a write up about their own contributions to the unit. I attached a rubric that was given to the students before their presentation so they could add any touch ups they might have forgotten.

I am lucky because the school I teach at allows me to teach the material the way I prefer as long as the correct criteria are being met. First I would need my lesson to be approved by the math department head. I would provide a copy of the lesson with the standards that are being met. Once he approved the only other gatekeeper I would need to convince are the parents of my students. Parents may question why I am teaching this material to high school students and see no relation to mathematics. As a result, I would send a letter home addressing the topic we are covering and what they will be learning (see letter below).

I encourage making learning a fun and enjoyable experience. Discovery is sometimes the best way for my students to fully understand a concept. I always enjoy new ideas teachers present to me and new units that can replace boring lessons. Students need to be engaged and we as educators need to engage them. If we as teachers can make our lessons related to their lives, we have a better chance of getting our students to know why it is important to learn the material.

Overall, teaching and learning mathematics though discovery and group interaction is more enjoyable for the class. As a teacher, I find it easier to teach the material when the class is intrigued and continually asks questions. A class needs to be in the correct environment for learning to occur. Through discovery, this is made possible.

Dear Parents/Guardians,

For the next few weeks the class will be learning about the income distribution in the United States. I wanted to keep everyone informed of what was going on in the classroom in case questions arose at home. The class will be learning the mathematics behind U.S. income distribution. We will be graphing the material in many ways, writing equations from the graphs and touch on the statistics of it all. Teaching and learning mathematics through discovery and group interaction is a more enjoyable lesson for the class. As a teacher, I find it easier to teach the material when the class is intrigued and continually asks questions. A class needs to be in the correct environment for learning to occur. Through discovery, this is made possible.

I am very excited to help the class discover new ways to learn mathematics, while also helping them see how mathematics is related to their world. After the class has completed the assignments, we will be having an in depth conversation about our findings. The groups will be responsible to present their material and unit to the class. All are invited to attend your child's classroom presentation. Thank you for your time and if you have any questions, please do not hesitate to call or e-mail me.

Sincerely,
Ms. Iaccio
Math Teacher

Kristen Iaccio

I am a mathematics teacher at Franklin Towne Charter High School in Philadelphia, PA. I am currently developing alternatives to the standardized tests in Pennsylvania. My idea is that students should be able to work in a small group or pairs to take a test made of open-ended questions. I researched different tests that other states administer and found that the tests are not properly evaluating the students. I am in communication with officers of the Pennsylvania Council of Teachers of Mathematics (PCTM) and contacts at the Division of Assessment in the Pennsylvania Department of Education.

MathWorlds 6:
Same Math, Different Metaphors

1. **Group Work.** In a group of three, work collaboratively on the following mathematics questions, assigning different roles based on the accompanying metaphors for how you will work:

Mathematics Questions:

a. Simplify: and Solve:

$$\frac{3x^2 - 19x - 14}{6x^2 - 11x - 10} \qquad -5|3x - 5| = -10$$

b. Find the equations of the lines that include the point (0,3) and are parallel and perpendicular to the line $8x + 7y = 40$

c. Describe three interesting aspects of circles that are not true about hyperbolas.

d. An architect wants to design a building that has no 90° angles and no curved walls. Can she or he do this?

e. Which of the following have the same cardinality, and which have different cardinalities? The Natural Numbers, the Real Numbers, the Integers, a circle with one point removed, a circle with no points removed.

Metaphors for the Workspace:

a. Internet Social Space (e.g., MySpace, Facebook)
b. School Playground
c. Fashion Section of an On-line Newspaper
d. Allergist's Office

2. **Social Issues Mathematics.** Find three resources on the Internet, three in print materials (e.g., newspapers, library reference books), and three "experts" (people you have contacted, and that have agreed that you could go to them for advice on the topic) for each of the following mathematics/social issues topics:

a. Global and local distribution of wealth.
b. Infectious diseases by age, race, geographic region, and one other characteristic of your choice.
c. Consumption of oil.
d. Number of deaths attributed to genocide.
e. Volunteers who contribute to their community, globally and locally.

3. **Social Issues in Curricular Context.** For each of the resources you have found in question 2. above:

   a. Describe activities that would help students to engage with mathematics if the resources were used:

      i.   at the beginning of a lesson/unit
      ii. in the middle of a lesson/unit
      iii. near the end of a lesson/unit

   b. Describe the kinds of mathematical skills and concepts that your activities would assume students are already capable of applying, and several skills and concepts that students could acquire and/or understand through the experience of your activities, depending on their placement at the beginning, middle, or end of the lesson/unit.

   c. For each lesson/unit you have described, identify the NCTM and State Standards that could be addressed.

   d. For each activity that you describe, identify at least one stakeholder in the educational process (e.g., department chair, principal, parent, yourself) who might have concerns about such an activity actually taking place. Write a proactive memo to this stakeholder (to be sent at least one week ahead of when you expect to implement your plans), explaining how the activity will accomplish the goals that you imagine this stakeholder to hold for your students.

seven
Places where People Learn Mathematics

Chapter 5 introduced youth leadership, voice, and participation, "nomadic alternatives" for organizing classrooms as places that support the teaching and learning of mathematics. Disparity, desire, and differences were three other nomadic categories discussed in that chapter because they can organize our reflection on our work. This current chapter will describe infrastructures of classroom work that can realize these goals. Of particular importance are the ways that we enter into new topics with our students, the uses of information that are gathered during a lesson or unit to inform further planning, and the criteria that are established for student products.

The question at the heart of all teaching is how much to say and tell at the beginning of work with students, and how much of the decision making process can happen as the work unfolds. The teacher may get better results by not telling everything or by not laying out clear and careful explanations; or, it may seem like students need more guidance than we are providing in order to engage in mathematical thinking. But this question hides more fundamental and crucial decisions that can be helpful for instruction. School curriculum is delineated in standards documents and curriculum frameworks as specific bodies of knowledge and sets of procedures that need to be taught to learners; this provides little or no advice to teachers about how learners might interact with the material. Instead of providing opportunities for apprenticeship in the arts and crafts of doing mathematics, bodies of knowledge bypass this apprenticeship in mathematics in favor of an apprenticeship in "studenting." Jean Lave (1997) contrasts school learning with apprenticeship by noting that apprentices are given landmarks to strive toward rather than training in decontextualized skills. Apprentices work with masters in two senses: Masters have truly mastered their craft, and they are highly respected for this, which gives value to the project of learning to be "like them." Also, apprentices know from the beginning that when they complete their period of apprenticeship there will be a legitimate field for the practice that they are entering. Now, as teachers we do not claim to be master mathematicians; and we can't assume that our students are committed to the life of a professional mathematician. Even if we chose another subject position (see Chapter 2) where students might be considered to be apprentice citizen-mathematicians, there are ways in which the given school mathematics curriculum simply does not support the notion that our students have chosen to enter this field of practice. That is, the apprentice model is not directly applicable to our work as we define it. But it does provide interesting ways to think about teaching and learning that we can bring back to our planning and assessment.

Lave's study of apprentice tailors helped her identify significant differences between apprenticeship and "studenting." There are no formal tests in apprenticeship. Instead, apprentices judge their own work as worthy or not worthy of being sold, and what price to set. They discover their own mistakes, because they have models of the products in the work of the master, and they always have the comparison of their own work with others who are striving to compete in selling their own products. We might take this to mean that teachers should demonstrate carefully how to solve particular mathematics problems, and that apprentices in the classroom would try to copy their efforts. But apprentices don't copy their master's work; they are instead enculturated into the profession so that they themselves can create their own original products. More important for Lave is that she identifies similar apprenticeship behaviors in studenting; these behaviors just are not focused on the craft of mathematics. Students become very adept at figuring out how to be good students. If we observe a classroom, the majority of students will be very busy at finding ways to produce the high quality results that are expected—the correct answers to mathematics problems, the correct facial expressions that communicate proper classroom etiquette, and so on. As any good apprentice would do, students work as efficiently as possible in perfecting the quality of their results. Lave uses the example of an elementary classroom where the teacher introduces specific techniques for multiplication and division problems. Small groups of students learn mathematics, but not the mathematics that the teacher is teaching. Instead, they rely on previous knowledge and their own discoveries about the multiplication table to develop successful systems of obtaining correct answers to division problems on worksheets. Because the teacher and the students are working together to produce the appearance of successful students, most members of this class are satisfied with the results—the correct answers—even though nobody in the class has learned to use the techniques that the teacher spent several days explaining in whole-class direct instruction. This example helps us to see that the ways of working in this classroom do not foreground the craft of mathematicians. What the class time is very good at is producing students who can think independently about how to please the teacher in terms of answers. If this teacher had spent more time listening to what the students were talking about in their small groups, she might be very disappointed in that "they missed the point" of her lessons. All too often, teachers will take the correct answers as the most important indicators of learning, and the independent thought of their students as a sign of their own flexibility. Don't we all want to embrace students' own inventions of algorithms? Doesn't their ability to find the answer using any way that they can show that they understand the mathematics involved? Maybe. But if we want to instill a sense of the craft of thinking mathematically, then we want students to be reflecting on how their methods are related conceptually to what the teacher is talking about when she introduces a new technique. Examining the technique for what it communicates about the mathematical ideas involved is in this sense more important than obtaining the correct answers. Even if our long-term goal is to make it possible for students to obtain correct answers, on tests, in everyday life, and so on, I want to believe that they will be more successful at this with more complex mathematical topics if they become more open to thinking mathematically and reflecting on the meaning of the mathematics.

There is good evidence that students learn actively, and through construction and invention of mathematical procedures. But there is no guarantee that in doing so learners will understand mathematical principles. Mathematics is taught most often as a body of techniques and facts that experts know rather than as a process of learning. Lessons are usually based on the decomposition of skills into little pieces in the name of preparing students to be able to "master" these skills. Lave writes that this "strips problem-solving activity of any relation to mathematical practice" (1997: 29). "The problem," she writes,

> . . . is that any curriculum intended to be a specification of practice, rather than an arrangement of opportunities for practice (for fashioning and resolving ownable dilemmas) is bound to result in the teaching of a misanalysis of practice . . . and the learning of still another. At best it can only produce a new and exotic kind of practice contextually bound to the "educational" setting . . . In the setting for which it is intended (in everyday transactions), it will appear out of order and will not in fact reproduce "good" practice.
>
> (33)

What we can learn from apprenticeship is how the tasks are named for the apprentices. Instead of specific procedures, apprentices are told to "learn how to make clothes" or "learn how to fix a leaky pipe." In the process, they can observe how the masters do these jobs in particular situations that call for different procedures. Apprentices observe and experiment until they achieve a first approximation; then they practice. The apprentice's curriculum shapes opportunities for being able to do these things, rather than setting out a series of training sessions.

When a teacher specifies the practice to be learned, pupils improvise on the production of that practice, but not on the practice itself. This is the key issue. Ironically, the more the teacher, the curriculum, and the text materials "own" the problems and set the agenda for what is to be produced, or the more the steps are decomposed in order to make the techniques more easy to copy, the more learners are pushed away from owning the problems and the harder it may become for them to develop the practice. Most school mathematics experiences that we have had ourselves required us to move around inside the problems that were presented to us. We needed to figure out what these problems were asking of us in terms of how we should react, behave, and in general respond. We would copy the teacher's behaviors as best as possible. In other words, we were not thinking about the mathematics but about what the teacher was doing. You might say, "But if the teacher was modeling thinking about the mathematics in front of the class, then in the end we were copying how she thought, which is what this chapter seems to be about." I can see your point. But compare this typical classroom setting with places of apprenticeship, where the relationship between learning and the practices of mathematicians would be "reversed." Apprenticeship makes the encompassing significance and meaning the central focus; apprentices have the opportunity to develop understanding about things that they are learning. In our student experiences the subject matter became our environment, enveloping us in a special world. In apprenticeship, the learner's understanding encompasses and gives meaning and value to the subject matter: "Because I need to create this thing I am working on, I need to understand such and such." As a long range goal of education, I suspect most of us would be more attracted to this latter development of learning about learning rather than the acquisition of specific skills. We want to imagine our students being able to pick up specific skills on their own when they need them, learning how to figure out what they need to know and where to go to find this knowledge. When we create spaces of learning that are experienced as places of doing mathematics, we have a better chance to realize this goal, and to make the classroom a more intellectually lively and interesting place.

Structuring Ways of Working

> **Box 7.1**
>
> **Where's the Assignment?**
>
> Throughout this book you've found textboxes with inserted suggestions for exploring the ideas presented in this book. By now, you should be ready to think up your own. As you read this chapter, take notes on the ideas, words, and phrases that strike you as important, surprising, difficult to understand, or otherwise noteworthy. When you are done reading the chapter for this time, go through it a second time and think specifically about the words and phrases that you noted on the first round. As you do so, think in the back of your mind about potential pedagogical experiments you could try based on what you have learned here. Design an action project that you can undertake with a colleague, with youth, or in a school. Carry out your project, and afterwards write a reflective summary of what you learned through the project, beyond what you learned in reading this chapter.

Much of the literature on teaching mathematics takes the daily lesson as the unit of instruction. Reys *et al.* (2004), for example, describe three options for structuring the ways of working in the classroom: investigative lessons, direct instruction, and review-teach-practice lessons. They reassure their readers that any good teacher is bound to use all three types of lessons at one point or another, depending on the teacher's goals and the students' previous experiences with the particular content at hand. Keeping in mind the basic goal of increasing student understanding and maintaining active student involvement, they suggest that you begin planning a lesson by deciding what sorts of tasks the students will be doing. "Students learn best," they declare, "when tasks are motivating and challenging, but not out of reach, and when tasks involve them in actually thinking about the mathematics at hand" (53).

Investigative lessons involve students in pursuing a problem or exploration on their own. The task may have been identified by some of the students or by the teacher, but the lesson itself revolves around ideas that the students generate through their investigations of the task. The teacher needs to guide the lesson, but the students are expected to identify their own approaches, strategies, and solutions. Reys and his colleagues base this lesson structure on an idea from Glenda Lappan and her coauthors (1996); if you, too, use this structure, then your lesson plan has three main pieces: 1) launch; 2) explore or investigate; and 3) summarize. In the launch phase, somebody provides motivation for the lesson and explains the task. The launch might come from a student presenting a problem that she or he has been struggling with, and include an invitation to classmates to work on the problem as well. Or, another classmate might share a newspaper article or present a puzzle. The teacher might provide motivation by doing either of these things herself or himself, or by reading a children's storybook, leading the class in playing a game, and so on. Once all of the students understand the challenge, students work on it, either on their own or with others in small groups. Students would be encouraged to use models, drawings, computers, calculators, or any other tool that would be useful in the exploration. As Reys and his colleagues explain it, this lesson structure seems to offer much in support of the feelings of apprenticeship that I discussed in the introduction to this chapter.

Note that in an investigative lesson, the challenge should be a true problem for the students, so it is important that you do not explain up front how to address the challenge. The students should not be simply mimicking strategies or skills that you have just shown them. Instead, they should be deciding how to get started and what to try. Your role as a teacher is to let them go. You should circulate around the room, listening in on conversations, observing what individuals or groups are doing, and occasionally interjecting questions or comments to help students recognize where they may be going wrong or to suggest a different approach (Reys *et al*. 2004: 54–55).

During the "summarize" phase of the lesson, the class comes back together as a whole group to talk about everyone's findings. Using information you gathered from observing while the students were working independently or in their groups, you orchestrate a discussion in which various groups or individuals report what they tried and what they discovered. Your job is to encourage sharing of ideas and also challenging of each other's ideas, while maintaining control of the discussion and trying to guide it in ways that advance your agenda (of focusing on the mathematical meaning rather than on mimicking what the teacher does). It is crucial that you make the importance of this phase clear to the students, by marking the transition (perhaps students all move to another part of the room where they can be more intimate, or perhaps you establish routines for presenting findings so that they are prominent and become the basis for future lessons).

If you follow the structure that Reys *et al*. call the investigative lesson, then your plan will include the following parts: objective, rationale, prerequisites, materials, lesson outline, gearing down, gearing up, assessment, and references. You would probably choose this structure, they suggest, if you want to emphasize developing problem solving skills, learning new concepts, or applying and deepening understanding of previously learned material. They contrast this with the *direct instruction* lesson, where the teacher's main role is to communicate knowledge, introduce new vocabulary, or teach certain specific procedures. In this alternative structure, the motivation is followed by the teacher demonstrating how to solve a problem, or by the teacher carefully leading a class discussion on how a problem might be approached. The third option, the *review-teach-practice* structure, is the most familiar form of mathematics lesson, and the authors note that it is likely to be the least effective, especially if it is used mainly to introduce different content from previous work. Because this structure makes it hard to focus on understanding, it is the least recommended, despite its prevalence in most of the classrooms that we have seen ourselves. You might say, "But that approach worked for me, and I liked it." Don't fool yourself. You may have learned in spite of this approach; and most of your students do not have the interest in mathematics that you have, so you can't expect it to work for them. Even if the review-teach-practice structure got you to be able to do math, take a moment to reflect on whether or not you see yourself as a mathematician, or whether you routinely find yourself experimenting with mathematics on your own for fun now.

Flow-structures. I find that a lesson plan often inhibits authentic interaction with my students. I am so busy trying to follow the plan that I can't really do the instruction and assessment that needs to get done! This is a real problem for many teachers. Once we have a script, we need to follow it. An alternative to the kind of lesson plan usually expected in courses on the teaching of mathematics is what I call the "flow-structure." In this structure, I pay attention to what I want the students to be doing rather than what I need to be doing. By putting all of my energy into maintaining the students' participation, I worry less about whether I am saying or doing the right things, and more about whether or not the students are involved in ways that are productive. The teacher's job in this type of lesson is to keep the energy going, and help students to find new ways to enter into the work of the classroom.

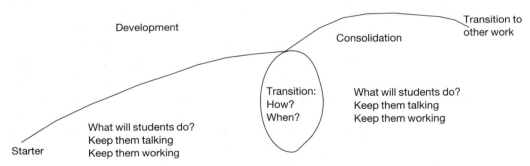

I draw a squiggly line in my planning notebook, like the one above. First I consider the learning time as made up of two parts, the first having mostly to do with development, the second having mostly to do with consolidation. In this first part, students are developing new ways of thinking, developing new ideas, and developing new vocabulary for communicating what they know and want to know. In the second part the group is consolidating what has been experienced, organizing information, analyzing ideas, describing patterns and relationships, and applying new vocabulary and concepts to their regular understanding of the world. What I do is consider a collection of possible ways that students can begin the development process, and I imagine potential outcomes of these experiences; based on my imagined outcomes, I think of several ways that students would be able to move into the consolidation phase. I need to plan what will happen in order for the activity to get started. Who will be starting us off? Myself, a student, a group of students, an email message, a visitor? How will students be asked to transition from the way we start off the experience into whatever they will be doing in the developmental portion of the activities?

For the developmental portion of the work, I plan what I expect students to be doing. I also plan questions and provocations that can keep them involved and help them to extend their work in new directions without losing momentum. One way to do this is to keep students talking about what they are working with. If they are talking about manipulative materials or pictures they have drawn, I ask them to explain what they mean using just words or symbols; if they have been using mostly words or symbols, I ask them to explain what they mean using a model, manipulative material, or picture. If they are not using charts or patterns, I challenge them to do so. I promote talking amongst themselves by asking them constantly to compare one approach to another, to figure out how one point in an explanation with one manipulative material corresponds to a moment in an explanation with another. If they are talking, I ask them to take a moment to write down their main thoughts and questions, then to organize them into a summary that they can share with others. If they are writing, I ask them to switch to talking. I approach the consolidation portion of the work in a similar manner, where my main objective is to maintain the momentum and constant probing of ideas and methods.

Where should the transition from development to consolidation occur? This is my most interesting decision as a teacher. I experiment with finding the moment when I can suggest to individuals and groups that they move into the next phase. If they seem to have developed a useful set of concepts for thinking about the ideas, and a vocabulary and system for communicating these concepts, then they are ready for consolidation. I do not have to wait until they are thoroughly understanding everything because I have found that the consolidation work can often help students to refine and modify their initial understandings as well, so that there is a blurred boundary between these two parts of the experience. Finally we run out of time; but before we get to that point, we need to have a planned way for us to stop what we are doing, take stock of what we have done, and make the transition to other work—other subjects in the curriculum for that day, or how we will pause until we begin again the next time we meet. One technique is to make a KWL chart, as discussed in the Prologue and Chapter 3. We record what we now know, and what we now want to find out. We have a brief discussion of which questions seem to be important at this point, and why we find them important. We also discuss possible strategies that we could employ to pursue these questions, and what we would need to do between meetings in order to be ready to start in with these strategies when we reconvene. Homework becomes the tasks that must be done in order for us to be ready to do our work together in class.

Units, not lessons. Places are created when the space has a life that lives beyond the immediate moment, so it is best to plan in chunks of meetings at a time rather than in individual lessons. Before the beginning of the school year, I consider what my students will be accomplishing during the year. I avoid specific lessons and topics in favor of the larger picture. "At the end of the year, what sorts of things do I want my students to be able to do, to be able to talk about and offer opinions about, and to be able to understand with guidance?" Most schools provide scope and sequence charts or curriculum frameworks that help with this. Most states now have grade level standards that are useful for this as well. But in the interest of detail, even these documents lose sight of the "big ideas" that will be the cornerstones of our work during the year. As I see it, this provides each teacher with the opportunity to personalize the experience through the important decisions about what the key concepts and classic problems will be. "If this particular year were based on a canon of essential questions that every student should be able to respond to in an intelligent way, what would these questions be?" In the states that I have worked in, the standards websites also include samples of such questions that I can use in organizing my thoughts about what will happen for the year. I make a list of these questions and add my own, along with the big ideas of the year and the year's goals.

The next step in planning is to figure out for myself what I think the "story of mathematics" is for the year. "What are the one or two essential messages that I want to convey about mathematics, so that if my students were asked to characterize what mathematics is, I would expect them to offer some variation of this message by the end of the year?" Curriculum is more than a pipeline through which facts and skills get injected into students. As the Canadian curriculum theorist Keiran Egan has emphasized in many contexts, teaching is a form of storytelling about the content, and about what it means to know and learn (Egan 1988, 1990, 1992). His poignant example illustrates well the sort of story that is often told about mathematics and its central purpose in the school curriculum.

Egan encourages teachers to compare their curriculum organization with the storytelling qualities of fairy tales, one attribute of which is the dramatization of binary opposites. For mathematics, and for place-value in particular, Egan suggests the magical drama of power versus powerlessness. In an elementary unit coordinated with Colonial American social studies, students are introduced to the theme: they hear about how pioneer families would help each other out when a crow needed to be removed from a barn. Crows are about as good as people at counting—they can recognize around five objects in a cluster. So, if a farmer went into a barn and waited for the crow, the crow would know that one person went in and would not fly in himself until that farmer went out. Again, if two, three, four, or five farmers went in, and one or two walked out, in an attempt to fool the crow, the crow would still know that a couple of farmers were waiting with shotguns to shoot him dead. Now, if a bunch of farmers help each other out, and a group of seven or eight go into the barn, one farmer can hide while the rest go back out. The crow will lose count, and fly in to find his nest. BLAM: the hiding farmer no longer has to worry about the crow eating his seed stock. Extensions of this story can move children into the study of different animals and their relative ability to count. The story would be consistent with the drama: whales and dolphins, who can count up to 12, can outsmart people and save themselves in various ways; counting is placed in the life cycle in terms of the power or lack of it that the counting ability enables the animals have. Back to arithmetic: The myth of the origin of troops in the military is conveyed and acted out by the children. A general calls his advisors and says, "We need a better system for keeping track of our soldiers, or we will never beat our enemy." One advisor after the next fails to come up with a scheme, until one genius suggests having each soldier drop a stone into a vase as they walk past the general: each time a vase is filled with ten stones, that group of soldiers is clustered into a team of ten men; ten vases are grouped into ten tens of men, and so on. The general then is able to plan the movement of his troops with such precision and creativity that the battles are won with great finesse. Thus is the notion of counting and place value intimately linked with the normalization of counting as a tool of power and the ability to take another living being's life.

Depending on the age of the students and content of our curriculum, the story is often continued. Even the Calculus, usually considered the keystone of school mathematics, is implicated through its etymological origin in the meaning of its name—the word is Greek for the stone or pebble used for counting. Taking the story a step further, Davis and Hersh (1981) string along Archimedes, scribes and astrologers, and the mathematicians of Napoleon, continuing on to the development of operations research techniques during World War II, the marriage of mathematics and physics in the atomic bomb projects, Norbert Weiner's controversial work in prediction and feedback that led to his later work against the "nonhuman use of human beings," the origins of the computer industry in the intensified cold war space race, and the futurist notion that while World War I was the chemist's war, and World War II was the physicist's war, World War III will be the mathematician's war.

In Chapter 5, I mentioned Christine Keitel's (1989) very different story of mathematics as power, in terms of its role as a "technology" or tool that people use to accomplish newly possible tasks. Ledger systems of accounting, for example, made it possible for a whole culture of mercantilism and a merchant class to emerge in Medieval Europe. As Keitel notes, however, this form of accounting and the use of columns for adding and subtracting numbers of and costs of objects also structured a form of culture that previously did not exist, to which people adapted in the slowly emerging assumptions of trade systems within capitalism. More specifically, Cline-Cohen (1982) writes a history of mathematics education as one of "calculation," pointing out that calculating the population to be governed and the idea of a calculating population (assisting the running of government and the emerging capitalist system) were intimately linked. Students in schools today are smoothly enculturated to this notion of being calculated and studied as objects of pedagogy and administration, writes Cline-Cohen, and they are similarly "prepared" to both calculate and be calculated. Valerie Walkerdine (1987, 1990), another author mentioned in Chapter 5, continues this story of power one step further, by recording and analyzing the ways in which bourgeois democracy was to be upheld, not by a coercive pedagogy, but by a "natural pedagogy" of love, in which reason would unfold. "Reason was to become the goal of a technology designed to provide reasoners who could govern, and those who might, at least, be hoped to be reasonable, not pushed to rebellion by repressive and coercive pedagogy" (Walkerdine 1990: unnumbered).

In Egan's storytelling, the teacher constructs a resolution of the story: mathematics should be learned and practiced for the power it yields. You, too, can count beyond stupid animals and uneducated people, so you too can get the jobs such skills promise. Of course, this story of mathematics could also be resolved in other ways—a less optimistic moral that focuses not on the power promised but the objectification of people through number: a libertarian fear of big government; a distrust for the numbers claimed by military and government statistics; a distaste for the anonymity and subsequent loss of community fostered by contemporary studies of "average" people and the loss of attention to particular individuals (Greene 1973). We might study how the school has turned each of us into objects of study, through calculations of various projective data about our "ability" or "personality," twisting each of us into a particular projective future. We might further examine the role of numbers in mystifying the public rather than communicating information: "termination units" in discussion of new weapons in Pentagon budgets; hard-to-understand units of radiation leakage that spring forth in local debates about the location of a new toxic waste dump or faltering power plant; the manipulation of testing data to satisfy local taxpayers that the schools are accomplishing their stated goals; data used in arguments about "raced ability" (Bachman 1996; Kincheloe *et al.* 1996).

In the construction of the elementary curriculum, we should note a common practice of using relatively "small" numbers because, as some psychologists would tell us, children need to learn about numbers that they can construct concretely. Big numbers are hard to see and feel and thus inappropriate for young children. The concept of place-value is introduced within this larger curricular context, in which big numbers are for the powerful big people, little numbers for the constructed powerless little people. The ageism within the curriculum is implicit but important. Walkerdine is relevant in this discussion as well (Walkderdine 1989; Walkderdine and Lucy 1989). In the studies of girls and mathematics already discussed in Chapter 5, working class girls were found to have intimate understandings of large numbers of abstract mathematical reasoning in their home life; middle class girls' lives were distant from such interactions with mathematics. School mathematics based on presumptions of middle class lives did not meet the needs of either female population. But the key point here was that we should not presume a certain universal model of development that psychologists could abstract from studies of children; in fact, such universal models are often flawed in terms of class, race, and other categories by which children might be grouped and clustered. Place value might not even be a concept to be taught when various family and life experiences are taken into account—like abstract concepts of good/bad, fair/unfair, tasty, fun, and so on, place value might indeed be a concept that many children bring with them to school along with comprehension of large numbers.

By "teaching" place value, however, we tell a variety of stories about the world and the idea of numbers in that world. For most of us, numbers are a natural truth that we can see and use to understand reality. They are basic skills essential for successful life experiences and jobs. Brian Rotman's (1993) work on a non-Euclidean arithmetic articulates how hard it really is to imagine that numbers are as socially constructed and context-specific as Euclidean geometry. The idea that there might be a more generalized notion of number and quantification for which linear series of counting is locally "reasonable" but totally absurd for other contexts is almost impossible to fathom. Indeed ethnomathematics has argued for years that "Western" mathematics is not as universal as we wish, and has told alternative stories of worlds incommensurable with this so-called universal truth of arithmetic (Pinxten *et al.* 1983, 1987; Fasheh 1989, 1990). By subjecting children to particular "models" of the concept of place value without first trying to understand what models they bring with them, we are denying even a range of cultural variants within this narrow cultural construct of "Western" mathematics (Lave 1991; Carraher 1989; Mellin-Olsen 1987; Ladson-Billings 1995). Instead of "teaching mathematics" we are teaching that in school one must understand what the adult tells you to do; a new layer of obfuscation is sometimes added, to neutral or negative affects, occasionally positive enrichment, on top of what might be brought tacitly with the child to school.

Most important, however, is that we recognize how teaching and learning evoke these stories about what mathematics is, and that the stories are incredibly powerful in determining the ways that we make meaning out of our lives, the moral choices we are aware of, and the kinds of relationships that we establish with others. If we design our year of mathematics based on textbook chapters or units, then we are mindlessly acting as cogs in an ideological machine, and we are depriving ourselves of the most fundamentally interesting and meaningful decisions that we make as teachers. Without designing the stories that we are telling through our teaching, we are leaving the experience of mathematics void of compelling purpose and cultural context, turning mathematics into a flaccid copy of someone else's story. The story is there but we don't know what it is. This would be like telling the story of Cinderella, a version that emphasizes how someone who is good and follows authority gets rewarded in the end, but always accenting the wrong words as we tell it. If we chose our own version, say one that dramatizes how women and girls are always at the mercy of men in the Cinderella story, or how class and privilege win out in the end (Cinderella is after all the real princess, unlike her social climbing step-sisters), we will know where to accent and emphasize the important words and phrases: the meaning is in the details. Similarly, when we plan with the story of mathematics in mind, we know where to dramatize the key questions and concepts, and how to interpret the story, because we are telling our own story, not one scripted by someone else. If you are concerned about forcing an ideology on your students, consider how telling one's own story of mathematics also makes it easier to leave the interpretation up to the students. I can present an unresolved story about power and mathematics, and ask my students to either believe in one version or another, or to write their own complex weaving of the forces and conflicts that are part of the history of mathematics and social control, military prowess, scientific enterprise, and so on. By composing a meaningful storyline for the mathematics that we teach, we make it possible for our students to consider how that storyline compares with alternatives; when we ignore the story, we and our students are all wrapped up in the invisible story.

Within the year's story are substories—units that build to form the whole. I think of these as chapters and sections of chapters in my big story about what mathematics is. What are the main topics and subplots of the story? What are the dramatic moments of each chapter, and how do they connect with each other to make the point of my story? Some schools require teachers to maintain a lesson-plan book in which you would note the objectives and details of activities for daily or period-based lessons. You are likely to be required to record the specific state or district standards that are being addressed in the daily lesson, along with how this standard is going to be assessed. This is your opportunity to organize the sentences and paragraphs of your story. I see it as an opportunity to explain to my supervisor or principal what I am doing in my classroom, using the language that they need in order to convince the powers that be that I am doing the right thing and doing it well. It is very helpful for me as well, because I can see if I am ignoring some standards, paying too much attention to others, and whether or not I am really addressing my assessments to what the state and local authorities are concerned with. But this lesson plan booklet does not replace what I need to do for myself in my planning. Without the year's story broken down into dramatic plots, important and minor characters, forces of evil and good, justice and chaos, patience and urgency, I would not have the slightest clue about how to teach or what to do in my classroom. Just as an actor researches and experiments with their character's motivation and how their background would influence the ways in which they would dramatize the script that they are provided with, a teacher needs to fill in the background story around the script of the standards and frameworks documents. At the same time, an actor needs stage directions—"I will move forward to stage left at this moment to dramatize my unwillingness to genuinely consider my father's anger . . ." And so we too need the stage directions and props that our daily plan book provides—"I need Geometer's Sketchpad ready for two groups to use, while I need to carefully demonstrate safe uses of a compass to a third group today in order for them to be able to realize Standard 2.17–a.iv regarding properties of parallel line segments; and I need to have index cards available to record my anecdotal records about how they are achieving this standard." But as I write this in my plan book, I also have in mind the grand picture about why people study properties of line segments in the first place, and how it fits into the larger scheme about why we are studying geometry and learning how to use various technologies in the process.

Entering into work. The most artificial aspect of teaching for me is the "starting point" for any group on any topic. Before we have begun to create that "It" that David Hawkins refers to (see Chapter 2), we have not yet established an I–Thou relationship, and therefore do not yet know how to interact with each other. We are still at that awkward moment of "entering." Here are some ways to handle the awkwardness: One option is to make the awkwardness transparent and part of the process. Simply tell the students that this is going on. "We need to find a way to start studying blah-blah." Try different approaches with the students, making it clear that they are each attempts to find the best way to maximize student involvement in the new topic. One time the teacher can present a problem to consider; another time, a student can read a story; yet another time, a collection of student questions collected from comment cards written at the ends of every week over the last four months can be distributed to small groups of students; still another time the class may play a game together, then begin an analysis of strategy.

Another entry option is to arrange a variety of activities as *centers around the room.* Either have students visit each center, or have them select the centers based on interest. At each center, present a different form of guided inquiry into one of several topics that will be covered during the unfolding unit. This can be organized in the form of an "arcade" (see Chapter 5). Or, you may wish to use another design element in forming the centers, such as learning styles or multiple intelligences. One of my favorite ways to organize the beginning of a unit is with literature circles. I typically use one of two approaches to literature circles. In the first approach, every student is reading the same chapter from a mathematics book, the same mathematically themed novel, or a common reading that a previous student has written for the course. This works especially well when I first use literature circles with my class because it allows for students to practice the skills of each role. Students form into small groups where each group member takes on one literature circle role. Groups of five work nicely for this, because you can get a good range of roles in each group. My favorite roles are "literary illuminator," "connector," "soundtracker," "illustrator," and "question collector"; you should invent your own to try out. (For more information on literature circles and roles, check out *Literature Circles*, by Harvey Daniels (2001).) Students prepare for their role as assigned or chosen within their group. Before the groups meet, however, I have the students in each role meet first in role-groups to discuss how they prepared for their discussion, what they prepared and brought for their groups to use in the discussions, and to generally plan how they will go about doing their role when they are the only one with their role in their discussion group. I move about the room helping those role-groups who are having the most trouble planning for when they join the literature circle. Then we move into our circles and discuss the reading. One of the main goals of Harvey Daniel's original ideas for this way of working is to develop self-directed readers. So if we use literature circles again in my class, I have students select their own readings and form groups based on their reading interests. The group is asked to define reading goals for themselves as part of each regular ongoing meeting. But in our first round with this, I usually provide my own set of questions for the whole class to discuss after the groups get their own chance, so that they have a sense of accomplishing something that was designed by me. When we form a whole-class discussion, we talk first about the questions and ideas that came up within their groups, moving on to my questions only if there is still time or if someone in the class found one of my questions particularly provocative or interesting. In this way, my questions provide a back-up "safety-net" in the beginning, in case students are anxious about finding something to talk about.

Chapter 8 has several further ideas that could be used for starting off with activities that are not directly central to the main topic of your unit, but which provide ways for students to practice mathematical thinking parallel to the regular classroom work. The activities described in that chapter are not necessarily best used at the beginning of class meetings, but I do find it useful to start with short warm-ups that are not part of the regular, more prolonged parts of class. Teachers have often begun class with short minute-long tests of how well students remember previously taught material, or with curious and challenging puzzles. This can work well if the class finds these starters engaging. But in my own experience, they are not tied-in in any meaningful way with the rest of our work, so that students do not receive them as important. They see them as time fillers rather than as fundamental learning experiences. So I try to make the warm-up activities central to our regular work even if they are independent of them. Short writing experiences and mini-discussions are always used as the basis of later long term projects and class investigations, so that students recognize that these short experiences are part of the building of inquiries and the development of essential understandings. Serious attention early on means more influence on how they are used later. Participation in these warm-ups is thus a way of creating community and forms of further unit planning.

Criteria for "Performance"

Because so much of what I am writing about encourages changes in expectations for what students do in the classroom, I realize in my own work that I and my students together need to be clearer on what the expectations for work are in my classroom. I am not referring here to classroom management issues such as rules of behavior or consequences for violating rules. I mean that sets of criteria for what is produced in the classroom help us to make explicit how mathematicians work. My students come to me with enough experience in understanding the criteria for playing the role of "student" in school—as Lave has described, they have been apprentices in "studenting" for as long as they have gone to school; but too few students come to me with any sense of what the expectations are for how to play the role of "mathematician" in school. Explicit criteria fill in the gaps and make the space of learning into a place of mathematical activity.

I start with my own set of criteria such as those in Appendices 2 and 5. As the year goes on, I ask the class to add to and modify the criteria based on what they feel they need to clarify and make even more explicit given their experiences so far. By making the "rules for working" more transparent than my students have experienced in other classes, I reduce the sense of secret or hidden expectations. Fewer of my students try to psyche me out or second-guess "what I really want" from them.

One by-product of lists of criteria is that it is easier for me to figure out what sort of feedback to provide. Products of work in our class need to demonstrate that the criteria for working have been met. Whole-class meetings provide the context for developing ideas about how to meet the criteria. How would or could a student show that a question they are sharing is provocative or interesting to them? What are indicators of things that are provocative or interesting in our everyday lives or in other classes? What would be mathematically provocative? What would it mean for a math question to be interesting? These are indeed questions that mathematicians must ask of themselves, and when our work really does answer these mathematician's questions, then the classroom is a fundamentally interesting place. I recently used the rubric sheets in Appendices 2, 3 and 5 to communicate with students my sense of how they were meeting such expectations in my class. We use such rubrics as part of "workshops" in class where students provide feedback to each other on how they can better meet the expectations in their work.

Good criteria are about the ways of working, not about the content. It is better to say that a student needs to have a section in their portfolio that includes weekly reflections on what they need to do in their studying in order to improve their performance than to dictate a set number of correct answers or properly used algorithms. That is, it is better to require that students log on their reading of the textbook (write a short response to what they were assigned to read or chose to read on their own) than to require that they correctly use a certain procedure for solving word problems of a particular type. When students meet for ten minutes in peer-feedback pairs or threes, they can use the rubrics to help each other strategize how to do the work of a mathematician. This shifts the emphasis from a space of learning mathematics to a place where mathematicians do their work. And, when mathematicians are at work in our classroom, I know they will be learning mathematics.

One final strategy for the use of criteria involves negotiating individual grade contracts with each student. When I use this approach, I begin with a basic set of criteria in order to give the students an idea of my expectations for their work. But then I hold short conferences with each student to discuss what they want to take on as particular goals for the assignment at hand; we consider other criteria as well as specific ways of meeting the criteria given their own choices of how to do the assignment so that they know exactly what they need to do in order to receive specific grades. The contract includes criteria for working as a mathematician (maintaining a problem-solving notebook, a reading log, etc.) as well as criteria for the final product (that it identifies a significant aspect of the ongoing inquiry, or that it includes charts and diagrams as well as words, etc.). Sometimes this is tricky to do at the very beginning, before their inquiry has progressed to the point where they know what the product will be. In this case, we discuss only the criteria for working as a mathematician who is formulating an investigation, and we include in the contract the stipulation that a later conference at a defined date will determine criteria for the product of the work. Such conferences support the development of student leadership, voice, and participation in the design of the teaching/learning space as well as skills of intellectual work.

Leadership, Voice, and Participation

Standards documents, curriculum frameworks, textbooks, and teachers' manuals are excellent resources for thinking about the content to be learned. These clearly are important things to use in planning and assessment. But as this chapter points out, they do not help you to figure out how to establish your classroom as a place of learning, that is, as a place where people are mathematical thinkers and actors. Places emerge in spaces where leadership, voice, and participation are powerful enough to make things happen. In the least productive mathematics classrooms leadership is entirely absent or invisible: the leadership is somewhere outside of these classrooms, in the hearts and minds of the curriculum developers, standards writers, teacher-evaluators, and standardized test scores. All of that leadership is attending to the content to be learned, and is ignoring to some extent the ways of being in the classroom. Teachers in these classrooms try to fill the gap with gestures of authority rather than leadership. True leadership moves people to act. This is important: leadership is not the same as authority. Leaders help people to change their lives, rather than force others to do something. In the classroom, leadership may be found at different moments in the teacher, in any one of the students in the room, or shared by combinations of people in the room.

My thinking about leadership owes a great deal to the legacy of Ella Baker's five decades in community organizing, and especially in youth organizing, as part of the civil rights movement. It was essential for Ella Baker that localized organizational structures offer women, poor people, and youth—the three forces that see-saw as the backbone of the movement—an important entry point into movement leadership circles. For Ella Baker, the single most important goal of community organizing was to ensure the leadership development of poor people, women, and youth to participate in and contribute to local political activism by initiating projects and influencing strategy. I do not want to minimize the significance of Ms. Baker's attention to disenfranchised members of society, that is, I am not equating youth in schools to former slaves and other members of the larger society who have historically been discriminated against. But I do see important parallels between the civil rights movement, which is primarily in my mind a movement of power and participation in democracy, and my desire as a teacher to make my classroom into a democratic community that fosters participation. Ella Baker led by listening to the youth whom she worked with, and by facilitating their own participation as leaders in the larger social movement. Indeed, Baker emphasized that the larger social movement was far more important than any individual's rise to power. So she fostered a group-based notion of leadership rather than a leader-centered group.

When a group has a moral commitment to the democracy of the group, and on the urgency of its work—that the moral balance of the universe is poised on the efforts of the group—then it is important to keep the movement democratic and to avoid struggles for personal leadership. If only we could set the tone with our students for a similar intensity of purpose in our classroom, grounded on the need to avoid struggles over the leadership in the classroom community. In the movement that Baker "led," it was further evident that desire for supportive cooperation from adult leaders and the adult community was also tempered by an apprehension that adults might try to "capture" the student movement, to take it over. The students showed willingness to be met on the basis of equality, but were intolerant of anything that smacked of manipulation or domination. I believe this sense of leadership is at the core of mathematical communities in our classrooms as well. We need to establish the purpose of our work on the sense of group purpose rather than on the adult's desires or the adult's carrying out of the expectations of others outside of this special place.

This inclination toward group-centered leadership, rather than toward a leader-centered group pattern of organization, was refreshing indeed to those of the older group who bore the scars of the civil rights battle, the frustrations and the disillusionment that comes when a prophetic leader turns out to have heavy feet of clay. Similarly, a classroom will falter and feel the scars of failure if one person—be it the teacher or an individual student—takes on the role of prophetic leader only to wither in their responsibility and fail the group. One of the ideas behind the group-centeredness in the youth movements of the civil rights era grew out of the fact that many schools and communities, especially in the South, had not provided adequate experience for Black youth to assume initiative and think and act independently; this accentuated the need for guarding the student movement against well-meaning, but nevertheless unhealthy, over-protectiveness. Similarly, in our classrooms, our students have not been given increasing opportunities to take responsibility for their own learning, and they come to us lacking in classroom-based initiative and leadership skills. This does not mean, though, that they lack these skills. Outside-of-school experiences often provide the space for youth to develop such skills. Just as we run into difficulties when we want to take advantage of informal and everyday life mathematics in the mathematical activity of the classroom, however, we also find it very hard to make use of students' initiative and leadership talents that they have honed outside of school. Ella Baker's inspiration was to give youth the freedom to work toward their own goals and desires; she provided them with the tools they needed to accomplish the work of the student movement. Similarly, teachers can provide students with the mathematical skills they need to apply mathematics in their efforts to realize their own dreams. This starts with listening to students and practicing leadership in small groups within our classroom, and continues out into the larger community.

Making sure that each member of the group has a "voice" is an important part of democratic places as well. Many of our students do not feel like they have a mathematical voice. Most of us probably would be at a loss to describe our own "mathematical voices." Fortunately, Ira Shor (1980) once came up with a ten-step plan for how to help students ease into rediscovering their voices, and for making these voices part of the larger group discussion. His book, *When Students Have Power*, tells of his experiments with a college English Literature course, not mathematics, but the basic ideas were of interest to me, and you might enjoy reading it yourself. Begin with a problem, question, or issue. OK, this works well in my classes because students are used to being confronted by math problems in math classes. Shor's next steps enable students to find their own voice within their work on this problem, rather than to listen to the voices of others and try to turn them into directions for how to solve the problem. Most relevant to this chapter is step 9, in which students read their work aloud to at least one other person (in his case, essays; in our case, essays, posters, solutions to equations, proofs of a conjecture, etc.). Once they have spoken to an audience, it is a small step to share their ideas out loud with the rest of the class, or even people outside of the class. Each member of the classroom community is *ready* to share their voice. Only after step 9 can step 10 take place, an assessment of the impact of their contribution on their audience, and it is only at this point that students can begin to look *critically* at their own voice: once it is present. When students can examine their own voice from a standpoint, then they are aware of this voice and are ready to act upon it.

Participation in the classroom space takes on a variety of forms depending on the nature of the place that is being crafted by the members of the classroom community. Here are several teacher techniques for promoting increased student participation in the classroom: First of all, the teacher should structure into class time periods where the teacher is not directing the experience, so that students take responsibility for their learning without the teacher hovering over them. One way to do this is to form small student groups that are required to produce something that will be used by the class, leaving what they produce and how they produce it up to the groups themselves. At first, the results may be disappointing to you or some of the students. This is natural, as there were no criteria for the results of the group. Based on what happens in these early uses of student-only groups, the whole class should have a discussion of the products of small group work and how they can contribute to the subsequent class learning experiences. As long as I never judge what a group member says, but instead focus on how well the groups have produced what the whole class has agreed is necessary for meaningful follow-up discussions, I can restrain the teacher influence on students' participation in these groups, and facilitate conversations independent of these groups on the kinds of interactions that promote successful group work.

When groups come back to the whole class for discussion, I carefully take notes on what each student says. Sometimes it is very effective to have each student report one by one on something that transpired in their group. At other times, this puts too much pressure on students. When I take notes on what they say, it helps me concentrate on what they are saying, it provides a written record of what happened in class, and it gives me something to study in preparation for the next class. Taking notes sends an important message to the students that their comments are indeed important. When they see that someone with authority listens carefully to what they say, they start to recognize the value of what they are saying. John Dewey once wrote that a democracy requires not just the freedom to speak, but more importantly, the right to be heard; when I attend to everything that is said by students, and then refer back to what they said, quoting directly as much as possible, students recognize themselves and each other as "stakeholders in and creators of the learning process" (Shor 1996: 49). This pedagogical technique uses the teacher's authority to "authorize" student authority in the classroom space.

My next step in promoting participation is to provide tools for analyzing and organizing what students have reported to the class. Given the different comments raised by students coming out of the small group work, we need to develop a meta-cognitive awareness of the themes and issues that groups had in common as well as ways in which their ideas differed. This is where my knowledge and authority as the teacher can further push students into increased participation. I can offer "experiments," things to try as the group seeks to find a solution to the problem at hand, or seeks to identify good questions to pursue. Such "experiments" may be based on mini-lessons, where I offer new techniques or concepts, facts that mathematicians have come to know, and so on. Since these lessons are very brief and because they are offered as potentially useful in organizing further work that has grown out of the students' own comments, their participation has made a crucial impact on what transpires in the classroom. My own mini-lessons do not detract from student participation; instead, my mini-lessons make further participation possible. The key is not to turn this into a series of lessons on how to solve the problems that the teacher wants everyone to solve. This approach can too easily be reduced to the traditional review-teach-practice lesson, and lose the sense of the investigation based lesson or flow-structures. To help me avoid such a reduction, I make sure that we are always organizing and analyzing the students' results, so that we are always working as David Hawkins (Chapter 2) suggests, as allies studying "it."

Ira Shor (1996) has another idea for promoting student participation. He offers as one assignment option among several the participation in an outside-of-class planning group. This group of students meets regularly—after every class, once or twice per week—to discuss how class went, and how it could be improved. The students in this group plan the next class with the teacher. In this way, if any class is awful or boring or has just plain silenced a student or a group of students, then the outside-of-class group can search for a way to change things right away. Shor begins each meeting of this group by asking what worked and what did not work in the previous classes; what did they like and what did they not like about class? Each student has the chance to make a statement, and then he leads a general discussion. He takes careful notes on what they say, so that there is a record of the meeting, and so that everyone is clear that what people say matters. He encourages students to talk with each other, not just to him, about what should be done to improve the climate of the class meetings. Since he works with college students, he can perhaps more easily support students' input into the content and nature of instruction. But I have had success with this approach with any age students as long as I do not expect very young students to use the same language as and to solve problems in the same ways as adults. In fact, my own experience with college students is that they do not really differ all that much in their experience with such reflection on classes from elementary, middle, and high school students; so the groups work out in very similar ways. Should material be repeated? Is this a waste of class time? How can we include those few students who have not yet found their voice in the class? Because any student can volunteer to be part of the group that decides these issues, those who opt for other forms of participation and contributions to the class are more apt to accept the way that the class is run. And those students in the outside-of-class group have greater opportunities for leadership, voice, and participation in a very direct way.

The Teacher's Way of Working

This chapter is mainly about maintaining a focus on what the students are doing, and how to value their ideas and participation in order to create a place of mathematics learning. As teachers, though, we always come back to what it all means for what we do. As much as I want to say that the flow-structure keeps my attention on what the students are doing, and not on what I am doing, I still find that I need to be able to tell myself what to do. Here is another irony in the work of teaching: in order to avoid thinking about ourselves and what we are saying and doing, we need to think about what we should be doing in order for that to happen. For example, in promoting student participation, the teacher can take careful notes on what the students are saying, and provide copies of those notes to the class for further discussion. This is a concrete thing that the teacher is doing. But the nature of what the teacher is doing is qualitatively different from the sorts of teacher behaviors that are part of review-teach-practice lesson plans. Instead of preparing for class by dissecting each problem into tiny pieces that can be demonstrated to the class, the teacher is editing and organizing notes from the previous discussion.

More generally, I rely on a working portfolio to help me do the job. I organize my portfolio into several sections that support my work. One section is assessment and planning. In this section I organize all of my information on my students, their participation and interests, as well as my observations of what has taken place in class. I use all of the assessment ideas in Chapter 1 to collect information, and then, in this section of my portfolio, I maintain a regular analysis of that information. At least once per week I ask myself Susan Ohanian's questions (page 32), and I try to document every answer with specific examples from the data that I have collected as part of my assessment. I then use my reflections on assessment to make decisions about what the class needs to do in the next week and month as well as long term for the rest of the year. These decisions constitute the bulk of my ongoing planning. Do I need to develop a mini-lesson on a skill that students still do not use effectively? Is there a critical aspect of a concept that students would benefit from addressing? Do we need to change the structures of working in our classroom? Is there something that works well that we are not capitalizing on enough? My planning may take any of these sorts of things into account.

My portfolio is not just a place for immediate planning and assessment. I also maintain sections where I log on what I am reading and where I pursue my own personal inquiries and research. In my reading log, I use whatever I am reading as a lens for looking at my teaching, my students, my school, and my life as a teacher in general. I use the broadest definition of "reading" to include comic strips, T.V. shows, videogames, and anything that helps me to think about an issue related to my teaching. In this way I force myself to always maintain an intellectual stance on my work; I do not allow it to become a meaningless routine or daily or weekly exercise because I am always applying whatever I am doing outside of my work to the analysis of that work. In my section on inquiry and research, I organize anything that I am collecting on things that I need to pursue for my career. This may include, for example, information on extending my certification, articles on working with autistic children, or silly websites that I can share with students. The important thing about this section is that I am always identifying a question that I want to pursue about my work. "How can I get better at working with the families of my autistic students?" "How can I provoke students to use statistics to understand on-line information?" I make sure to visit my inquiry and research section at least once per week so that I can reconsider my questions, reformulate them in light of the work I have done on my inquiry in the last week, and set for myself the new questions that I now have. By working this way, my inquiry takes on a high degree of focus, even as this focus may change from time to time.

My portfolio is my refuge from the confusions of teaching. It is the place I go to figure things out, to be away from the work in order to examine it, to be working when I miss the work . . . in general it is my professional "home." The point of my portfolio is not to impress anybody but to help me to do my own work. It is not an evaluative portfolio. It is a working portfolio. The audience for this portfolio is myself. But the portfolio also serves some nice professional purposes. I always have careful records of assessment that I can refer to when meeting with a supervisor, principal, or family member. So I can always point to specific examples of how my students are meeting my expectations and the standards that others are concerned about. I always have ongoing records of my work in general, so that I can look back over what I have been doing and make informed decisions about what my class should be doing next. And I already have a portfolio started if I want to seek further certifications such as National Board Certification, which require a professional portfolio; I can take materials from my working portfolio and reorganize them for these other evaluative portfolios. The use of a working portfolio supports my efforts to create a place of learning in my classroom by structuring the ways I go about doing my job. Instead of seeing my role as planning and carrying out lessons, I see the daily lessons and longer units as part of what I need to do to maintain my portfolio. I need to collect assessment information so that I can sit down every week and organize that information to be used in planning the following week. I need to log on my reading in order to use the quotes I copy into my portfolio, the reactions I write down, and the ideas that I want to try out in my teaching. I need to revisit my inquiry and research section in order to identify what to do for my professional inquiry in the next week. My portfolio is where I decide these things.

Within my assessment and planning, I spend a lot of time challenging my own assumptions about what it means to be developing mathematically in my classroom. The next chapter explores this in detail in terms of students who are not learning. The distinctions that I inherit as a teacher of mathematics run the risks of defining what is going on as non-mathematical, or inappropriate, when it may be precisely what is needed in order for students to be learning in the first place. For example, we in mathematics education seem to value conceptual understanding over what we pejoratively describe as "mere calculation." Yet many of our students focus on calculation over abstract representations of the concepts. We sometimes refer to this as an adherence to procedural knowledge when we are promoting conceptual knowledge (see Chapter 1). But I need to consider the implications of this system of values as I apply them to my classroom in light of the fact that subgroups of the population may have a tendency to approach mathematical knowledge more in one of these ways than another. To label students who prefer calculation over conceptual reflection "non-learners" is to deny the very learning that is taking place. In the next chapter we consider whether or not these students are "resisting" learning, and why we want to call it not-learning. My concerns about this issue arose in my reading logs, as I read the work of Valerie Walkerdine (1987) and Thomas Popkewitz (2004), and subsequently changed the ways that I go about using my assessment information to inform my planning decisions. What you now have before you is an example of how the working portfolio can push a teacher in new directions. As I am pushed in new directions, my students and I together can try out new places where mathematics is learned.

Response to Chapter 7:
The Classroom is Always Changing

David Scott Allen

My current batch of students provides me with some good illustrations of the ideas in this chapter. All of my students are brighter than ever—brighter, that is, at knowing when and what to parrot to me on demand. They are like machines bent on jumping through every (easy and useless) hurdle I put in front of them. They have shown much resiliency toward my efforts to thwart their roteness, and I in turn have learned much about what causes this, or at least, what the effects are. I'm in a position to critique neither the individual elementary school teachers nor their curriculum. What I can say is that "the system" is failing the students by not instilling the ability to think about a problem independently. They learn all too well the types of artificial problems presented, but when presented with other problems to solve, they fall back on this stilted repertoire of strategies. If it looks good, add. If the numbers fit together, divide. No thinking is required, just blind number crunching. As we discuss "new" topics in class (I'm finding that they are exposed to more topics by seventh grade), almost all hands shoot up immediately before I even ask a question. Even the advanced eighth graders do not read directions and do not answer the question asked. But these are good problems to have. I tell them that part of my job is to turn upside-down everything that they have ever known about math so that they slow down and really think about what it is that they are doing. They don't like me in September, but I look forward to June, when they have gotten the point. By midyear, we can do some pretty interesting mathematics.

So just how can teachers provide interesting ways for our students to interact with the material, accomplish the standards, and avoid the apprenticeship in studenting? This is at the heart of all my struggles with this approach to teaching. I can see that some things are not working well using the old methods, and in some cases I can even identify them, but the changing is hard, and so above all it is slow. It also seems risky when we are under pressure as teachers to get our students to perform well on those standardized tests. We learned math a particular way, and so that's the way we teach. Maybe students won't be professional mathematicians, but maybe we can give them that experience in a small way early in their schooling so that they can decide based on knowledge of, rather than assumptions about, what mathematicians do. Being a mathematician and being a math student in school are usually two very different things.

As stated previously, I believe that the State Standardized Tests and other tests are easy targets for my students to hit. They just can't do justice to what my students should or could learn. Johnny can distribute and Jane can add fractions, but to me, that isn't the point. So I'm left with the incongruity of having to teach them to jump through hoops that are not at the heart of mathematics. Take properties, for example. It is unclear to me why it is important for my students to recognize $(x + y) + z = x + (y + z)$ as the associative property. Maybe they can pick this out from three other expressions, but do they understand why it is helpful for them to know? Do they know for what operations it is true? For that matter, can they recognize it if I substitute other letters or numbers for x, y, and z? This is why many teachers feel that these new standards are adding to their curriculum rather than transforming it. If the tests we use are the same, how can we teach different material? I still strive to "integrate" both approaches, but there are certain areas which don't seem to lend themselves to new methods. We are still teaching to the old tests. It can be done, but it takes an enormous amount of thought, time, energy, and experimentation before it is coherent. It is easier to do it badly than well. I keep my eye on my own children's teachers in this area, and suggest and supplement when I can.

In order to do investigative lessons, you must have rich problems in hand for use. Textbooks typically do not provide this. Finding them creates more work for the teacher, but greater benefit to the students. In discussing this with one of my colleagues, we have decided that we would try starting at the end of the lesson, or with the most complex material, in order to intellectually motivate students to sit up and take notice of the sum of the parts and perhaps the parts with which they are unfamiliar. Students' minds are so full of technique that they need to slow them down to actually think about meaning. If we give them easy or trivial problems, it reinforces to them what they already think about mathematics from past experiences. This is not mathematics, but recipe following. Any other problem is considered by the students to be too hard, but what they really mean in most cases is that it is longer by a few steps.

I am inclined to think that the unit is more critical than the daily lesson. The beginning of a given unit is important. While there is something to be said for packaging lessons into bite size chunks, the philosophy must meet the goal halfway. Students need to wrestle with unresolved issues. Some things just need a day to be slept on before they work themselves out. I'm not always going to teach by the clock, and I teach by it less all the time. In order to make concepts coherent, good problems are very important. Perhaps we can use them as a base to get the students to be thinking about these issues. Why are textbooks so cookie-cutter? Can't students handle more? Why do the publishers need to baby students this way? Clear thinking will lead to better results.

One area of concern for me is the balance between the individual, the small group, and the class as a whole. I like the idea of using small groups to generate discussion and endorse positive social relationships, but I am concerned about the unevenness of progress that can occur in this setting. It is another risk! Homework is a substantial part of this dilemma. Some students might have no idea how to do certain things on their own, but others may need more than the simple and mundane. I am truly thinking of (gasp!) not assigning homework, but rather pouring my energy more into what happens in class, and truly giving the students the idea that it is not the number of problems that they did, or even how many they got right, but what they were thinking about and how and what happened next that matters. By reputation, my middle name is homework, but that may change sooner than I ever thought possible.

Still, one of my greatest frustrations is a lack of time just being able to talk to my students about what they are learning. I see the only solution as my stepping down from the stage and letting them do the hard work of mastery with my guidance, not my teachering. I don't want my students to think that the ultimate in mathematical achievement is page 270: 1–22 even!

The truth is that I laugh at many of the stories I hear about storytelling versus good teaching. That is, until I look at my ten-year-old daughter. Today is her tenth birthday, and we went to see the latest family movie in the theater with nine of her friends. I believe that stories can be powerful influences, and so I hope my children are not too influenced by Hollywood. So, how do I organize the attractiveness of the topics? Variety is important, and I will keep doing little things for classes who cooperate. One day I read one of my children's books to my students and challenged them to find why this story is mathematical. This was uncharted territory. I had one idea, but they saw math everywhere in the story. I also find class warm-up activities all too ignorable, so I sometimes struggle to get students to focus at the beginning of class. I try to let students relax, because most students will do most of what I ask most of the time. This is where the discussion and off-the-wall problems come in. My students like them as long as they are not literally impossible. They need each other to play with potential answers. Eventually we can get to the point where students can independently critique one another. For now, a comment from a group member is enough. But we do make projects and grade each other on the products. There needs to be different sets or levels of rules because this is such a foreign experience for them.

I believe that mathematics can train students to be leaders, and that it is my responsibility to lead by example of leading, not force-feeding. But in order to do that, we must risk that they might lead enough to take charge. We can't force them to do this. It is complex, indeed. Group-centered leadership is something I need to ponder long and hard before I implement. I like this idea, it is just not natural to me. Much of this discussion may seem very far from your classroom experiences at school or at work, but rest assured they are far from mine also in many ways. Much of change involves just doing it. Maybe a starting point would be choosing a unit or topic to transform your classroom into what it once was not. You will experience pedagogical discomfort, and may fall on your face more than once. But it can be done, and you can share in it. The paperwork alone can seem overwhelming. Keeping track of students' progress can be daunting. I find it saves me time in the long run once I get a handle on a system that works for me. Much of this is like the mathematics: it's in the use of it that we find the meaning we seek.

You may want to attempt to have students have a say in their classroom atmosphere. I have done this in the past, and I always have gotten a certain sector of the larger class to participate. We cannot really change the class to include an excluded group if no one from that group attends. The more popular people use it as a platform for self-serving purposes. Some students need encouragement to use their voice. Some students sell themselves short.

I do find myself thinking in math teacher terms at this point in my career. I think about everything in terms of how I can use it in my teaching, even things that are non-mathematical. All of my lessons ask this question, and all of my other endeavors do the same. I don't do it to show off, but rather because this helps me with my voice. I enjoy writing and it gives me an outlet for organizing my thoughts and life. What do I need to cut out of my life that I may not want out, but which I know must go or else my schedule will explode? How can I get my daughter to memorize her multiplication tables? How can I encourage my Kindergartener's budding geometric competencies? The writing is where I collect all this and see connections between concerns in many areas of my life, and so serves an integrative function.

Action Research 7:
Empowering Students Who Don't Learn

Petal Sumner

Barbara H. Wagner (undated website) states in the last line of her "Educator's Pledge" that her educational goal is to empower her students. Most teachers feel that they empower their students by simply transmitting information to them and having the students receive knowledge through repetition of the information. At the end of the year, the teacher's job is done when the student passes their class. Although students have passed the class, most of them will not remember what they were taught. Can we say that students have learned anything in the class if they do not remember the material? I don't think so. This problem can be seen in most classrooms where teachers are frustrated because they feel that the student should have learned certain material in one of their previous classes.

As I begin to make the transition from a working professional to a mathematics teacher I have been analyzing my motivation for becoming a teacher. Before I read Wagner's Educator's Pledge, I listed my motivations as being the ability to share my knowledge of math and work with minority students that have been labeled as remedial and unable to be successful in a math class. The things that I listed all lead to one goal: empowering my students by helping them to compete above and beyond others when they leave my class and go out into the real world. I believe that students who know and understand mathematics are on their way to empowerment.

How can I achieve my educational goal? How can my students become empowered by learning math and become successful when they have been told from the beginning that they would never be able to do so? At first I believed that meeting the learning needs of every student that was placed in my class would be the way to accomplish my goal. I began to look at Howard Gardner's (1999) theory of multiple intelligences to see how I could adapt his ideas into the math curriculum. I believed I could develop lesson plans and activities based on his theory to help each student get a better understanding of what they are learning. Gardner's theory defines intelligence as an ability to solve problems or create products that are valued in at least one culture. He suggests that each individual possesses at least eight mental abilities or intelligences which they blend in various ways in the course of creating something that is meaningful or in performing a meaningful rule or task. This all sounded like it would work well in creating empowered individuals, but I felt that something was still missing (see Appelbaum 2004 for a good critique of Gardner's theory). As I went through the semester and read the chapters from this book I began to recognize that a lot of these theories were being practiced in the classroom, but were still not producing students who were skilled in mathematics. Students are still sitting in our classrooms and not learning the material we are teaching. This is the basis of my action research, to find out why students are not learning in our classrooms and what we can do to bridge the learning gap.

I began my action research project by volunteering to work with a ninth-grade female student at Parkway High School in Center City who was performing poorly in her mathematics classes. I looked at some of her standardized test scores and progress reports from previous years and it showed a pattern of below average scores. I spoke with her principal and mathematics teacher and they both noted that they believed she had the potential to do better, but it seemed that she was "just not getting it." They were hoping that personal attention would help her to perform better in math.

In Chapter 1, Appelbaum discussed ways to help students embrace math. I began to look for ways to change the experience she was having in school by creating activities that would be interesting to her. I began to incorporate weekly free-writes into our sessions. I noticed how well she wrote and saw that this was something that she enjoyed doing which could be the beginning of her success in mathematics.

I then began to conduct interviews to look with and through the student to know what she was interested in so that we could bring those interests into our sessions and interweave them into the lesson. I was excited to use interviewing because I felt that she would open up and I could gain additional insight into other activities that she enjoyed and could possibly incorporate to make her math experience better. During the interviews, I discovered that besides her love for writing, she also loves music. She told me that she writes poetry and has been playing the violin for the past six years. She enjoys writing music, playing music, and listening to music. Her favorite style of music is classical music and she aspires to be a classical musician when she finishes school.

This really surprised me, because research shows that instruction in music stimulates early brain development and helps children develop intuition, reasoning, imagination, and dexterity into unique forms of expression and communication (Consortium of

National Arts Education Associations 1994). With music's mathematical relations, she should have the ability to transfer those skills into success in the classroom. One of her math teachers should have been able to use this information to create an environment conducive to her learning abilities.

This is why it is important to be informed on the psychoanalytical self of our students. We can do this by interviewing them and promoting an environment that inspires critical thinking. It seems like teaching should be enough, but students walk into the class with a lot more than an open mind ready to receive the information we want them to learn. I would never have been able to tap into her musical interests based only on progress reports and test scores. Now that I had some background information on her, the goal would be to use it to help her be successful in math.

Although success in math is important, helping her to think like a mathematician would produce long-lasting results. I tried to use the information I discovered along with Polya's four-phase plan to begin to help her develop mathematically. As I tried to do this, it did not seem to work as well as it looked on paper. The hardest part of using the four-phase plan always seemed to be the first phase, understanding the problem. The student would start by saying, "I'm not sure what that means" or "I don't understand what they are talking about." I addressed this problem by incorporating a writing task where I would go over the chapter and have her write a poem or a song that would include the important facts and terms that would help her complete any assigned tasks on the topic. I thought this would work well and would help her see how her interests could help her excel in math. This worked one time and after that it seemed like the wall was back up; this became another unimportant task that was pushing her to do something she really was not interested in. It did not matter how I tried to make her see that doing this would work to her advantage: she was not willing to try any more.

Now I feel that nothing is working and I need to try to find a different approach to the situation. None of the critical thinking pedagogy seemed to work with this student. No perspective, no pattern, no story seemed to work towards creating a skilled mathematician. Her test scores and grades did seem to slightly increase, but this was probably due to the intense drill-and-practice activities that we had to do to keep her working and to keep her teacher happy. I was getting impatient with the slow progress and became another aggressive seller of mathematics, trying to make her see its importance and relevance to having a successful career. I felt that I was failing at helping her believe that her everyday experiences represent mathematical situations.

I asked the student what she thought of our sessions and how she felt she was doing since we had been meeting. She began to tell me that she never understood math and this made doing math hard. She did feel that she was doing better and that she understood math more than before. This still was not enough for me because I could not see the results of successful learning during our sessions. I began to realize that her success was based on standards that I had set. I then tried to use the metaphor of a "birthing center" where she would feel comfortable and at ease to present her own ideas and standards. I would support her experiences by connecting other mathematical ideas and show her that her explorations had meaning. This still was not working and I began to see myself acting like a preacher during our sessions (this was a metaphor one of my former classmate's came up with when asked what metaphor could be used to describe teachers in some classrooms). I think if I asked her she would also agree that our sessions felt more like mini sermons where I was constantly trying to convince her that math was the key to life and without it she would die.

As I near the end of my research, I feel like I have run out of time. I don't think that I have come close to finding a solution that would help me empower the students who refuse to learn or cannot learn. Time plays a major role when you are trying to model and apply the many theories you have learned into your classroom. In her eyes, she gained confidence, increased her grades, and did better than she would have on her own. In my eyes, I would have liked to have more time to explore the ideas I have learned with her.

My initial concluding analysis of her overall attitude was that her grade in her mathematics class was not important as long as she could pass the class. This is the attitude that I have encountered with many minority students; passing the class is more important than the grade they get or what they have learned during the year. Yet, this initial analysis changed during our last session. She gave me a card signed by her and her mother and it made me realize how much she did care and how happy she was with the progress she had made during our sessions. She did want to learn, and did learn. But what do we do when a student does not learn? Students are learning that they cannot do many things and math may be one of those things they cannot do. I believe that a student's classroom environment is an important factor in helping them learn.

How can we change the classroom environment to enhance students' learning? Although there are many facets that will enhance students' learning, I will focus on self-regulation, motivation, and support. The first component, self-regulation, is termed by Myron H. Dembo (2004) as the ability to control the factors or conditions that affect one's learning. He says that this can be done through the use of different learning and motivational strategies. Most classrooms focus on telling students what to learn instead of telling them how to learn. It is important for teachers to provide students with the necessary tools that will enable them to learn what we are teaching. Electricians cannot complete their assigned tasks without all of the required tools, so we cannot expect our students to learn without all of required learning tools. Each subject requires a different learning tool and having these tools will be the deciding factor of success or failure throughout their academic life.

Motivation in the classroom is another important component of the student's learning process. Many teachers use positive reinforcers such as money, stickers, and other benefits to motivate their students to participate and excel in their class. These types of motivations do not work among all ethnic and social groups and tend to create biases against the students who are not able to achieve these rewards. Some researchers (e.g., Kohn 1993) suggest they even work against our goals in the long run. We should not motivate our students to outshine other students in the class. We should motivate our students to understand the material and recognize their improvement in the class. This will lead them to set individual goals and not compete with others to see who does best.

Supportive classroom environments also help students to not feel intimidated when they are not doing well and do not understand what is going on. Teachers must celebrate their students' successes and support their weaknesses. There are no dumb answers or stupid questions. We must turn those questions that may not make sense into probing questions that will make our students think critically. I could see my student try a little harder to complete her tasks near the end of my project. I believe that she saw my dedication to her success and that I was not going to give up on her. Students are willing to participate when they see that the teacher is willing to work with them.

I believe that these three components will move our students in the right direction. As we get students involved in the learning process and make the material relevant to their diverse backgrounds and needs, they will feel more confident and become capable learners. We will then begin to see successful math students who are knowledgeable and able to compete at any level.

The next task then becomes the ability to qualify a student's success in the classroom. Success is defined as the favorable outcome of a desire or attempt; this means that we must know what the student desires to accomplish based on their beliefs and goals for education. Our students' commitment to achieve success in the classroom will not be the same because each student will not be working towards the same goal. We cannot use the same measuring stick to conclude the success of all of our students. Our job is then to take time to address the needs of each student in our class and judge them based on their individual goals for success.

Petal Sumner

I am a high school mathematics teacher for the Baltimore City School System where I have taught mathematics for two years; I am currently a third-year doctoral student in mathematics education at Morgan State University in Baltimore. The chapter included here is the result of a research project I carried out for a course at Arcadia University in Glenside, PA, in the spring of 2005. Since then, my interests in alternative assessment have led me to pursue research methods that may help my students gain mathematical knowledge in non-traditional ways. I also teach developmental mathematics for Baltimore City Community College and hope that my research can be used with adult students.

eight
When Students Don't Learn

Gus was in my college math class last spring. He was really interested in our discussions, and always participated in them. Gus seemed to have a knack for zeroing in on the crucial questions, and was not afraid to ask them. He was curious about other students' ideas, too, and eagerly listened when they spoke. When we were studying rational and irrational numbers, Gus impressed me with his ability to develop an investigation into their relative densities on the real number line. His tactic was to generalize the proof that there are infinitely many primes, which he found intriguing given that we also explored how the primes get further and further apart, so that he could look at the distances between rationals and irrationals for bigger and bigger numbers. When we moved into our unit on infinity, Gus was naturally drawn to the paradoxes. He enjoyed Zeno's paradox, where the hunter shoots an arrow at a lion, but the arrow never hits its target; even though we know that hunters seem to hit their prey more often than never, Zeno described how at any given moment the arrow always has half the distance left to go. Gus reveled in trying to remove an infinite number of ping-pong balls from a barrel taking out ten at a time, while always replacing just one; it seems like you are always removing more of the balls than you are putting in, eventually taking out all of the balls, yet the thought experiment allows you to do the ten out and one back faster and faster each time, strangely resulting in a barrel full of balls. He offered clever insights and provocative questions during our discussion of Hotel Infinity, where, because there are an infinite number of rooms, even if they are all filled, you can always find a room for another guest by telling the current guests to move to the next room. Then suddenly Gus was not in class every day. Suddenly, his investigation was minimal and poorly executed. He got a "D" on a quiz. But he still wrote an amazing essay on power sets—the sets that contain all of the subsets of a given set.

When Gus first stopped coming to class every day, I asked him if anything was wrong. He told me absolutely not. He really enjoyed our class, but he had other things to do right now, like travel with his band, finish a major project for his painting class, and spend more time with his girlfriend. I was bewildered. Here was possibly the strongest student in the class, someone who so easily caught on to every idea in our course, and he was simply opting out. I pushed him on this, and he told me it had nothing to do with me, and he hoped to take the next semester of mathematics with me as his instructor, but his grades in the first part of the semester were good enough to let him coast right now and not end up with lower than a "C" in the course, which would be good enough. But Gus should be a mathematics major, I thought! *This* class should be the first priority for him, because he is so good at it. Gus, though, was happy as an art major, and his painting was his first priority.

Herb Kohl (1994) writes about a student, Rick, who refused to learn algebra. He notes that there was nothing wrong with Rick's mind, his ability to concentrate, or his ability to deal with abstract ideas. He could read, and he did read the books he chose to read. He knew how to do very complex building projects and science experiments, he enjoyed playing around with athletic statistics and gambling odds; he just rejected the whole idea of being tested and measured against other students, and though he was forced to attend school, there was no way to force him to perform. I immediately remembered Rick when I had that first discussion with Gus. Rick had refused to learn and through that refusal gained power over his parents and teachers. As a free, autonomous individual, he chose to not-learn, and that was what his parents and the school authorities did not know how to deal with. Gus, on the other hand, had found a way to avoid the strong hand of power; by doing enough to get a "C" he made it possible to control his own time. By missing class, he could spend more time on his painting; by doing just enough on the investigation, he could devote more energy to the essay (he liked to write).

I went back to Herb Kohl's book, *I Won't Learn From You*, for inspiration. He writes,

> It's interesting how stuck parents and school authorities are on a single way to live and learn. Any youngster who refuses to perform as demanded is treated as a major threat to the entire system. Experts are consulted, complex personal or family causes are fabricated, special programs are invented, all to protect the system from changing itself and accommodating difference. People like Rick then get channeled into marginal school experiences and, too often, marginalized lives.
>
> (1994: 11)

Kohl's story about Rick is similar to my story about Gus in that Rick, too, devised ways not to learn, as opposed to being unable to learn. Rick, like Gus, found the mathematics intriguing; he described the abstract representation of mathematical relationships interesting him as much as chess did. In order to force failure upon himself, Rick found ingenious ways to dissolve equations into marks on the page by creating visual exercises that treated the equations as nonmathematical markings. Kohl shares the example of reading an equation from the equal sign out in a number of steps so that, for example, $3a + 2b = 12a - 32$ would become the sequence =, b = 1, 2b = 12, + 2b = 12a, a + 2b = 12a −, and so on. Rick would sometimes even memorize the sequence. When the teacher asked him what he was doing, he would explain exactly what he was doing, which would infuriate the teacher even more than if had merely said he didn't understand the problem.

Resistance

Gus and Rick are overt cases of not learning. Kohl distinguishes not-learning, which takes active effort on the part of the student, from failing to learn and unlearning. This essay and other works by Kohl have convinced me that it is surprisingly rare that a student fails to learn mathematics. In most cases, I now believe, the student has been forced to unlearn the mathematical ideas, strategies, skills, and facts that she or he exhibits in everyday life outside of school. In some cases, school has taught students to unlearn what they could do mathematically outside of school as well, instilling a severe lack of mathematical self-confidence and trust, although most of the time school simply presents "school mathematics" as a different kind of beast from that other stuff that one might do in life that a mathematician could label mathematical. In other cases, school experiences have resulted in learning that might be characterized as "mislearning." An example of this is the student who learned in early primary education to follow directions and memorize facts in order to produce a paper that looks like the teacher's, and to turn off conceptual understanding in order to do this; such students have unlearned their natural mathematical reasoning in order to create what is called mathematics in school, only later to find that they no longer know how to think when they need to.

A psychoanalytic framework would ask what is at stake in not-learning. Instead of assuming that a student has failed to learn, we ask why they have not learned. In a sense, not-learning for any reason is a kind of refusal of mathematical knowledge. The examples of Gus and Rick are overt refusals. Other students' refusal of mathematical learning may be more complicated. As teachers we assume that knowledge is liberating, all learning is "good." But there are at least two ways in which new knowledge may lead to resistance to that knowledge. The first considers ways in which new knowledge disrupts students' prior self-understandings; while we would hope that such disruptions would lead to feelings of empowerment, these disruptions are experienced as producing new and often debilitating forms of helplessness and isolation. The second way that resistance may emerge from the experience of learning is when our efforts to create learning produce social inequalities and conceal a pedagogical will to dominate. The way we think about not-learning, and resistance to learning, must engage us with two things: the ways in which acts of resistance to learning are attached to the social experience, and the ways that the teacher and the students are attached to resistance itself. Being attached to resistance might account for some of Gus' and Rick's actions. It is not so much that they are not-learning mathematical knowledge, but that the resistance to learning accomplishes something for them.

For psychoanalysis, resistance occurs when someone meets the otherness of their own unconscious knowledge. We always resist confronting the unconscious. We try to meet the unconscious, and instead, the unconscious is "forever creating itself anew from the bits and pieces of everyday life that remind us of wishes that are taboo and experiences that are too painful to confront head on" (Pitt 2003: 55). Learning might be said to deal with, on the one hand, the curious ways in which ideas and affect organize and reorganize each other and attach themselves to new experiences; on the other hand, learning has to do with the method of approach—how to "go there." To "go there," one passes through resistance. When Freud first started working in psychoanalysis, he thought that resistances to learning were like barriers that the unconscious constructed to protect what was hidden there. Later he refined his theory. He decided that the barriers were constructed by the ego as a defense against such knowledge. What might it mean for teaching mathematics to consider resistance to learning as a mechanism of defense against unconscious knowledge? The French theorist Lacan called unconscious knowledge "knowledge that cannot tolerate one's knowing that one knows" (Felman 1987: 77). Learning poses dangers to the learner: she or he must pass through resistance in order to learn. This insight—that knowledge may be refused because it may be threatening to one's sense of self—is painful to us, but our desire that the learner just accept and understand the value of mathematics is also partly a symptom of our own struggle to master our own difficulties. Mathematical knowledge provokes a crisis within the self when it is felt as interference, or as a critique of the self's coherence or view of the world. These crises call forth a "crisis of witnessing" in which the learner is incapable of an adequate response because the knowledge offered is dissonant on the order of trauma; the response can only be a working through—a mourning—of belated knowledge. The confronted self vacillates, sometimes violently and sometimes passively, sometimes imperceptibly, and sometimes shockingly, between resistance as symptom and the working through of resistance. When the movements of affect and idea are in conflict, varying forms of aggression also can be staged as the self struggles for elusive mastery through strategies such as the discounting of an experience as having nothing to do with the self and the freezing of the event in a history that has no present. These mechanisms of defense—undoing what has already happened and isolating the event in a time that has long past—are key ways the ego attempts to console itself. But the cost of consolation is severe.

I don't want to say that teaching is the same thing as acting like a psychoanalyst, but there are many parallels. During the course of every psychoanalytic therapy, the patient will behave in ways that interfere with the progress of the treatment. This interference is called resistance. Because psychoanalytic therapy helps the patient achieve freedom of thought and action by talking freely, the negative emotional forces that caused his or her symptoms manifest themselves as obstacles to the talking therapy. Similarly, we should not be surprised if students exhibit behaviors of resistance throughout every educational curriculum. In mathematics education, too, we want students to speak about their thoughts, to share their ideas and beliefs and opinions, and the difficulties of confronting new knowledge will manifest themselves in behaviors that delay learning. Like a patient undergoing analysis, a student may exhibit any of the following: becoming unable to talk any longer; feeling s/he has nothing to say; needing to keep secrets from his/her teacher; withholding things from the teacher because s/he is ashamed of them; feeling that what s/he has to say isn't important; repeating him/herself constantly; refraining from discussing certain topics; wanting to do something other than talk; desiring advice rather than understanding; talking only about thoughts and not feelings; talking only about feelings and not thoughts (Beck, undated). These and many other forms of possible resistance keep the student from learning about himself or herself, and from growing and becoming the person s/he wants to be. A patient and analyst study together the meaning and purpose of the resistance and try to understand the key to unlocking it and allowing the patient to continue growing. Many therapists recognize that a patient may *need* to resist, and use a relaxed approach as an aid in overcoming the problem. Similarly, a teacher and his or her students must together study their resistance to learning, the meaning and purpose of this resistance, and try to understand the key to unlocking it and allowing the students to keep growing. And teachers recognize that resistance to learning is a part of learning itself; students need to do this in order to learn.

Our first interpretation of resistance is often to label it as behavior problems. We ourselves feel angry, annoyed, or helpless, and we want to do something about our own feelings. The trick is to remind ourselves that these behaviors are just as possibly signs that important learning is taking place. If we move directly to punishment or correction, we may be inadvertently teaching the student to avoid the hard work of learning, to not-learn, since trying to learn is a punishable act. How ironic! A critical response to resistance behavior requires that we internally monitor ourselves. "I feel angry . . . so stop thinking of myself and focus on the student: talk to the student about the mathematics, backtrack with her or him to the point where the resistance can be identified; help the student look at the moment in time when the knowledge was lost."

Losing Interest

Less directly oppositional to learning are boredom, disinterest, or losing interest in the task at hand. These, too, may be manifestations of resistance, and therefore indicators of a person confronting the serious work of learning. But there is also the sense that students need guidance in order to find what is interesting in the task at hand. In *Wrestling with Change*, Lew Romagnano, a professor of mathematics education at Metropolitan State College in Denver, writes about changing the content of the mathematics taught in general mathematics classes when co-teaching with Ms. Curry. As the expectations changed, students disengaged from the activities. After playing a game, when the teacher leads them to explore the mathematics of the game they begin to disengage and wait for the teacher to tell them things. Romagnano's expectation that they try to figure out answers for themselves only accelerates their disengagement. At that point, he doesn't know what to do either, and that is when he, too, begins to disengage from the mathematics and think more about controlling classroom activity. After three days of investigating how much rice cereal boxes of different dimensions could hold, students still had not produced the box size that could hold the most, the original question they were given, and students were beginning to say that the class was boring. Despite his intentions, this problem just wasn't interesting to his students. His response was to introduce a variety of novel activities, to move toward entertainment: stations around the room for comparing and computing volumes; classes in the computer lab simulating the building of boxes using Logo procedures Romagnano had written; collecting all of the data and graphing the results. Ms. Curry felt even at this point like half of the students simply had lost interest. A few students would give answers when prodded. Most seemed really bored and just did not want to participate. A fair number of the students seemed to just play around with the computers, as opposed to really focusing on what they were doing.

For Romagnano, this was an *interesting* question! If *he* were given the chance to simulate boxes on a computer, he would take advantage of the opportunity. But for the students, this was not a question they cared about, and they had trouble motivating themselves to look for or even imagine that there might be a functional relationship between the size of the squares cut out of each corner before turning up the sides and the volume that the box would contain. The situation was repeating an earlier experience in exploring how the length and weight of a pendulum affect the period. Ms. Curry finally confronts the class:

> What I really feel is, as soon as you get frustrated, you just want to stop. . . . A large part of the way you get through life is you run into a problem and you figure out how to solve it so you can move on. If you run into problems the whole time and drop out, you're never gonna get anyplace. You're never gonna move from the point where you're stuck.
>
> (Romagnano 1994: 61)

Of course, this only made matters worse.

For Romagnano, mathematics is about "good problems." He has found a number of such good problems that lead to the study of functional relationships, an important central concept in mathematics. For the students, school mathematics prior to this class had been a collection of clearly defined tasks to be completed either during class or during daily "resource periods." The tasks involved basic arithmetic skills, to be completed at their own pace, which meant that if they finished them early they could spend the rest of the period controlling their own space, mostly socializing with friends. If they needed the time or wanted the help of the teacher they could expect to get it. Feedback was immediate, based on the number of correctly completed tasks. If we are to shift to new ways of working in our classrooms, we have to facilitate our students' understanding of the new expectations. We also need to help the students to see that these new expectations will help us together to create a place of mathematics learning that is worth moving into; we have to help them to see how it will be worth their efforts to leave the old place and enter the new one. In this new place, activities are neither clearly defined nor short. Students are expected to search for answers to questions without being given carefully defined procedures for finding them. The procedures devised rather than the answers they find generate the feedback received. The feedback comes in the form of another question.

Students are expected to think, talk, and write about unfamiliar mathematical ideas for the entire class period instead of until they can get the tasks done. Sometimes, they have to remember stuff from one class to the next, and activities may take more than one class period. Problems may never actually be solved. Students are expected to ask questions that might help them figure things out for themselves instead of shutting down until the teacher shows them what to do. The implications are that every day of school prior to the day they entered this new type of classroom has set them up to behave in ways that are counter to the ways they are now being asked to act. For example, in the new classroom, when the teacher asks students to offer ideas for how to proceed in a new and confusing situation, "good" students in the former sense of "good" will sit and wait to be told what to do, because they can't offer a correct answer, and the proper behavior in a class when you can't use the already-given algorithm is to wait for the teacher's help. When the reward for a correct answer is suddenly another question instead of time to socialize, a "good" student might really wonder what she or he has done wrong here not to get the proper reward. Two fundamental things are undermining successful school mathematics in this situation: first, the task assumes motivation for good problems when the problems are only good if they are posed by the person trying to solve them rather than imposed upon this person; and second, the structure of classroom participation, the space of learning, is so radically changed as to be unfathomable. I do not believe that the question is between being directive or indirective as a teacher; instead, I believe the question is between working with the students to create a new place or not.

Romagnano calls the first issue the "good problems" dilemma. Problems and topics that were intrinsically interesting for his students—such as when Ms. Curry did a unit on owning a house and setting up a budget—did not generate the same kinds of mathematical problem solving and attention to what he considered important mathematical concepts (such as functional relationships) that he hoped to accomplish with experiences like studying the probability of a certain game or the design of a box that maximizes volume. Students easily seem to lose interest in a teacher's "good problem" before the mathematics is really studied. Another aspect of good problems that he identified was that a problem raises different mathematical topics for different students, which means that the teacher has a complicated task ahead in figuring out lessons based on one student's topic that another student would be interested in as well. How does a teacher build a lesson out of one student's construction of a problem that addresses other students' needs? And, if a teacher responds to some students' disengagement by moving on, what does that teacher do with the few students who are still engaged? Most of these issues grow out of the assumption that everyone in the place of learning mathematics must be doing the same things at the same time, or at least working on the same questions. If we free ourselves to imagine small groups of students moving ahead to different questions, or continuing to work on an engaging question, some of these issues are resolved. But there still is the central dilemma of how to engage students in each others' work. Prior school experiences work against this as much as any other aspect of a mathematical classroom community. It is a rare class that has had any experiences with caring about the ideas of others in the room. This is where the establishment of a purpose beyond the immediate goal of being able to do what the teacher wants you to do can work to achieve community. If everyone has a need to support each other because there is a public event next week, or because what they find out will help you with your own investigation, or because you can see how to help them figure it out for themselves, then this class really has successfully established the expectations of a community.

But what do we do about those students who always lose interest? Or those students who cannot find something of interest to them? If we allow some students to move ahead while others deeply probe the mathematics of problems that these students leave behind, will we be failing those students who never get to probe the mathematics of our "good problems"? Let us assume that we respond to most forms of resistance by taking them as signs of seriously trying to learn, yet we still have a few students in our class who are not able to become interested in the material. The investigation is boring, so they stop. When asked to decide on their own what to pursue, they draw a blank. If they do ask their own question, it is so simple that it really is not worth pursuing, and you can tell that they too know this about their question. Such students range in behavior from complete disengagement to engagement with the need to please the teacher, but each has in common a lack of attachment to the mathematics. What I usually try next is to recognize with these students that disengagement is part of the process. We just need to find ways of working around it. Writers get writer's block and don't have anything to write; what do writers say about how they get past it? Scientists get paralyzed by not knowing what to do next, not being able to find a good set of experiments to pursue; what do scientists do to get around this? The Polya questions and Mason and colleague's specializing and generalizing are ways to continue working or begin to work on a mathematics investigation when you do not yet know what to do next. Sometimes just the act of working mathematically will generate ideas (see Chapter 3).

Remember, too, that engagement is related to membership in the community. Disengagement is an overt sign as part of ongoing assessment that such a community has not yet been established in the classroom. Rather than focus on an individual student or the specific content that is currently being covered, I will often reflect on our ways of working and the perception of purpose for what is being done. Something that is not engaging today might be engaging next week if we change the place in which we are working. This means reflecting on the issues of space and place from Chapter 6. It also means reflecting on the roles of exercises, problems, and puzzles and how they are currently being perceived by the students in the classroom, as discussed in Chapter 3. Boredom is a response to mathematical exercises; one would never be bored by something that one takes as a problem. Puzzles can be boring if one is not yet ready for them, as they are designed for people who are already enculturated, not for novices; introduce a puzzle too soon and it may seem like a dumb question. We can also distinguish between mathematics exercises and practicing the skills of a mathematician, something for which we sometimes also use the word "exercise"; but in these other kinds of exercises we mean something very different. Even a boring exercise can be something we do seriously if we have already committed to the skills the exercises promise.

It can often be the case that we simply need a break from the investigation at hand in order to come back to it with fresh eyes. I will regularly suggest this to my class: "Everyone is getting bored with this investigation of boxes. We're not able to see what is interesting about it any more. We need time away from it so that we can look at it with fresh eyes. Why don't we have some of us continue to work on this, while the rest of us pull out our ongoing projects and use the rest of the week to make progress on them. If you are still interested in trying to find a pattern in our data, let me see your hands. . . . O.K. the three of you can report to us tomorrow on what you find out. See if you can identify a relationship that would help someone build the box with the greatest volume. Everyone else: while they are doing that . . ." When students report on what they have found, others are regularly interested in what they have to say.

Practicing Thinking Mathematically Outside of the Regular Work

Here are ways to practice and games that mathematicians can play when they are not getting anywhere, or to break out of a boring routine:

Free-math. One just starts doing anything mathematical—making a chart of something, drawing a picture and then adding things to it. Try special cases. Do this for a specified amount of time, such as five or ten minutes, or a half hour. Do not feel compelled to share this with anyone, but save it because you might want to use some of the ideas at a later date. This is similar to "free-writing," when writers start writing even if they have nothing to write about. A writer may write for five minutes, even if what they write down is "I can't think of anything to write about so I will simply say this right now. I am promising myself not to leave a blank page but to write continually for the next five minutes . . ." Similarly, I may write down, "I have nothing to think about today so I will write down these facts: $2 + 3 = 5$, $3 + 5 = 8$, $5 + 8 = 13$, $8 + 13 = 21$, . . . hey, that's starting to look like a pattern I think I might remember from somewhere." Or, "Here's a triangle, there's a square added on the side of the triangle, here's a half-circle with one of the sides as the diameter of the circle . . ." Just as actors refine their skills by improv exercises, so do mathematicians practice by math improvisation. In math class, we can free-write with the requirement that we write about arbitrary things, such as how circles are related to the number 13. The mathematician John Conway sets up a series of roadblocks to working every day. His computer sets a puzzle before he can use it, and so on. These little daily exercises that at first prevent him from working keep him limber and ready for his work, indeed anxious to get started.

Connection Cards. Each student writes on five or ten index cards mathematical ideas, questions, or facts that have been part of class discussion in the last week or unit. The teacher adds a card or two that s/he wants everyone to have to think about. In small groups, students mix up their cards to make a deck, including the extra teacher cards. Variation 1: Play like dominoes. Deal the cards to each player in the group. First person selects a card and puts it in the center of the table so that others may see it. The next person can place a card next to this one if s/he can explain to the others how the two cards are related or connected; any explanation goes in this game, as long as each player understands the connection. Play continues in the same way, but a player can only place a card next to an open card, and if the card needs to touch more than one card that is on the table to be placed in a certain location, then there must be some connection with each card it touches. If no match can be made, the player forfeits his/her turn. First person to place all of his or her cards wins. Variation 2: Shuffle all of the cards together and deal out the first twelve cards onto the table in three rows of four cards. Players simultaneously look for pairs of cards that they can explain a connection between. If they see such a pair they say, "connection." As soon as someone says connection, play pauses, they show the two cards to everyone playing, and they explain the connection. If each other player can understand the connection, the player who found it gets to keep the two cards and two new cards are dealt to replace these. If at some point nobody can find a connection, deal out four extra cards, and play with these extras on the table; do not replace pairs until you are back to fewer than twelve on the table. Play until there are no more cards to deal and nobody can find any more connections. The person with the most cards at the end of the game wins. Variation 3: Same as variation two, but play to find sets of three cards that have one connection among them at a time.

Write a Book. Take time out to write a comic book or graphic novel, using words and pictures, to "teach" either younger students, or older students, about something you have recently learned or figured out. Pick the audience so that you have specific kinds of readers in mind. Both audiences present issues having to do with the content being something seemingly too hard or too easy for them, so the writing decisions are not so much about the mathematics, but about the representation of the ideas. Share with students samples of comics and graphic novels, some of the "beginner's guides to . . ." books (such as to post-modernism, Einstein, feminism), and look at children's books together to get ideas for how to use the pictures to help the reader think about the material, or to make the book more entertaining. Variation 2: write a story in which the mathematical idea is either the main decision in the story, or the punch line, but do your best to avoid making it sound like a word problem. Variation 3: Create a poster without any words allowed that shows someone how to create one mathematical object out of others (e.g., how to make a copy of a triangle using only strips of paper).

Math Talk. Students routinely write questions that they normally can't ask in math class on a large sheet of paper posted on the wall of the classroom. At regular intervals, perhaps once per week or every other week, the class holds a "math talk": the person who asked the question chosen facilitates discussion. The one rule is that you are not allowed to raise your hand when you want to speak. Instead, the group has to learn how to know when the person speaking has finished, and whether or not what they have to say is best said at that moment. Before and after math talks, the class should discuss what they have been learning over time about how best to speak, and what sorts of questions and comments are the best sorts of contributions to the talk. As discussed in Chapter 1, the teacher's notes on such classroom conversation make for excellent assessment material.

The strongest argument *against* using these exercises is that they take time away from the regular curriculum. But they really save time for the curriculum in the long run because they train students in ways to get a move on in thinking mathematically. Also, the content of these tasks is usually the curriculum itself, so you are not really pushing the curriculum aside when you are using them. One approach to finding time for these is to take just a small amount of class time, perhaps once per week, to work on them. A student can work on a book or students can play connections cards for ten minutes every Thursday, for example. For most of these exercises, the main focus is not on the mathematics content, even though a lot of mathematics is learned during them; attention is on ways of interacting, rules of the game, or ways of presenting the math content. This can help students to positively concentrate without feeling pressured to be mathematical—even though we know they are developing important skills of mathematicians.

Even these exercises may not grab a student's interest. One approach is to accept that this may be the case and just not worry about it. Over time, each student will find the best exercise for themselves. Another approach is to allow students to opt out of any of these exercise options if they have something else related to class that they would rather do; the downside of this is that a student may never learn to appreciate what the exercise has to offer, something that can take some time to become evident. Yet another option is to have students suggest their own ideas for games and exercises like this. This last option can be good except that it often leads to trivial games that they enjoyed in the past, like math jeopardy, which you may feel do not get at the essence of what is important mathematically. In the latter situation, you are back to Romagnano's dilemma of the good problems, where what you know can lead to important mathematics does not capture the students' attention, and where you feel like the class is avoiding the very important concepts of mathematics to be learned. Each of the exercises I suggest work well if they are leading to some public sharing of what the students have accomplished. If there is a date set when students need to perform a math improv, read from their book, or facilitate a math talk, then they have a form of external motivation to work on it to the point that their part of the event is ready for a public audience. Of course, there may still be students who are not motivated by a public performance. As we will discuss below, we can't control everything; all we can do is try a variety of responses and it is really surprising how often just trying leads to positive results.

Concretizing Academic Literacy

My next level of attention to students who are not learning, whether it be from lack of interest or something else, is to examine with them the skills of being a successful student. The field of academic literacy has helped me to understand the depth of knowledge that a student must have about the arts of being a student that some students simply have not had the opportunity to learn, and that others have not been ready to learn before they work with me as their teacher. The skills of being a student can be explicitly named and learned. What do readers do when they read a math book? What do teachers expect students to be thinking about and attending to when they are completing homework problems? When a teacher asks you to work in a small group, what is the teacher expecting each member of the group to do? I started thinking about these issues a great deal when I had the opportunity to visit the Academic Literacy Group at the University of Cape Town in South Africa. After Apartheid, as part of the "new" South Africa, UCT, a traditionally white institution, seized upon its commitment to being a school for the new democracy, and developed programs to support the success of all students, especially those new to the university from rural, previously non-white "homelands." Students admitted to the university scored well on matriculation exams, so they were clearly qualified for university study according to established criteria; yet they were often unprepared for some of the work at the university. Smart students do not always perform well, and the courses offered by the Academic Literacy Group can make a big difference. By extension, in my classes back in the United States, students who are not performing are, I should assume, often smart students who would benefit from academic literacy work; since there is no special academic literacy group at my school, I look for ways to build in such experiences as part of my regular classes.

In academic literacy studies, the teacher helps the students to examine the concrete ways of working in academic study, and to adopt these ways of working. For example, successful students often socialize with each other and make studying part of their socializing. They may go out for pizza together, and while they are eating talk about the homework problems, how they each did them, what is important to remember about them, and so on. Some very smart students work for extra hours studying by themselves, redoing the homework, testing themselves on things to remember from the textbook; this studying alone is useful, but the social studying is crucial to performing well in the end. So I need to help my students to make this way of studying part of their approach to school. It can be quite a challenge, since social time away from school for many students would only be ruined by talking about homework. But I can make the homework be social. "Tonight, you must telephone or IM at least two people and explain how you approached each of your assigned problems. Turn in tomorrow a print-out of the IM discussion, or a picture that you drew when the other person gave you directions over the phone." "For homework, decide with one other person what you will do in class tomorrow so that your group is ready for a presentation on Friday; just decide on how you will work together in class, you should not do the work until you come back." "Compare the methods your group invented with what two adults demonstrate as how they were taught or have figured out on their own." "By next Wednesday, meet with two people and come up with a list of those problems that you still can't do on your own. We will talk about them in class." I can also create social events centered around mathematics class—parties, field trips, and so on—where students have the chance to develop friendships with others in their class, so that it is easier for them to plan social events on their own.

Other aspects of academic literacy involve skills related to the languages that are used in the classroom and in the texts that are utilized. What goes into a complete explanation of what you were discussing in your group? Many students will benefit from role-playing and experimenting with possible responses to teacher assignments for in-class and outside work. If three volunteers discuss a math problem in front of the class, and then the class notes what they did well that helped the group to move toward a solution, or toward the identification of strategies they could use, then each group can try these group techniques out on their own. The point here is that performance is not always tied to conceptual understanding of the material. Also, not understanding the ways of working academically can interfere with producing what the teacher wants. I couldn't help wondering about this as I was reading Lew Romagnano's wonderful self-reflection on how so many fantastic ideas for exploration fell flat in his class. Would it have helped if he made the expectations explicit, instead of wondering why they were losing interest? We can only guess. But I personally believe that he might have been happier with the outcomes if he described the kinds of things that he wanted them to have ready after cutting paper and making boxes to fill with rice. I know why he didn't do this: he had good pedagogical reasons for thinking that too much information would spoil the surprise. And the surprise was the punch line for the investigation: there is a functional relationship between the length of the square you cut out of each corner of the paper before folding up the sides and the eventual volume that the box can hold. But in this case, the punch line and the purpose of the activity were too intertwined, so that holding back on the punch line meant holding back on the purpose, and students felt aimless.

I strongly suggest that you use "classic good problems" like this box volume problem as examples for academic literacy work in your class. I would say, "This next week we are going to work on understanding what you are expected to produce in this class and in other mathematics classes when you are presented with a certain kind of ambiguous and open-ended situation instead of a formulaic task. First, you have to find a way to collect enough data to form the basis of an analysis that could produce a pattern. You may find this aimless and you may feel like you are losing interest. This is only natural, and part of the work of being a mathematician. Sometimes the collection of enough special cases gets boring and you forget why you are doing it. But keep in mind what we are trying to find: a pattern or relationship, which means first a range of data that we can organize into a table." As the students worked, I would keep the purpose in the front of their minds: "Remember, we are collecting data to see if there is something worth exploring. I could tell you what to pay attention to but we are practicing what mathematicians do, which is figure out what to look for themselves. Let's not spend too much time redoing what everyone else is doing. If one group is trying small cuts for the squares, can another group try large cuts? What about in-between?" Fifteen or twenty minutes later: "Do we have enough data to start suggesting a pattern? If we want to make the biggest volume box, should we use small, large, or medium cuts for the squares? We can't go on to make these boxes for our packages that we're sending to our partner school in Managua until we figure out how we want to make the boxes." Rather than getting frustrated with students' disinterest I would just not let them fall into that state: "O.K., so after working for one day on this, what can we say that we know? What can we say that we still do not know? Remember, as mathematicians, we are expected to find a pattern, to write an equation if we can, that fits our data, so what might we do tomorrow in collecting data, to look for a pattern within the first fifteen minutes of class?" Then tomorrow: "Did our plan for today help us within our goal of fifteen minutes? Why? Or, Why not?" Now, in order to concretize the literacy skills: "Look at how Mr. Romagnano's class used their table in studying the coin toss game; can we use a table in the same way for our reports on the box problem, or do you think we should organize our data differently for this investigation?" If we put the table at the top of the page, how will we explain what we are saying to the reader who hasn't looked there yet? (Write a note to the reader to look at the next page?) How should we start a paragraph that is going to introduce an equation for the relationship between the length of the side of the square and the volume? How can we explain to the reader why this is so surprising, or otherwise interesting? What is useful about this result? Why should the reader care? Each of these things must be communicated in our report. And so on. Finally, because I find this important, I will ask students to identify several new questions that they want on some level to know the answer to, and to which they agree ahead of time to be willing to work on with others. I will then put them into groups based on their questions and ask them to pursue their own investigations. Before we go ahead with this, I will again orient them to the idea that interest is not always automatic, and that mathematicians use clever tricks to keep going and maintain their attention long enough until they get absorbed in their work. We will have regular, brief whole-class chats about what works for various members of the class in maintaining interest, creating interest, and sustaining interest, because, as we know, even when we care about the answer we sometimes find our mind wandering to other things. In this way, losing interest and finding a way back in are both part of important contributions to the class.

Codependency

This issue of academic literacy comes up in another form in Romagnano's book. He is discussing Ms. Curry's teaching of the coin toss game, where the goal is to collect percentages of wins and losses in order to make conclusions about the game. In this lesson, Ms. Curry is proceeding in ways similar to what I suggest above: she tells them they need to produce a chart together, and to move along with this task so that the interesting analysis can take place. She instructs them to copy the table into their notebooks, which I would say is what mathematicians do. Then Romagnano shares a compelling story with us. One student asks where in the notebook to put the table, and Ms. Curry tells her to turn the page and start a new page with the table. Romagnano reflects, "I wondered if it was possible to get students like ours to 'do mathematics,' when some of them needed to be told when to turn the page" (97). He marked this observation with an exclamation point in his notes, and brought it up with Ms. Curry during their next planning meeting directly after class. He admitted that this student was an extreme case, but decided this was an evocative moment in their teaching, given that students were always expecting clear instructions about what to do in order to complete assigned tasks. He would challenge students to "figure out the percentage of the time you won," and give "no instructions on how to do that initially, assuming they would know or would remember with a little prodding" (99). Instead, the students would become frustrated and this frustration would lead to other forms of resistance. In response, he would even more rigidly resist telling them what they wanted to know—exactly how to figure out the conversions. Ms. Curry decided to try a more directive approach and received much less tension from the class. But this does not give us an "answer" to what we should do as teachers. The struggle for all of us is in deciding what to tell directly and what to not tell, to ask them to figure out on their own. As Romagnano writes, "a big part of doing mathematics is struggling with what to do and why" (100). He wondered what he would have done if someone had asked him where in the notebook to put the table—probably ask *them*, "Where do *you* think it should go?"—leading, he suspected, to more frustration and distraction from the primary goal of analyzing the table. But here's my thought on this: what if Ms. Curry comes back to this question at a later time, outside of the task at hand, and makes it a central question for discussion. Perhaps the next day, before the regular work begins, we can ask the class to discuss where they usually put tables when they are about to use them, or where each person actually *did* put their table, and how the location made a difference? The students' seemingly distracting and seemingly disempowering questions can be made the focus of serious discussions of how work is done. The location of a table in one's notebook can make a difference, and needing to think about that can prevent you from moving ahead with your work. What are some tactics to use in making this decision in the future?

When we work with students who are used to being told what to do, clearly assigned tasks have a better chance of engaging our students. But, asks Romagnano, "in what?" (101). Much to the teacher's frustration, the students are better engaged, but they are engaged in simply following the outlined procedures, and in the process succeeding in not doing any mathematics at all. What a complicated thing it is to be a teacher! We want our students to figure something out for themselves, to make decisions on their own, to internalize skills of self-monitoring. But our students seem to be able to do this better when we carefully organize the tasks for them. Yet, when we think about this, organizing it for them means that they focus on the directions we give them, and following these directions; rather than growing as mathematicians, our students enlist us as accomplices in their efforts to resist learning mathematics. They enlist us as accomplices out of reasonable and responsible goals of being good students who can do what they are supposed to be learning how to do. We become "codependents" in their persistent resistance to learning because we, too, need to feel like they are learning. Every teacher goes through a phase where they decide to ride out the resistance. What if I just do not tell them what to do for the next week, then month, then two months? Some teachers cycle in and out of a self-promise not to give in to their students' demands for clear expectations, because these teachers recognize that learning requires confronting this resistance. When students say a certain teacher is a bad teacher, they may be describing a teacher who is working in this way. And they are in a sense right that this teacher is not doing their job, since, from these students' perspective, the teacher is not making it possible for them to succeed. From the teacher's standpoint, such teachers are doing exactly what is necessary for the students to truly learn.

Planning Questions Versus Directions

Facilitate small group investigations of a mathematics problem in two different ways:

a) Start by identifying the "big questions" that make the problem interesting from the perspective of how the problem illustrates an example of a key mathematical concept. Pose the problem for the group and encourage them to work together.

b) Work through a problem solving it using two different strategies. Then create a worksheet in which you give students step-by-step instructions for how to understand the two different ways of solving this problem.

Discuss with each group what they experienced and what they felt like during the group investigations. How did the worksheet ease or increase students' frustrations? How did the open-ended experience increase or decrease frustrations? Why do some people prefer one way of working to the other?

Note that it is not merely a matter of "learning styles" or preferences, but that different kinds of learning take place. Even if students like one way of working more than the other, the teacher may have good reasons for using the less desirable approach from the students' perspective. List reasons for requiring students to use their less preferred method in either case. Discuss how you would explain to students why you are doing this.

In Romagnano's story, Ms. Curry prepares for class by making sure she can anticipate every possible student question or difficulty, and by being able to break down problems into all of their parts for the students. Romagnano writes persuasively that when teaching becomes clear explanations of every step involved in completing tasks, then there is no longer any dilemma about what to tell students and what to ask them to figure out on their own. Providing unambiguous explanations of procedures reduces student anxiety and helps them to learn how to complete those tasks, leading to the possible conclusion that this is exactly what teachers should do. However, the best teaching may not be the easiest or clearest and may require a lot of self-doubt on the part of the teacher.

If a teacher believes that students need to make sense of mathematical ideas by constructing their own explanations for perceived problems, the instructional decisions become more problematic. Explaining the procedures for completing tasks places the focus on the tasks and not on the mathematical ideas embodied by them. Removing the ambiguity inherent in mathematical problems (as distinguished from mere exercises) makes students feel less uncomfortable, but then there is no longer anything conceptual to make sense of.

(Romagnano 1994: 112)

This is why I suggest the academic literacy approach: we need to turn some potentially open-ended experiences into exercises that illustrate how to do the ambiguous investigations and how to communicate about them.

We become codependents for our students because we too feel anxious about what is happening in the classroom, but over different things. We want our students to feel good, to participate, to learn. We know that students thrive on risk-taking, but only when they feel safe and know that they cannot fail. When their anxiety is reduced, so is our own. We also know that real learning involves confrontation with resistance, so that our attempts to avoid codependency themselves lead to codependency. It feels like a double-bind, as if there is no way out. We, too, are learning—about our teaching, which means that we too are confronting resistance if we are to learn. What are we doing to prevent our own learning? We need to look at ourselves and what we are doing whenever we are not learning how to help our students through resistance in order to learn. The moment of looking at our mechanisms of resistance is the moment of psychoanalytic progress. But of course we are always frustrated in our efforts to look at our own resistance, because, well, that's just the way psychoanalytic progress works: knowledge is the purloined letter, right there in front of us until we need to find it, suddenly lost when we know it is there (Chapter 2). Yet we can identify some pointers to finding it: when we feel angry, annoyed, or helpless as our students demand step-by-step instructions, or when they shut down when we want them to figure something out on their own, what exactly are our actions and behaviors? These actions and behaviors are perpetuating our own need to feel secure in our own knowledge of what is best for our students. As we dig in our heels or completely give in to their demands for instructions, we are codependents and they are resisters; but we can also say that *they* are codependents for us as *we* resist learning. We are using them in our quest for certainty, a certainty that is always doomed to be undermined.

Separate and Connected Knowing

Think about how you are reading this chapter. Have you been reading this with a critical eye, insisting that everything I have written be justified? Have you been examining every argument, playing the "doubting game," looking for flaws in my reasoning, considering how I might be misinterpreting the examples I provide, whether there might be alternative interpretations, whether I might be omitting evidence that would contradict my position? Or, have you been reading with a receptive eye, playing the "believing game"? If something I have written seems absurd, have you asked, "What are your arguments for such a silly view?" Or have you asked, "What do you see . . . give me the vision in your head . . . you're having an experience I don't have, so help me to have it?" The critical eye is what Blyth McIver Clinchy (1996) calls "separate" knowing, while the receptive eye is what she calls "connected knowing." In connected knowing, the focus is not on propositions and the validity of inferences but on experiences and ways of seeing. Instead of asking "Why do you think that?" a connected knower would ask, "What in your experience has led you to that point of view?" In separate knowing, you take an adversarial stance to ideas; in connected knowing you try to understand ideas. Most people are combinations of both kinds of knowing, of course; separate and connected knowers are stereotypes. But before we dismiss the distinction, it is worth noting that we approach teaching mostly from the perspective of a separate knower, that is, we approach theories about teaching as if it is a science of right and wrong techniques that we can challenge, even when the ideas seem intuitively appealing. We are quick to find fault with any suggested teaching strategy. I believe teaching can never be a science of correct techniques, because there will always be immediate examples of when a certain pedagogical strategy fails. People are far too complicated and do not control their actions in the ways that such theories would want them to.

We can use the distinction between separate and connected knowing to think about both the way we approach students who are not-learning, and as an academic literacy that our students can use to think about learning. As teachers we can actively shift from a stance in which we are applying techniques of teaching to the listening stance where we are trying to understand our students' experiences of learning. This involves both their experiences of being in a classroom, in which they are asked to learn mathematics, and the ways that they understand the mathematics itself. What do they think we are asking of them as students in terms of student behaviors, and what meaning are they making of the mathematics that they are using in the classroom? How are their interpretations of what it means to be a student affecting what they think the mathematics means? For our students, we have to make the differences between these two ways of understanding what they are being asked to do and understand explicit. Most of their prior experiences have required that they practice separate knowing, both about being a student in a classroom and about the mathematics they are learning. Connected knowing expects that they attempt to think from others' perspectives—the teacher and their peers. Connected knowing also asks them to listen to the mathematics. "How is this concept or this method of solving a problem changing the way I think about mathematics, knowledge, the world I live in? What are things I can do to get better at answering these questions?"

Both separate and connected knowing are procedures for obtaining information and developing understanding. Each creates a way of being a teacher or of being a student. In other words, both forms of figuring out what we know are "objective" in the sense of being oriented away from ourselves and toward the objects of our knowledge—our students, or the mathematics, or the experience of being in a classroom—in order to analyze or understand things. The image is altered from invading another mind toward opening up to receive another's experience into our own mind. When we read a book, we open our mind to the point where we see what the author was all about, see the issues of what the author was trying to say. You let the ideas pass into you and become part of yourself, rather than treating them as something outside of yourself. You search for connections between your own prior experiences and the experience of the person whose ideas you are trying to understand, not to better understand your own prior experiences at this point, but in order to better sympathize with the author of the ideas you are trying to understand (Belenky *et al.* 1986). Languages other than English make it easier to make the distinction between these two kinds of knowledge. German's *kennen*, French's *connaître*, and Greek's *gnosis* imply a personal acquaintance with an object (not always a person); understanding involves intimacy and equality between self and object. *Knowledge* implies a separation from the object and mastery over it, as in German's *wissen*,

French's *savior*, and Greek's *sabor*. Both teaching and learning mathematics have suffered from the knowledge orientation when they ignore the understanding orientation. We have been taught to seek mastery and control over both teaching and mathematics, as if they need to be tamed, and we have been taught to fear the ways that understanding can change us, make us into new people, when we allow the experiences to enter us. When we accept things in all of their openness and ambiguity, avoid evaluation in order to understand, we risk not knowing the results of how we ourselves will change. We will become a different teacher, or we will no longer see the world in the same way thanks to being changed by the mathematics. We no longer know ourselves?

A separate self experiences relationships—with our students, with mathematics—in terms of reciprocity, considering others as it wishes to be considered. The connected self experiences relationships as a response to others in their own terms. A separate self approaches teaching based on impersonal procedures for establishing justice. Separate-self teachers tend to make sure that everything is objectively "fair" in their classroom. But a connected self bases the structure of the classroom on "caring." Separate selves approach knowledge as a quest for truth; connected selves believe that truth emerges through care. Nel Noddings, a philosopher and mathematics educator at Stanford University in California, says that we can apply these notions to the intellectual world of ideas. "In the intellectual domain, our caring represents a quest for understanding" (Noddings 1984: 169). Relationships with ideas are harder to grasp than relationships with people, so it will take a lot of time and effort for both us and our students to act on Nodding's notion. The relationship between an idea and a person "seems doomed to be one-sided since an idea cannot reciprocate the care lavished upon it by a thinker" (Belenky *et al.* 1986: 102). But, as Noddings suggests, when we understand the idea, we feel that this object-other has responded to us; we hear it speak to us. We know "the joy attendant upon intimacy with an idea is not so different from the joy we feel in close relationships with friends" (Belenky *et al.* 1986: 102). Connected-self teachers embrace mathematics and model this for their students, accepting how understanding leads to new truths. They also develop procedures for gaining access to the knowledge of others in their classroom. Since knowledge comes in this orientation from experience, they value and validate the experiences of their students, and try their best to share other's ideas by sharing the experience that has led to the idea. They also recognize that their understanding is always and forever partial, since they will only gain limited access to the experiences of others; therefore, they must accept the tentativeness of what they know and understand at all times. Connected-self teachers make a strong effort to avoid judgments, and they request from their students that they, too, refrain from making judgments until they understand the topic at hand. At the same time, they encourage some students to make judgments more often, as they attempt to understand the origins of these students' passive stance lacking agency.

Connected knowing has the potential to lead more smoothly than separate knowing into a personal, integrated knowledge. Separate knowing has the more immediate feeling of power and control, a selfless knowledge. Yet it also leads sometimes to alienation when the knower is always separated from the objects of knowledge. Perhaps our attachment to separate knowing is part of why so many students find a lot of mathematics purposeless. They no longer feel any personal involvement in the pursuit of knowledge.

a) Go back to one of the investigations from the Prologue and work on it some more yourself. Can you identify acts of separate and connected knowing within your investigation? Try to make a list of problem solving strategies that fit into each; avoid placing a strategy in both unless you can describe it with two new names for when it fits into each type of knowing.

b) Design a mini-lesson for teaching the differences between separate and connected knowing to your students. What are the main conceptual goals for this lesson? How will you start it? What questions can you ask during it to make sure that students are comparing, analyzing, synthesizing, and evaluating during the lesson? How will you use the end of the lesson to segue into other topics?

They feel as though they are always answering other people's questions, and they have trouble making themselves care about the answers. Connected knowers, in contrast, are attached to the objects they seek to understand. This makes it easier for them to make the transition beyond purely procedural knowledge into a personalized, integrated knowing: instead of echoing the voices of others, connected knowers end up making their own personal experiences part of what they know, since they use these experiences to better understand the experiences of others. When one's own experiences are part of how one understands, one comes to new knowledge of those experiences and oneself; one is also better able to reflect on how those experiences have led you to become the person you are becoming.

As procedural knowledge, both separate and connected knowing have a lot in common. They both objectify knowledge even as they differ from each other. A more personal, integrated knowing requires the knower to move in and out of the variety of voices of knowing, separate and connected, authoritative and questioning, and to move in and out of oneself as the source and others as the source of knowing. Again, we may quickly say this is obvious, but if we dwell on it a bit we can see that this is saying more than what is apparent when we first hear it. Such knowing appreciates the tentativeness of all knowing, but places this tentativeness in contexts: What we see and understand depends on the context. Ideas and values, like children, must be nurtured, cared for, and placed in environments that help them grow. Integrated knowers seek the opportunities to make this happen. As teachers we need to understand what our students are experiencing, not necessarily the experiences as we have designed them and interpret them; we need to put students in spaces of learning that allow them to make their own place. As teachers and students, we need to understand mathematical ideas not as we see them, but as they demand of themselves, in order to place them in new contexts and see how they grow.

The Myth of Instructive Interaction

If only it were the case that things were simple. If only it really were the case, as so much of our actions assume, that our nervous systems receive an intact message directly from the world outside our bodies. Then the teacher could just tell the students what they need to know and there would be no misunderstanding or loss in fidelity. Adults and children would understand each other completely, and students who are having problems could drop them off for the teacher to solve, picking up their solutions a week later. "Instructive interaction," this belief in little packets of information directly transmitted in an efficient, computer-like manner, bit by bit, owes its origins to the scientist Claude Shannon and early work in the information management sciences, stemming from the design of telephone networks and later in the crafting of computers. It became a basic model of perception and communications transmission in general, where we imagine that neurons work this way, passing on bits of information, and that larger systems of communication also work in ways analogous to telephones and computers. What is put in, bit by bit, is passed along as communication; other than the noise that distracts us from the information, there should be no disagreement, no confusion, and no misunderstandings.

Yet we continue so much our lives as if this model of communication really is the way the world works, leading to many of the perplexities of interpersonal relationships.

> How could she have done that to me? Why don't they care? Who do they think I am? Didn't he know how I felt? What's the matter with them? Where are their heads at? These are the kinds of questions that tumble out of people's mouths when communication doesn't follow a lock-step progression – when what is "received" isn't necessarily what was "sent." If we understood that in communication what is "heard" is *never* what was sent – that each person's hearing is unique – then we might begin to take these everyday misinterpretations less personally. Each person is marching to a very private drummer, but thinks the drumbeat is loud enough for everyone else to hear.
>
> (Efran *et al.* 1990: 71)

The "other side" of connected knowing is to realize that the classroom we believe we are creating is not exactly the classroom that everyone else in the room is experiencing. I may have written the words you are reading right now, but you are not reading *my* book; you are using the words I have written to create your own book, the one that speaks to you. I control neither the way you interact with the pages (such as whether you go chapter by chapter, or skip around, whether you take notes or highlight passages, whether you tell someone else about what you have written, try out some of my suggestions, and so on) nor what you take from the book as understanding. If you disagree with something I write, you may have missed the point, and so, too, might those who agree completely. Writing and reading are conversations that include audience participation. This is sometimes referred to as "reader response theory." Classrooms are much the same. As a teacher you have a certain class activity in mind; what that activity is and what it means for each of the students in the room has a lot to do with what you do and what you plan, but is really only what the "reader," the student, makes of it. Each student is reading a different classroom. Efran and his colleagues (1990) describe life as analogous to a collection of people each writing their own scripts for a play that is performed in the same theater at the same time, using each other as cast members. There might be moments when the scenarios fit together well enough to create the impression that only one coherent story is being told. Indeed, from our own individual perspectives, we try to fabricate a coherence out of what transpires.

> However, since many divergent scripts are involved, sooner or later discrepancies in plot become painfully obvious. At this point, each producer-director-actor is apt to get huffy about how the others are sabotaging his or her masterpiece. Of course, that's not what they are doing. They are only trying to advance their own productions. Sometimes, bargains can be struck: writer-producer A agrees to incorporate part of B's third act, if producer B will reciprocate by amending his or her Scene 2. However, when each performer keeps insisting that the other plays are of inferior quality or – worse yet – doesn't realize that there *are* any other plays, bedlam results. Moreover, in the realm of relationships, rarely is there a recognized representative from the producer's guild or Actor's Equity to mediate disputes.
>
> (71)

The telephone metaphor for information is misleading because of the myth of instructive interaction. But it is also misleading because information is represented as coming from the outside in the form of "inputs." In fact, the nervous system is a closed neural network: it doesn't import information from outside but generates its own information (the word information is a combination of "formed" and "in"). People are brought up to believe they perceive the outside world. Our eyes supposedly provide direct and immediate access to our surroundings. However, while our eyelids open, the neurons of the retina do not. We attribute our experiences to an idea of light that is fashioned entirely within our own systems of creating information and meaning. That we are "fooled" into believing that we "see" the outside world dramatizes how well coupled we are with our environment. But because our nervous system is closed, it is incapable of distinguishing between perception and illusion. The exact contributions of the environment to internal changes of state cannot be discerned. Distinctions between reality and illusion are therefore social ones, not perceptual ones. It is pointless to strong-arm delusional individuals into giving up their reality in favor of our own. Such an approach creates bitterness, suspicion, and alienation. (It would be bad manners to go to a foreign country and chastise the natives for doing things oddly. Instead, we would watch and listen, attempting to find meaning in their customs and commonalities between their experiences and our own. We would work to bridge the cultural gap.) Sometimes the differences between our versions of the classroom or mathematics are so fundamentally different from a student's that we might want to label their version "delusional." But to tell them so directly would probably lead to fewer opportunities to work productively with the student.

Efran and his colleagues describe their work with a delusional client who thought others were inserting secret messages into her newspapers and broadcasting special thought-controlling signals over radio and television. The client needed her perspective validated rather than denied, yet this was something her therapists could not do in good conscience. She was angry at those psychiatrists who had attempted to convince her that she was talking nonsense. The key step was to recognize that this was not nonsense for her, no matter how disruptive and disturbing her behavior was. Fortunately, they all had seen a demonstration of the "Ames room illusion" at a local museum. This illusion is created by having a person peer through a window of a room at another person walking around inside; the person inside appears to change size as they walk back and forth. Observers looking down at the room from a catwalk overhead can see that the room isn't rectangular (as it appears through the window), but is actually trapezoidal. Because the walls are set at odd angles, they cause anyone seen against them to look like they are shrinking or growing depending on their location in the room. The delusional client recognized that a person who only had access to the top view would never understand what the person looking through the window was reporting, and vice versa. Each would be convinced of the validity of the perceptions produced by his or her frame of reference. She also understood that there are times when life is just like this—in fact, at some level, it is *always* like this. Two people are unable to confirm each other's perceptions, because things look very different from each one's perspective. They disagree not because they want to be a problem or create trouble but because they don't see any other option. In our classrooms, there is rarely a catwalk built above the "room" to help observers obtain a meta-view from which to mediate between the perspective of the non-learner and the teacher.

Efran and his colleagues describe a success story, in which their client begins to grasp that although her perceptions do not concur with what others believe, they are nevertheless legitimate. This led to her agreement to try medications and to read about schizophrenia. Up until that time, she responded angrily to anyone who hinted that she was "not in her right mind," and had assumed anyone trying to label her "mentally ill" as working to delegitimize her perceptions. I think non-learners may feel the same way when we try to label them as problems. Only by validating their own perceptions of what is happening, and simultaneously making it possible for them to see things "from the catwalk" can we move them into a place where they may be ready to try alternatives to their current actions. At the same time, teachers need to create "catwalks" of their own to help them see that students have different perceptions of what it means to learn in their classroom. However, the purpose of the catwalk is to recognize the particularities of your own perspective, not to obtain a "god's-eye view" that would help you discern the ultimate truth. Each story we tell about what we see ourselves, each explanation we give, and each purpose we attempt to identify is as much a part of the stories we are telling from our own perspectives as anything else. Teachers, students, parents, counselors, and others who meet to talk about a student who is not learning may want to label what is going on, assign an explanation, or prescribe a solution based on a notion of what is causing the situation. What they usually find is that things that help do not fall into simple patterns, nor do they necessarily "work" for the reasons they were chosen.

Box 8.4

Considering Cooperative Discipline

A popular theory of classroom management originally formulated by Linda Albert (1996, 2003) suggests that you can better respond to student behaviors by understanding the "need" or "motivation" behind the behavior and responding to that instead of the behavior itself. The following table includes common motivations/needs, the reaction that the teacher tends to feel at the moment, and suggested teacher responses.

Power	Feels alive only when controlling others	Teacher feels angry or manipulated	Provide power to choose between different options; avoid immediate power struggle; offer choice of participation or consequences
Retribution	Feels alive only when hurting others	Teacher feels threatened or anxious	Avoid conflict at the moment; discuss options with the student in private, including alternative motivations for participation
Attention	Feels alive only when others are directly attending	Teacher feels annoyed, pestered	Provide a lot of attention when it is not requested
Fear of Failure	Avoids working in order to avoid failure	Teacher feels helpless, at a loss	Provide small incremental tasks that can build confidence through success

a) Think of a management situation that you were involved in yourself as a student or teacher. Use the table above to decide on a positive approach to the situation.

b) Role-play situations in groups of threes: make small cards that have teacher goals, student goals, and math content; shuffle each stack of cards; the "teacher" looks at the teacher goal card and the math content card, the "student" looks at the student goal card; have a third observer watch what happens and offer their perspective after you are done role-playing. (e.g., the teacher may want the student to complete a homework assignment on ratio and proportion; the student may be seeking attention; the teacher may want the student to share a strategy with the class on how to factor a polynomial, while the student is seeking power.)

c) Theories are both enabling and constraining. This theory gives you ideas about what to do, but also moves you very quickly into a small amount of information about the student as proscribing what you might do in response. Discuss both how this theory is useful and how it might get in the way of productive interaction with some students.

Stories and Explanations; Causes and Treatments

So much of teacher education has prepared us to apply labels and categories to students and pedagogical solutions to particular learning "problems." But then we enter the classroom and little of this preparation helps us to figure out what to do. Mostly it "works." Students are able to get answers to the questions we pose for them, to use correct methods on tests or to apply explanations that we ourselves would use to make meaning out of given situations. But we are forever plagued with failure when we think about those students who are not learning. When we open ourselves to connected learning about our teaching and the lives of students in our classrooms, we have a chance to create places where even these not-learning students may find a space of learning. Yet we also open ourselves to the labels and explanations that may interfere as much as enable new solutions. People create "causes" and purposes by subdividing phenomena into parts, by trying to control the situation in order to manipulate it towards their own purposes. We want something to blame for what is happening, so that we can focus on that thing to blame. This makes sense to us in general because we interpret so much of our actions in terms of purposes that appear after the fact. "We work hard in order to be able to reap the benefits of our work." "We learn about common denominators in order to be able to add fractions." "We study mathematics or go to school in order to . . ." "We assign homework so that our students can practice skills." "We do homework in order to reinforce what we learned in class."

It is only natural that we would therefore assign causes and reasons for not-learning. Our explanations take on lives of their own as complete stories that rarely exist until we tell them, but which then propel further confirmations that they must have been accurate in the first place. "So many years of failure have caused me to be afraid that I will fail again; besides, if I show the teacher that I can do this simple problem, it will cause her to challenge me with problems that I will not be able to do." "This student needs attention and gets it by always preventing the class from having a discussion of the mathematics." "That other student needs attention and therefore always asks for step-by-step instructions because he won't get the attention if he figures it out on his own." According to Efran, "Explanations and stories help to sustain traditions; they orient and organize action. However, no matter how compelling they seem, they are still just stories" (Efran *et al.* 1990: 88). The trick is to use the stories without getting overly attached to them. There are always other stories that are told; they may not even have a basis in what actually happened, if they feel true; and if you don't like today's stories, there are always new ones coming along.

If not-learning is a problem, then it is established by a story regarding what is happening, what is causing what is happening, and what people need and desire. To be kept alive, problems have to be talked about. Psychologists suggest that this is why young children seem to get over problems so easily; they don't rely so heavily on language. As we get older, we linger over problems, conversing with ourselves repeatedly about every conceivable possibility and outcome until our entire conversational existence is problem dominated. We tell and retell the stories over and over again; there is no way to distinguish between the story and the problem. Problems can also disappear even though the circumstances associated with them remain essentially intact. The story has changed. At other times a problem might have resolved itself on its own if we had merely moved on and not dwelled on it. This all makes it even harder to know what to do.

As handy as labels for problems can be, they short-circuit our observations. They shut down detailed examination of the phenomena to which they refer. (Efran and his colleagues note this is why some artists refuse to attach titles to their paintings; they know it will prevent you from seeing the canvas.) A person labeled attractive or unattractive can no longer be seen for who he or she actually is. The moment we describe someone as "dyscalculia" or "math anxious" or "unable to process sequences of ideas" much of their behavior gets swept into a category and is no longer available for alternative appraisals. For example, I have often found myself reminding "math anxious" students that most everyone gets anxious over standardized tests. It is not a syndrome that is causing them to experience anxiety, but a routine part of everyday existence.

Another more subtle point is that the labels we use to create stories about ourselves and others are applied from *outside* (Efran *et al.* 1990: 90). We only use these special terms when we are describing others or when we are describing ourselves from another person's viewpoint. After a night of IM-ing, playing guitar, talking on the phone, and watching TV, a student can look back on what she has not accomplished on the math homework and label herself a procrastinator. But each individual thing she did was not an act of procrastination at the moment that it occurred. Even if she "knows" she is postponing her work when she reaches for the remote, it is only during that instance of self-appraisal that procrastination is born. When another student says he is bored and wants to be told what to do in class, he can only later use the teacher's word to describe himself as a "misbehaving distractor." When he says he is bored he is not shirking responsibility for his actions; it is the description applied after the behavior that makes it into misbehavior. Because of the potential confusion between the two domains of acting and observing the actions, both teachers and students are likely to make the mistake of construing a poor-performing student as having a "learning difficulty," another as "having a problem with authority" and yet another as "unable to think conceptually." The problem is not in forgetting how to do a problem, not being able to identify the key words in a word problem, or being unable to represent arithmetic patterns in the symbols of algebra, but in the realm of appraisals and comparisons that produce the simplified explanations and labels.

Suppose a student is so afraid of mathematics that they freeze every time they come into the classroom. She or he is not displaying abnormal behaviors but responding in the same way that anyone would act when under severe threat. Any supportive encouragement can do little to help her or him "let go," "take a risk," or "relax and focus." Can the teacher use a time when he or she felt similar fear as a resource for relating to this student? Imagine yourself attempting to let go of an airplane for your first parachute jump. The most helpful approach would be a full acknowledgment of the legitimacy of the person's fears. There is no such thing as irrational fear in such situations. Instead of having to deal with both the feared object, the mathematics or the mathematics classroom, and its associated issues that are part of the story (Why am I such a coward? Why does the teacher always put me on the spot in front of everyone else? Why are my parents making me take Algebra II?), all of the resources can be focused on the first, the mathematics and learning mathematics.

We want to give advice, but we really don't know what works. We share what worked for us when we needed to deal with a certain issue, but all we really know is the story, not exactly what worked. Our stories change over time, so that their connection to reality is only partial. We often don't even follow our own advice. We sometimes give advice that we have never tried ourselves. Advice tends to mirror the segmentation of the world into the realm of experts who are authorized to speak on special topics, but the "problem" may be biological, not social, or emotional, not content-based, or with the structure of interaction in the classroom, not the difficulty of the material. (Who actually studied for a math test using the techniques they now recommend?) I say this not to make it seem impossible to figure out what to do, but to help you realize that your job is not to figure out the causes and explanations so much as to make it possible for your students to shift their focus, to change their lenses from time to time, first using one level of analysis of what they experience and then moving onto a larger or smaller or different unit of examination. Each analysis of the circumstances opens up fresh opportunities for intervention. If they change the definitions of events, retell the stories using different languages, they can see and think about different things. Each story leads to new paths of discovery about oneself as a student, teacher, member of a family, and each lens can lead to new conclusions about what is happening. The teacher can make this a part of the learning experience in the classroom. One goal will be to help students see that purposes are usually invented after the fact as part of a story. A description is turned into a purpose which in turn is asked to account for the description. Stories are tautologies.

Behaviors seem neurotic or maladaptive because they are repeated. We want to find the cause of the repetition and do something to break the cycle. But human beings too easily repeat what they have already done. This is part of the story, and part of human beings as closed systems who make meaning internally, rather than through "instructive interaction." Whether a student cannot do the mathematics on their own or attempts to manipulate classroom activity for attention or power, they and we can tell a story later about why this happened. The point I want to make, though, is that the story constructs a belief in purpose or reason, yet is not necessarily right; in some cases, there may not be a reason or purpose, in other cases the reason or purpose may be unrelated to the story. Efran and his colleagues quote the psychologists Erhard and Gioscia:

> Paradoxically, the experience of helplessness or dominance results from the attempt to locate responsibility outside of self and sets up a closed system out of which it is sometimes very difficult to extricate a valid experience of self; since the self which might otherwise be responsible has been excluded in the attempt to protect it from guilt, shame, burden and fault.
>
> (in Efran et al. 1990: 109)

Unless we take the responsibility upon ourselves to avoid descriptions that lead to prescriptions, we run the risk of perpetuating an unproductive illusion; hidden in the descriptions are value judgments as much as objective analyses. When we tell the stories we avoid philosophical choices and turn people into "good" and "bad" students. Better is to construct with our students places where we and they together can confront learning as the subject, resistance as a necessary and essential part of learning, and the moment of searching for the resistance that always hides itself further as the method of learning. Diagnoses are not meaningless. They provide a framework for understanding. But they are often wielded as magic wands when they really just obscure and slow down new interpretations, new stories that lead to new action.

All too often our story about why students are unable or unwilling to do something assume that they do not know or do not understand. A fine example of such a story can be found in an article by Frances Van Dyke and Alexander White (2004), two professors of mathematics education at American University in Washington, D.C. When they gave students entering Calculus a test of their understanding of graphs and functional relationships, they found that students' knowledge of fundamental properties of graphs is in general very limited, so that using graphs effectively is impossible. The way they framed their assessment, they had to assume that the "problem," if there is one, is a lack of knowledge. Surely it would indeed be difficult for students to use graphs effectively to talk about functions if they do not have the skills to do so. But once we represent the issue to ourselves in such a way it is impossible for *us* to understand what is going on in the minds of our students in any terms other than a lack of knowledge and skills. These were all students who had taken pre-Calculus and were therefore presumably ready to study Calculus. So why do Van Dyke and White describe their test responses as a "reluctance" to use the techniques of visual interpretation of graphs? And why is these teachers' response to set up ways of pointing out and highlighting such techniques for students? When we describe "the problem" as a lack of knowledge or understanding, then we end up prescribing a solution to the problem that is very much like what has already been done: the solution to this particular problem is more teaching of the fundamental concepts of graphs and the visual interpretation of graphs. The solution is made up of "reminders and simple exercises that emphasize the . . . connection" (117) that students presumably are not making on their own. This article is valuable for its ideas on how equivalence and variation are at the heart of the meaning of "function," on making a strong connection between a graph and its equation, and for its examples of questions that can help a teacher to learn how well his or her students interpret graphs. It also nevertheless reminds us how easily we slip into a "deficit language" where we assume that reluctance is related primarily to missing knowledge. Given the tremendous amount of attention to graphs throughout the school curriculum, a more interesting inquiry might lead us to wonder what else other than a lack of experience with graphs would lead students to a reluctance to interpret them or to think of graphical interpretations as a method of mathematical understanding and problem solving.

Should we say that entering Calculus students in general have not learned enough about graphs? Similarly, should we say that entering third graders have not learned enough about place value? That entering sixth graders have not learned enough about fractions? That entering seventh and eighth graders have not learned enough about algebraic representations, or the order of operations? At every level of mathematics education, we find students "reluctant" to apply previously taught material in new situations. And our first response is to re-teach with exercises. From the students' point of view it is the same old stuff again, nothing new, and nothing interesting. And we say, "They still don't get it." Perhaps they never will until there is a reason to use "it" toward something that matters. Perhaps *they* would be demanding "its" use if they were in places where mathematical practice supported leadership, voice, and participation (see Chapter 7); that is, perhaps places that are grounded in desire, disparity, and difference would help us to think about resistance, reluctance, loss of interest, and other forms of not-learning in new ways.

If we use the ten questions from Susan Ohanian (Chapter 1) to reflect on how well things are going in our classroom, then we automatically identify students who are *not* learning as we had hoped. In other words, by constructing goals we automatically create students who are "left behind." Our expectations for what should happen in the class-room are defined and refined by noting students who are *not* doing what we expect. The ultimate irony is that most of our language for thinking about this very issue is designed to include the excluded students, to support the growth of students who have not yet learned what we had in mind for them, to enable students to become the people who they "should" or "could" be. Yet in the process of using this language, we create the very categories that label certain students as "excluded" or "included," a "successful problem solver" or a "non-learner," and so on. The differences between and among students congeal as types of students that we need to "deal with," which perpetuates and rigidifies our categories so that we see the world as if these categories were "truths" about reality.

So, you might ask, what is my point? I recommend Ohanian's questions for assess-ment, and then later in the book I worry about how their use is bad? I just want us to question our assumptions about what we think we are doing. When students are not learning, please ask yourself, according to what criteria and whose expectations about what learning is? Please recognize that the "truth" about a student is not in the label we apply based on some theory we know. Theories create the labels that define the stories we tell about our students. Some are "fast learners," some are "slow"; some are "critical thinkers," some seem to resist posing questions for themselves. Our practices create our modes of understanding and describing the learners in our classroom. No student is "not-learning" in general; not-learning is particular to what a teacher is looking for. Every student is learning *something* at the moment; it just might be that they are learning that they can't do word problems, or that they are very good at distracting authorities like teachers from their goals. For other students, not-learning has more to do with a resistance to understanding themselves as learners or with not appreciating that another person's idea may have something to offer them. Teachers run the risk of not learning themselves, when they fall into routines that do not challenge them to reflect on what is happening in their classroom, or when they apply readily available labels to their students without considering the specific context in which the not-learning is taking place.

Response to Chapter 8:
We Are All Students

David Scott Allen

This chapter hits very close to home right now, and will probably do so as long as I am teaching. Even the best students with the best teachers want control of their time. Even if we assume that all people are mathematicians, we must inevitably conclude that all mathematicians do not mathematize the same things because not all people enjoy the mathematization of previously designated mathematical objects. There are so many objects to mathematize in life that one lifetime is not enough to explore them all. We must choose with care what we pursue, because the way we spend our time is the substance of our lives. You can spend time spending money, or choose to hoard it, but money can't buy time. Time is the ultimate exhaustible resource. The truth is that mathematics as we teach it is a small fraction of the essence of its usefulness in the lives of real people. Any field worth pursuing has many fine choices within it to make. How does one go about choosing the right size nail, the right type of yarn, the right friends? The often bewildering array of choices in today's world, coupled with the proliferation of more and more sectored opportunities to make those choices, leads us to pause if not ponder who we are, what we believe and why we believe it. This set of "givens" then will drive the human spirit to be the person defined by that future. So I am not surprised by Gus' choice to opt out of the opportunities at hand, even if he seemed to be a math natural. You may be surprised, however, to know that I am not troubled by it either. He already had a purpose, and perhaps his conversations with Peter will yet return to mean something to him in the future. It is our responsibility as teachers to reflect back to students where they stand in the grand scheme of things with respect to our discipline. Mathematization is a part of playing with a band and painting just as much as it is a part of writing essays on infinity. Gus must pursue who he is in this way just as much as I must pursue who I am by writing. Everyone must become who they are.

Having said that let me turn more directly to the subject at hand through my everyday experience. In conjunction with the above statements, I am deeply troubled by the number of twelve year olds who mentally opt out of school. I can see why they do it. Not all teachers are engaging and interesting every day, including myself. But students this young have no idea what life is going to hold, nor what they will need to learn in order to get and keep a job and do it well. To put them in the position of deciding such things for themselves would be irresponsible on my part. I used to take much of this very personally, because I came from a modernist home where responsibility was laid down and enforced. I learned my lessons well, but don't have as much savvy with transferring it to others as I could have obtained under a more flexible rule.

What I can do as a teacher is make sure that it is clear what the short term and long term consequences are for doing or not doing x, y, or z, then informing the student and the parents about this. At some point they need to take up the work and learn. There is nothing that I teach which is so difficult that my students cannot understand it and do it.

It is more a matter of them choosing to do so or not. If I am doing more work than my students in trying to get them an education, then shame on me. There are some students whom I need to keep from others, and some I need to try to save from themselves, but it's not always my choice in the end. I can provide the context and environment for ownership of the material at hand, and maybe a little bit more, but I can't make them open a brain cell. I can only encourage and direct. And maybe inspire a little. So while I believe that we should not curve students' grades to fit the mold that there will be As and Fs and everything in between, I also believe that there could be all As in one class just as easily as there could be all Fs. I have had a range of success over the years at both ends.

I'm not convinced that all learning is good, but I agree that learning disrupts the old self. The process of maturity is painful, but the cost of staying immature is higher. Students do choose resistance just to choose the opposite of the expected original value. This is a good reason to instruct students in such a way that they themselves drive the learning. If we give them less to resist, they must fill the void created by a lack of direction. Choosing resistance can become addictive in itself, until it breaks the confines of the classroom and finds itself being acted out in the real world, where people get arrested for "just joking around"-type behaviors. People are not being independent by choosing the opposite of the rule, because then the opposite of the rule becomes the new rule. Anarchy comes from all people making intelligent, informed, and truly independent choices, with no regard to anyone else. So laws really are for social reasons. So it is in the classroom. We do have to get along as a group. The rules should be to promote the social structure rather than to superimpose a given set of behavioral norms for the sake of the norms. If we tell students what to do less often, they must choose for themselves what to do more often. If we present them opportunities which intrigue them, they might fill the void with those opportunities, and in the process shift who they are and who they will become.

I don't want my students to become parrots or photocopies of me or my techniques and ideas, but I would take that over them choosing other activities, thinking they are getting one over on the system. Thankfully, my building does not feel threatened by students who do not choose to succeed. Though 90 percent of the teachers at my school want to do so much more than go through the motions, some students sadly force the teachers to just focus on the State Test scores so that the school doesn't look bad, and at the age of fifteen, the student will leave no more mature than when they entered at twelve. There are so many changes occurring at this stage of their lives that it is difficult for some students to keep up. This brings me again back to making choices based on resolved values, not on choosing the opposite of what is given by authority or peers.

No matter how we deliver or package our learning, or what exactly is encompassed by that learning, there is a certain amount of it that is crystallized and stable. Less of mathematics actually is this way than students think, but the perception is that most of it is this way. This can be threatening, and drive students inward for answers to deep questions. So we need an outlet for the resistance. I have a student now who is perplexing me along these lines. This is a student who is in an advanced eighth grade class and is a nice kid. I want him to come out for the Math Team, and he would make the official competition team easily. He is also the type of person whom I would want to represent me and my school. But he won't come out to practice. I ask him why, and he says that he isn't getting a lot of the problems right. I remind him that in this competition, even a score of fifty percent is considered extremely good. He shrugs and indicates it doesn't matter. So it seems like he has defined success in a certain narrow way, and if he cannot succeed according to that definition, he does not want to try. School needs to be a place where learning is success, and where not learning for a time before breakthrough is normal. There are times in my classes when I hear everyone mumbling "I don't get this!" and the like, but I always stop them and ask if they care. I know they don't know how to do certain things or they wouldn't be on my class list, but moaning about it won't help them learn it. There is a lot of parenting in teaching.

Just as alarming to me are the students who are genuinely bored by anything at all. Has this generation seen so much so soon that nothing is interesting anymore? We are producing a race of zombies, no better than wild animals which spend their days looking for food (social interaction) and shelter (a place to get away). We as teachers are called to show students that they are capable of better than this. These students can be workable, unless they are using the guise of boredom to achieve the noncompliance of resistance or the semblance of cool. So math class is indeed about good problems. As we have discussed, they are neither self-evident nor easy to find or construct. Good problems are desirable to be completed as tasks, and are attractive at a social, intellectual, and/or cultural level. They must elicit more questions than they answer. This is truly counter to students' experience. They must learn how to act all over again, including interaction with students and teacher and mathematics.

The idea of disengagement being part of the process is intriguing. I can imagine some very interesting circular conversations with students about their non-learning. But if I assume that whatever they are doing is attempting to learn, the student must eventually learn something, even if it's not something on my list. They must be confronted first by themselves and their behaviors and attitudes, but not by me: Is that really a question you find interesting? Go for it! One really must just press on when a stop is at hand. And on top of this, the idea that students are doing different things at the same time can be uncomfortable for some teachers. I haven't found a good way to implement this, but I am close. It would be almost comical to challenge some of my students to do some of these things just to break out of mathematician's block, especially if they are just not interested in doing anything. Past, present, and future work is the question from me at every point. How could my students write a book to illustrate a concept? I have many good artists, could they use their skills to produce the next great textbook?

In Peter's classes, we bring snacks. Why can't I do this in my classes? Fear that it would lead to chaos? Some days I have chaos anyway. What if we set it up so that you must bring a snack but you may not eat your own snack? One of the key aspects of the students' transition from elementary to middle school is organizational mastery of both time and materials. Teaching this is essential for the students' survival at higher and higher levels as the road gets harder. Technology integration is a key here with today's tech-pervasive generation. My students also ask where to put things in my room, from found items to completed work. I ask them on the spot to pause and decide what would make sense. Similarly, our students are masters at getting us to help them in their efforts to resist our efforts to teach them.

I have also been accused of not teaching. Again, I used to take offense at this and point out what I in fact did do, but now I accuse them of not learning, a far greater offense.

There is a certain point where I cannot teach a student the distributive property, he or she must learn it. I can give geometric visualizations, I can use constants and variables, I can give repetitive exercises, but until a student gets the connection between $3(x + 2)$ and $3x + 6$, they can go no further in that direction. They must be able to (forgive my sensory metaphor) see it for themselves. But this is not teaching; this is conceptualization of a new idea. This is a metaphor unique to mathematics, like a mathematical currency.

There is a fine line between separate and connected knowing. In law, if you use evidence to win a case, then you have proven your point of view. In mathematics, the proof depends on the conventions of reason. But it is only a short but critical leap between the evidence and the conventions. And in some cases, the evidence can persuade a proof of an incorrect assumption or goal. So it is with mathematics. Mathematicians can construct proofs only with reference to a system that exists outside of their evidence. So separate knowing focuses on the proof, but connected knowing focuses on the evidence.

The teacher does not have control and the student does not have mastery. It is all an ongoing process by which we achieve definitional and functional familiarity with the objects of mathematics. Evidence of this can take many forms. But proof can only take the form of something within a given context. As mentioned above, we cannot move the student to connected knowing by ourselves. They must choose to go there. Until then, we can only assume that they are trying or that they are stuck. We need to get better at using their metaphors for objects rather than trying to convince them that ours are appropriate and comprehensible, and then explaining what they are and then making sure that they understand them the same way we do. In other words, we need a metaphor in order to package our perceptions. Trying to open two packages is more difficult than only opening one. And while insight and experience is good, every student is an individual. This is why we go to work every day. This is why we have our own children.

Our students' problems are hard to separate from the story behind them. My perception of what my students are trying to see also enters in. In other words, if I have a label for what is happening, then I might just try to find evidence for my proof, when in fact the evidence would more naturally lead to a different conclusion. This is what we could call self-fulfilling prophecy syndrome. We named it, so now we see it that way.

I am going to introduce an Anything Project this year. The students must take a topic about which they are passionate and mathematize it. They must go and find a way to analyze it mathematically, and also find out how people inside the discipline analyze it in any way. Comparing their ideas to the professionals' is important. So is sharing them with the group.

It is said that you are what you repeatedly do. I would argue that for middle school students and for teachers of mathematics, you are becoming what you repeatedly do. There is time to change, but by definition things will be different afterward, and so it will feel scary for a while. At a time when there is already enough change going on internally, my students need a safe place to experiment with their possible futures. Not necessarily mathematics, but the mathematization of mathematizers, which they are. One of my students called me a "professional mather doer" as a reason for why I could do something and she couldn't. I told her that the only difference between the two of us was that I had repeatedly practiced the mathematics which she was now learning until it was internalized by the process. I owned it, but I was willing to share.

epilogue
Becoming a Teacher and Changing with Mathematics

Teaching is an *ethical stance* one takes with the world (Block 2004). "Ethics" is explicit reflection on moral beliefs and practices. The difference between "ethics" and "morality" is similar to the difference between musicology and music: ethics is a conscious stepping back and thinking critically about morality, just as musicology is a conscious reflection on music. As I write this chapter, I am listening to *Lost and Safe* by The Books. I could listen to the album, I could write a review of it, or I could create my own album. Another option, of course, is to have the music playing in the background as I attend to something else. Each of these actions with their accompanying uses of the work of art, the music, might be an analogy for teaching mathematics. Listening to and creating music are two forms of imaginative, interactive action that foster the development of relationships between people. They are useful images when applied to teaching and learning—for example, we could ask ourselves when teaching and/or learning may be a form of interactive listening or artistic creation, ideas we discussed back in Chapters 2 and 3; we should further explore ways to make teaching and learning more like this, and in the long run better able to foster the development of relationships among people. Music playing in the background is a less positive metaphor. I do not look for ways to make teaching or learning the background for something seemingly more important. This book has been about the opposite kind of goal, about foregrounding mathematics teaching and learning.

Listening to and responding to the call from the stranger, "with reverence," is how Hongyu Wang, a curriculum theorist and teacher educator at Oklahoma State University, describes education (Wang 2002). She draws on the work of Dwayne Huebner to directly question the "tyranny . . . of the language of purpose (not experiences) and learning (not understanding)," and to suggest "alternative curricular languages" (289). We can use Wang's ideas to think about the mathematics curriculum, which is dominated by sequences of objectives, rather than provocative or engaging experiences, and learning goals, rather than understanding. Surely we can accept school, district, and state standards as one guide for our work. As a minimum barometer, I personally welcome lists of content objectives, process goals, and so on. I can use them as part of my overall assessment of my and my students' mathematical work. Yet, as Wang points out, education is rarely a "spiritual journey" when the primary language that we use to talk about it involves the controllable calculation of objectives that obscure the newness embedded in lived experience. Furthermore, if we use languages of learning that privilege observable (changes in) behavior, we cloud students' understanding of who they are, what they can do, and what they might become. This echoes the ideas of Stephen Brown discussed in Chapter 3, regarding the potential of mathematics posing and solving experiences to help mathematicians learn more about themselves and the kind of mathematician they are or hope to become. For Huebner and Wang, "the language of purpose and learning locks out the uncertain, the unknown, and the mysterious, which, in turn, drains out the spirit of life and puts everything under control" (Wang 2002: 289). When everything is under control, change is inhibited, including evolving relationships among mathematicians in the classroom community.

An ethical stance calls us to confront our fantasies of control, our dreams that we will be able to predict what will happen in our lesson plans, our desire to carry out educational processes according to predetermined procedures. An ethical stance further seeks to work with the "potentiality of the present" so that the "transcendent can be illuminated" (Wang 2002: 290).

David Jardine (2006) turns to the philosopher Hans-Georg Gadamer to make a similar point. Living in our current social and cultural age we have come to expect matters at hand to become matters of method. For example, "methods" of teaching particular mathematical content are marketed and sold for their efficiency and likelihood to lead to increased test scores; at the same time,

> isolated, anonymous, disembodied, clear and distinct, methodologically reproducible and assessable math facts become understood as more "basic" than the troublesome, toiling, ongoing, irreproducible, ambiguous, highly personal, and bodily engaging conversations we might have with children and colleagues about living mathematical relations.
>
> (275)

We are living out a logic of fragmentation and isolation that is centuries old and that is being worked out in our own lifetime, a logic attributed by Gadamer and others to René Descartes, the towering genius of the sixteenth and seventeenth century who left us with powerful ideas about logical thinking, Cartesian coordinate systems that link algebra and geometry, and, to the chagrin of people like Gadamer and Jardine, the reduction of thinking to a limited form of logic.

We have seen this suspicion of Cartesian systems applied to social spaces before, in Chapter 6; there we wondered about the efficacy of dimensions of social location (race, gender, class, ethnicity, age, religion, etc.). We considered the possibility that there might be more useful, "nomadic concepts" that cut across these dimensions—such as youth leadership, voice and participation—that is, useful for our current pedagogical purposes. Under the hood of Cartesianism, writes Jardine, "living conversations blur and despoil and contaminate and desecrate what is in fact objective and certain and self-contained"

(275). Gadamer sees all of this as wrapped up in modernism; he advocates placing what might have seemed isolated "math facts" back into the sustaining relationship that make them what they are, that keep them sane, and that make them rich and memorable. There is a recent development of serious consideration of mathematics in the post-modern moment which an ethical stance requires us to familiarize ourselves with, if only because it would be unconscionable to ignore something that might have implications for our work (Rotman 2000; Howell and Bradley 2001; Walshaw 2004). The idea is to replace the methodological with the substantial,

> full of smells and names and faces and kin, full of ancestral roots and ongoing conversations and old wisdoms and new, fresh deliberateness and audacity and life. It is also necessarily and unavoidably multifarious, contentious, ongoing, intergenerational, and unable to be foreclosed with any certainty because, for example, as a *living* discipline, mathematics endures. Therefore, . . . "understanding mathematics" means going to this living place and getting in on the living conversation that constitutes its being furthered.
>
> (Jardine 2006: 275)

Understanding is *not* method. That is, understanding is not a set of steps toward an answer. Understanding is not a recipe to follow. Understanding is not "knowing the right thing to do in a given situation." It is *not* decoding keywords in a word problem that dictate the mathematical operations to use to get the answer; *not* memorizing "FOIL" in order to multiply polynomials correctly; *not* regurgitating that "d for down" means the "denominator is down below"; *not* spitting out multiplication addition facts; *not* lining up the places neatly in a long division problem . . . Understanding is: asking yourself when 2 + 2 *might not be 4*, and finding an example or two of this. Understanding is: turning a simple calculation into an exploration of patterns. Understanding is "learning to dwell in the presence of this river edge, or learning to dwell in the presence of Pythagorean proportionality, and, under such witness, becoming someone because of it" (Jardine 2006: 275). In this sense, looking back on Chapters 6 and 7, dwelling in the actual places that people create will help us understand more about social dimensions such as race, class, ethnicity, gender, and so on, than these dimensions will help us make decisions in our classroom. But even this distinction has its problems for mathematics education, as does any dualism someone might set up; there are times when recipes lead to understanding, even if following them does not require understanding; there are times when dwelling in the experience entails serious practice in skills and procedural application. An ethical stance requires the post-modern move away from dualisms toward nomadic concepts that coexist with the modernist distinctions even as they enable other forms of action.

So, on the one hand, teaching as an ethical stance means stepping back and consciously thinking critically about the educational encounters that are ongoing in one's classroom. On the other hand, to step back and remain outside of the educational situation at hand would make it impossible to dwell in the moments of meaning and relation-building. We create a "critical distance"—like Descartes—in order to ground our logical conclusions in certainty. In doing so, however, we are taking ourselves outside of the situation, changing it so that we are no longer dwelling in it, and so that we are no longer understanding the situation as it is. So we need a post-modern update of the ethical stance itself. We are comfortable and secure in our clutching of control from outside even as we jump into the moment and lose ourselves in the muddled, delightful, challenging, and exhilarating confusion of it all. Some people criticize post-modernists for what is perceived to be a lack of commitment: everything is up for grabs, while the thinker looks at the ways of looking rather than simply looking and acting. Yet, teachers do experience each and every moment as truly up for grabs; we rarely can predict what will happen ahead of time, even though policy and standards have directed most of us to act as if we know precisely what we should and should not be doing.

The Books capture this feeling on their latest album, *Lost and Safe*, in "Smells Like Content."

> Balance, Repetition
> Composition, Mirrors
> most of all the world is a place
> where parts of wholes are described
> within an overarching paradigm of clarity
> and accuracy
> the context of which makes possible
> an underlying sense of the way it all fits together
> despite our collective tendency not to conceive of it as such
> but then again
> the world without end
> is a place where souls are combined
> but with an overbearing feeling of disparity,
> disorderliness
> to ignore is impossible
> without getting oneself
> into all kinds of trouble
> despite one's best intentions
> not to get entangled with it so much

"Most of the world is a place where parts of wholes are described within an overarching paradigm of clarity and accuracy." This is the dream of method, the fantasy of every student teacher: that there is "an underlying sense of the way it all fits together," despite our collective experience in classrooms that do not fit this dream. The world of the (mathematics) classroom is "a place where souls are combined," where disorderliness reigns, and where ignoring the complex unpredictable character of the experience is "impossible without getting oneself into all kinds of trouble despite one's best intentions not to get entangled with it so much."

Habituation, Conceptual Construction, and Enculturation

It's not just the nature of teaching and learning that makes the tangle of mathematics education. It's also the hidden conflicts in our professional field that underlie people's assumptions and hopes. The *Standards* movement and the increasing awareness of the National Council of Teachers of Mathematics of the importance of a unified position on reform for influencing government agencies' and the broader public's understanding of mathematics education have resulted in a vague acceptance of the *Principles and Standards* (NCTM 2000) and the *Curriculum Focal Points for Prekindergarten Through Grade 8 Mathematics* (NCTM 2006). Indeed, the subtitle of the *Focal Points* is "a quest for coherence," in response to the apparent lack of consistency and even logical sense in the mathematical topics covered in the first years of schooling across schools, states, and curriculum materials.

> With current mathematics curricula, students are expected to become acquainted with a wide range of topics in a short period of time, keeping them from developing deep mathematical understanding and connections. These curricula typically include long lists of concepts and skills at each grade level but never answer the question, "What are the key mathematical ideas or topics on which the others build?" The focal points are an example of how to answer this question. NCTM is offering a focused framework to guide states and school districts as they design and organize the next revisions of their expectations, standards, curriculum, and assessment programs.
>
> (NCTM 2007)

The Council is wearing the hood of Cartesianism, advocating purpose and learning. The pull toward clarity and accuracy is dramatic. Perhaps the *Standards* and the *Focal Points* have made it easier for the mathematics education community to affect policy. I am not sure, though. For example, the Office of Educational Research and Improvement once profiled exemplary and promising mathematics and science curricula, a number of which were created through National Science Foundation grants. In 1990, twenty-nine mathematics programs were highlighted by the Office of Educational Research and Improvement, and available in a free report through the ERIC Clearinghouse for Science and Mathematics Education (Mizer 1990); the programs were featured on the Department of Education website. When the winds of change swept a different president and legislators into Washington, these programs were no longer featured on the website, and the ERIC document was no longer available to the public, even though the featured programs continued to be consistent with the recommendations of the National Council. It was not the clarity of the Council's message that mattered at this point in history, but the content; a very different policy—associated in our mind for a long time to come with the No Child Left Behind Act—now sought to promote curriculum materials that were *not* designed with the clear and accurate NCTM message in mind. The hood of Cartesianism is so heavy, though, that the response from the Council, our professional association, representing mathematics educators in North America, was the *Focal Points*, as if the only explanation for this sudden policy shift must be a lack of clarity—where clarity had nonetheless reigned supreme. The dream, I suppose, is that an even clearer message that accurately details the topics that should be covered by schools up through eighth grade (and there are plans for a similar document directed at the high school level, likely to be available to you by the time you are reading this . . .) can help us all avoid getting entangled in the mess of politics and public rhetoric. I fear that, "despite our best intentions," the disorderliness of both the public sphere and the classroom community will remain at the center of our professional experience.

Despite the "disorderliness," the mathematics education community continues to act in general as if "reform" is a coherent, common goal; the "reform movement" in mathematics education is treated as a real and tangible object that we all understand, even though it is nothing more than an idea. Treating something abstract as if it existed as a real and tangible object is called "reification." David Kirshner (2002) has detailed both the reification of our "unitary reform vision," as well as the "crossdisciplinarity" of discrete notions of learning that underlie the conflicts among mathematics educators today. The conflicts are hard to describe, because we all espouse a commitment to either "reform mathematics" or its deterrence. At least, this is the story that we hear; even if we see ourselves as thinking in more nuanced ways, others apparently place us into one camp or the other. In the press, this is known as the "math wars."

The Learning Principle articulated in the *Principles and Standards for School Mathematics* (NCTM 2000) reemphasizes the well known distinction between "procedures" and "understanding" that characterize the supposed camps of traditional and reform approaches to mathematics; this echoes Huebner's "purpose versus spirit" as well as Gadamer's "method versus engaging conversation." The dualism keeps popping up! In the context of our profession, the "math wars" polarize while forcing any efforts at mathematics education reform to *conform* to the distinction set up by the Learning Principle. The "reform vision" is in this way reduced to a slogan, "understanding over (meaningless) procedures." Slogans remove personal agency from action. As Kirshner points out, the creation of the "vision" has meant all too often that teachers see themselves as implementing it rather than authoring it themselves. Sadly, that is, teachers are de-professionalized by the existence of this "vision." Their job is simply to conform to it, rather than to experience teaching and learning as the ongoing creation of their own vision.

The vision of the *Standards*, clear and accurate in its simplicity, obscures a range of languages that we might choose to use in describing the learning of the Learning Principle. It turns out that we need to get entangled in the disorderliness of the classroom if we want to take an ethical stance and really think about what learning "is." Kirshner suggests we try out at least three languages, each of which is based on a metaphor for learning. The first is *habituation*, informing behaviorist and information processing theories; the second is *conceptual construction*, informing constructivist learning theories that have developed from Piaget and so on; the third is *enculturation*, informing socio-cultural theories. Habituation is based on the idea that repeated practice of routine problems gradually leads to consistency between performance and task constraints. A curriculum based primarily on the metaphor of habituation is usually arranged topically with slow incremental coaching on homogeneous exercises for simultaneous consolidation of skills and concepts; teacher explanation of principles is reduced in favor of repetition over time.

In general, a constructivist teacher engages students in activities or tasks designed to cause perturbations in their current structures of knowledge, leading to conceptual restructuring. Such a teacher strives to obtain a model of the student's knowledge, and to apply theories that promote restructuring. Kirshner raises two interesting aspects of the constructivist classroom: the need for a trusting relationship between the teacher and student, in order to make it possible for the student to engage fully and deeply in the *teacher's* agenda; and the related need for a strong hierarchical relationship, in which the teacher often has to exert his or her power in order to direct a student in particular ways, for the restructuring to occur (Kirshner 2002: 52).

In contrast to habituation and constructivism, enculturation is the process of acquiring cultural dispositions through enmeshment in a culture. Kirshner defines dispositions broadly, as inclinations to engage—with people, problems, artifacts, or oneself—in particular (cultural) ways.

> Enculturationist teaching involves identifying a target culture and target dispositions within that culture, and working gradually to shape the classroom microculture so that it comes to more closely resemble the target culture with respect to the target dispositions. Students "learn" (in this sense) from their participation in the cultural milieu of the classroom rather than from other students or the teacher per se.
>
> (Kirshner 2002: 53)

Enculturation has a lot in common with the ideas of "apprenticeship" and "working as a mathematician" that have come up throughout this book. In general for enculturation, the reference culture (to use Kirshner's language) would be a mathematical culture; hence the idea of working as a mathematician. However, we also saw in Chapter 3 that working as a scientist may be more pedagogically effective for learning mathematics than working specifically as a mathematician. And, in Chapter 6, we noted the power of metaphorical play in the ways that people work in a classroom, leading to a variety of cultural analogies that can inform our planning (pedagogical design): the mathematician's work may be more analogous to that of a video-game player, roller-coaster-rider, or midwife than that of a typical "mathematician."

Untangling the constructivist and enculturationist strands of supposed agreement on "understanding" within the mathematics education community has, according to Kirshner, been a major challenge. Few teachers, administrators, parents, community members—even research professors in mathematics education—present themselves as aware of the muddled confusion. It is unusual to find educators targeting mathematical dispositions through development of a classroom microculture. I suggest we can use the "crossdisciplinarity" Kirshner describes to move through a rhetorical debate about learning toward the ethical stance: you need to examine your own values and beliefs, and reflect on where these values and beliefs came from, in order to think about the underlying premises of the activities you are wielding in the classroom. I use the word "wielding" on purpose, because, if you think about it, your actions are constant salvos in the "math wars."

Let's begin with the three types of learning. Kirshner's example is the typical secondary school mathematics lesson: review homework, lecture on incrementally new material, provide practice. Critics of "traditional" mathematics education, that is, advocates of "reform," would be expected to trash this type of lesson as merely serving students' acquisition of facts and skills. Yet, the lecture component of a lesson is often conceptual, at least in the teacher's mind, whether effective or not; and the grouping of the practice problems usually serves to support both conceptual and skill development. Stereotyped "critics of reform" are "right" when they note that practice problems do more than drill on skill. Crossdisciplinarity allows you to consider how each activity is grounded in habituation, constructivism, and enculturation at the same time. It also enables you to consider how these theories conflict with each other, and may undermine each other. You can't just say, I'll do a little of each, so that I get the best of all possible worlds. The practice problems may promote conceptual restructuring, but only if used in a constructivist way, which is unlike the coaching on drills that habituation requires. The hierarchy of teacher over student so essential for constructivist restructuring may counter the microculture of mathematicians necessary for the enculturation of particular dispositions, as well as the ally status of a coach at the heart of successful habituation.

Youth Culture and the Smell of Content

Our discussion of leadership, voice, and participation in Chapter 7 and our attention to ethnomathematics and consumer culture in Chapter 5 can help us here. Youth come to school having experienced the informal curriculum of everyday life for every single day of their lives. They have experiences with habituation, conceptual restructuring, and enculturation that they bring with them to the school mathematics encounter. Now, it is often the case that previous school experiences have "taught" our students to disconnect from their everyday lives and the funds of knowledge that they bring with them. But we might seek to reclaim these ways of learning and funds of knowledge. One way to think about doing this is to create classrooms as "spaces of mutual work" (Gustavson 2007: 151). We're working here within an enculturationist model. Students are more likely to live as themselves in a community that enables them to experience authentic relationships developed through the work of the community, to live mathematics, in Wang's terms, as a "spiritual journey." "Instead of the traditional 'detached spectatorship'," writes Gustavson, "where the teacher observes, interprets, and evaluates the learning of students, we need to shift to a classroom space of actors – with both teachers and students engaged in the challenges, frustrations, and benefits of real work" (Gustavson 2007: 151). This shift requires us to see ourselves as mathematicians, as writers of mathematics, as historians of mathematics, as comedians of mathematics, in our own everyday lives. It requires of us that we have mathematical passions that we share with our students. It further demands of us that we be interested in the mathematical passions that our students bring with them, that we ask students questions about their lives and their interests, questions we really want to ask, because we really want to know. Of course we want to know! We are mathematicians ourselves, and mathematicians want to hear about other mathematicians' mathematics!

The shift also demands of us that we use our ethical stance with the world in designing *real mathematical work* with our students. One way to do this is to integrate mathematics with issues that are literally affecting our students and their communities, right now, and to take action so as to impact upon an audience, to make a difference. This would indeed be helping our students to read and write their world with mathematics (Gutstein 2006). It would also be enculturating students with critical mathematical dispositions that support students' own ethical stances with their worlds. As The Books declare:

> and meanwhile the statues are bleeding green
> and others are saying things
> much better than we ever could
> as the quiet becomes suddenly verbose
> and the hail is heralding the size of nickels,
> and the street corners are gnashing together
> like gears inside the head
> of some omniscient engineer
> and downward flows the garnered wisdom
> that has never died

This vision of students and teachers using their critical competence to develop the mathematical skills and concepts necessary to affect their communities in some way that they together design is new educational terrain for most of us, even though the ideas have been around for centuries. If we live mathematics with our students in this way, as an ethical stance, we will not know at first what to do: there's no method for this.

> when finally we opened the box
> we couldn't find any rules
> our heads were reeling with a glut of possibilities,
> contingencies
> but with ever increasing faith
> we decided to go ahead and just ignore them
> despite tremendous pressure to capitulate and fade
> so instead we went ahead
> to fabricate a catalog
> of unstable elements
> and modicums
> and particles with non-zero total strangeness
> for brief moments which amount
> to nothing more than tiny fragments of a finger snap

. . . But it will "smell like content." It may not look like content, if you are looking with our past school mathematics experiences as your lens, but it will definitely *smell like real mathematics* to you and your students.

Now, I don't mean to be painting a new dreamscape. What I am describing will be fraught with resistance, will be painful and confusing at many times, and should be for all concerned. We'll have to figure out what we mean here, because there aren't a lot of examples that we can copy. It's as if our map of teaching and learning mathematics is tearing along its creases from overuse, when we never needed the map in the first place. Yet, we don't know how to find our way because we were trained to find our way with a map! "The statues are bleeding green." Still, it is worth it, because "real mathematical work" is satisfying. It meets all ten of Susan Ohanian's foundational questions (from Chapter 1). And anyway, we like peeking into our mathematician's notebooks.

We'll know we're onto something when we start getting confused about ourselves, start feeling that unsettling, slip-slidey, queasy feeling that we are changing, becoming someone new. Maxine Greene once wrote,

> To engage with our students as persons is to affirm our own incompleteness, our consciousness of spaces still to be explored, desires still to be tapped, possibilities still be opened and pursued. . . . We have to find out how to open such spheres, such spaces, where a better state of things can be imagined. . . . I would like to think that this can happen in classrooms, in corridors, in schoolyards, in the streets.
>
> (Greene 1986: 429)

The Books update her words:

> and meanwhile we're furiously sleeping green
> and the map has started tearing along its creases
> due to overuse
> when in reality, it's never needed folds

. . . and they also raise my one anxiety in writing this final chapter:

> expectation leads to disappointment
> if we don't expect something big, huge, and exciting
> usually, uh
> i don't know, it's just not as, yeah

. . . big expectations here. Big expectations require persistence, the pursuit of teaching as an ongoing project, and a firm commitment to our ethical stance. I am not fantasizing that everyone who reads this book is going to jump into a classroom and transform the world today. I *am* expecting that you will try things from this book over and over again until they seem to be working. And I further expect your ideas about what it means for something to be "working" will change by the time you get to that point.

David wrote in his preface that "education requires us to change our fundamental ideas of how the world is structured." I have found this to be true most every day that I teach. I am beginning to understand that this might be the core of my own personal commitment to teaching. It's addictive because it's always changing the world for me. On the other hand, maybe everything looks like it's different because it's me that's changed?

What will You Write in *Your* Chapter?

David Scott Allen

Why is anything interesting to anyone? This question underlies all of these discussions. So many of the differences of opinion which fuel the differences in mathematical/pedagogical perspective come from valuing rote memorization versus the ability to think through a problem: Facts vs. Processes; Traditional vs. Modern. Teachers don't need to choose, but rather should enjoy the best and richest ideas from both poles, finding our own mathematical/pedagogical equator. If you are at all familiar with the artwork of M.C. Escher, you will recall that in many of his pieces, there is no background or foreground. There is only space. We take it upon ourselves daily to fill the space/time of our classes with a similarly seamless mathematics.

Many of the answers to the questions we explored here come down to values. Each teacher should define his or her values on an ongoing basis and communicate them clearly to their students. They lay the foundation for what happens in the classroom. Values are hidden everywhere and underlie everything we do. If you didn't value mathematics (education) you wouldn't be reading this book. As in any other human endeavor, the more informed you are, the better decision you can make for yourself and your students. The bigger the picture you see, the more you can select the parts to focus on for that moment, knowing that the focus will need to change, just as your focus changes while looking at an Escher print.

Perhaps the greatest challenge we face as a professional is to make these ideas implementable and age-appropriate within our teaching contexts. Add to this the fact that each September brings a new batch of students with different group and individual dynamics depending on the maturity level of the group or of certain "leader" individuals. Good teachers do not teach mathematics, but students. They are our future. And that is not a metaphor. Some days I think all I do is remind students to put their name on their work and answer questions like, "What should I do if I find a book under my desk? Where do I put this paper?" But I love my students and I love teaching. The students I teach are at a very impressionable, yet more and more pre-impressioned, time in their lives. It is an exciting age group to teach, and this is why I chose it. Students never stop to ask the proverbial questions about math when they are interested. If we can replace our students' increasing apathy and hopelessness with passion for anything, ours will be a life worth living. So, teaching is the closest profession to parenting, the most worthy profession of all.

Now you go begin your journey, write your story, and help your students write theirs.

Part 1	Part 2 through Part 4	Part 5
Opening *Creating the Issue Finding the Question Generating the Interest*	**Doing the Investigation** *Three weeks devoted to active engagement in student designed investigation around curricular themes, issues, conflicts, problems*	**Archaeology** *Making explicit the knowledge gained*
Open ended activities to elicit student generated questions about issues or problems related to discipline concepts, curricular topic, or theme	Three parts devoted to mathematical investigations. Class time devoted to discussing work done, strategizing next steps, organizing mini-lessons and workshops on ideas generated by students, and putting the work back out into the world.	Time devoted in class to look back at the work done, to name what has been learned, and to extend it into new areas and directions.
Materials needed: Mathematicians' Notebooks; Criteria; books; Center materials; films; speakers; field trips, to help stimulate interest	**Materials needed:** Mathematician's Notebooks; Criteria specifications; conference forms; peer feedback; teacher feedback; center materials; new tools and materials as needed for investigations; assessment vehicles.	**Materials needed:** Mathematician's Notebooks; tests & other evaluation instruments; manipulatives; new problems
Envisioned activities: Quickies; Center work modeling Polya Phases, Specializing and	**Envisioned activities:** Interviews, experiments, debates, in-class writing and work time, peer-review of work in progress, reading discussions, mini-lessons and workshops developed by teacher, initiated by students, guest speakers, planning sessions, etc.	**Activities:** Quickies, Improv; Core curriculum and Standards based conversations where investigations are linked to
	2. Developing the Investigation	
	Activities: Quickies, Center Work, Polya Phases, Problem-Posing; Improv; Reflecting on their work; Discussions	

Generalizing, and Problem Posing; Improv warm-ups; Discussions; lists of questions; experiments, background information

Culmination: Each student identifies mathematically interesting and potentially significant ideas that they have been working with at a center (on posters, in discussions, in their notebooks, etc.); Students identify the center they will return to for their own investigation.

Assessment: Student work sample analyses; Center observation notes; Targeted interviews

Culmination: Peer strategy session

3. Doing the Investigation
Can start this part sooner if students identify their investigation.

Activities: Quickies, Center Work, Polya Phases, Problem-Posing; Improv; Reflecting on their work; Discussions of student work on large posters; mini-lessons and workshops as needed

Assessment: Student work sample analyses; Center observation notes; Targeted interviews

Culmination: Students identify a mathematically significant idea coming out of their investigation.

4. Putting the Work Back Out Into the World

Activities: Quickies; Improv; Writing about the idea; brainstorming in groups; getting ideas up on big sheets of paper; practice meeting with potential audiences; actually doing the work of putting the work back out into the world.
Questions guiding the critical activities:
What do you want to do with what you've learned? What *should* you do? Do something that impacts on you, or that impacts on other people.

Assessment: Student work sample analyses; Center observation notes; Targeted interviews

Culmination: Taking the action of work back out into the world.
Debriefing of the experience

school, city and state expectations; challenges presented by teacher to show students they can utilize skills and concepts developed in their investigations; activities that encourage students to transport the skills and knowledge they learned to other areas.

Culmination: Class addresses these questions: What should we do next? Starting a new project Leave taking, goodbyes, and plans for a reunion.

2: Criteria for Working as a Mathematician

Example 1 Portfolio Criteria

Problem Posing
Document you are posing at least twice per week

Brown & Walter's techniques
- What-if-not
- Attribute-listing

Polya's Understanding the Problem
- Have I ever seen a problem like this before?
- How is what I don't know related to what I do know?
- Can I define any variables?

Polya's Looking Back
- What's the meaning of my answer?
- What have I learned?
- Can I see an easier way now that I have done all that work?
- What new questions grow out of my work so far?

Mason's Generalizing
- Turn patterns and categories into conjectures
- Identify monster cases that can be barred from a conjecture, and ask what that tells you about the conjecture

Problem Solving
Document you are solving at least twice per week

Polya's Four Phases
- Understand the problem
- Plan what you will do
- Carry out your plan
- Look back

Mason's Specializing & Generalizing
- Try special cases
- Organize your cases in a logical way; try a chart
- Look for patterns &/or categories
- Look for general ideas that are true for all cases or for categories of cases
- Make a conjecture; now find a proof for your conjecture

Brown & Walter
- When does a crazy way of doing a math procedure end up working anyway?
- Which what-if-not's change the problem significantly and which only in relatively unimportant ways?

Example 2 Mathematics Folder Criteria

My folder includes:
- work that I am not proud of as well as work I am proud of
- three charts that I have used to further my investigation
- notes and suggestions from at least two other students who have helped me with my investigation
- work on two questions that I invented myself by asking what-if-not
- a copy of the letter I wrote to another student about their work, including what I thought was most important about their work so far, ideas for how they might do more, and suggestions for making their folder easier to use
- two mathematical equations that I created to further my investigation
- my plans for putting my work back out into the world
- my reflection on what I learned from putting my work back out into the world.

Example 3 Mathematician's Notebook

For the first part of this course, the main form of work will be to maintain a working mathematical notebook that will enable you to pursue particular mathematical investigations. These investigations will help us do several things: it will help each of us to work through a provocative mathematical problem; develop an understanding of what it means to be a mathematician; and show us how we can design mathematical inquiries for our students. Your notebook will include things like: drawings, writings, and reflections; first, second, and third drafts of charts; and notes to yourself in the margins. The contents of your notebook are what you will be evaluated on in the first five weeks of our course.

This notebook may not be like other notebooks you have kept in the past. Think of it as a sort of lab notebook where one keeps systematic records of experiments, or the notebooks that photographers keep about lighting, lenses, and settings. Anthropologists keep notebooks; so do cooks and architects. These notebooks all help these professionals keep track of, study, and improve their practice; they simply help them to do their work. You are developing a notebook that supports your working as a mathematician and your learning as a teacher of mathematics.

The goal is to create a useable notebook of your own work as well as our class's accumulating understandings and investigations, conjectures, and arguments. Your notebook will be a place for you to track and record your thinking about issues and ideas. It will also provide an additional avenue of communication between you and your professor. You will use your notebook to record the work you do within the initial group mathematical investigation and in the design, implementation, and reflection of your own investigation tied to a Core Curriculum unit.

Here are some specific qualities that distinguish work in notebooks:

Notebook Grading Criteria

	Satisfactory	Accomplished	Needs to improve
Mathematics Investigations and In-Class Mathematics Work	Organized to keep track of your own reactions and thoughts, solutions and experiments. Includes all entries. Records of representations and examples. Records of trial and error.	Consistent and repeated use of Polya's Four Phases; specializing and generalizing; and problem posing. Reflects practice of looking back to systematically understand what has been done and what could happen next. Clear records of others' thinking and work, solutions, and experiments.	Insufficiently organized. Difficult to read or understand. Missing some aspect of any entry.
Reading Reflections	Addresses issues that each assigned reading raises, highlights insights, raises questions or makes connections. Organized and clear. At least four reading reflections of at least 250 words.	Analyzes, synthesizes, and applies issues from reading. Connects different readings with one another. Connects deeply to work to be done in the field. Organized for useful future retrieval and reference. At least six reading reflections of at least 250 words.	Missing elements of reading response. Very brief. Lacking depth of record or comment. Poorly organized. Fewer than four reading reflections of at least 250 words.
Activity and Class Reflections	Comments thoughtfully on class activities, or ideas or experiences. At least two reflections of at least 250 words about issues other than the math investigations.	Comments connect to other experiences. Awareness of own and others' progress. Compares own thoughts over time. Reflects both on learning as a student, and insights as a beginning teacher. Pursues new insights or puzzles about a question or an idea in a new way. At least three reflections of at least 250 words that address issues other than the math investigations.	Missing some elements of reflections. Thin reflections. No evidence of reflecting on class activities.

continued

continued

	Satisfactory	Accomplished	Needs to improve
Problem Solving Strategies	Reflects complete cycle of Polya's Four Phases at least three times per week. Consistently uses pictures, charts, patterns, and application of similar problems.	Develops mathematical importance over time. Reflects repeatedly on use of different strategies. Pictures, charts, and patterns lead to more sophisticated versions of themselves that push the investigation toward a sustained focus.	One or more of the phases neglected consistently
Specializing and Generalizing	Record of at least six uses of this technique. At least one reorganization of special cases into a chart to look for a generalization.	Use of these techniques leads to mathematical generalizations that lead to mathematically significant results. Explains significance of generalizations.	Special cases have been tried; however, no record of reorganizing them in order to try to find a generalization can be found.
Problem Posing	Record of at least six uses of either *what if not* or *changing attributes*.	Questions have a purpose behind them that deepens the investigation or leads to a mathematically significant focus.	Fewer than six uses of problem posing. Problem posing is too disconnected, resulting in lack of mathematical direction or focus for investigation.

3: Mathematician's Notebook— Checkpoint

Polya's Four Phases

Work reflects consistent and repeated use (at least twelve times) of Polya's Four Phases (including the practice of looking back to systematically understand what has been done and what could happen next). Notebook documents repeated reflection on use of different strategies, and integrated use of pictures, charts, patterns, and similar problems.

no evidence of criterion   *beyond expectations*

1   2   3   4   5   x5_____

Specializing and Generalizing

Notebook documents at least six examples of this technique, and includes at least one reorganization of special cases into a chart to look for a generalization. Notebook contains evidence of at least one generalization.

1   2   3   4   5   x5_____

Problem Posing

Record of at least six uses of either *what if not* or *changing attributes*. Questions have a purpose that deepens the investigation or leads to a mathematically significant focus.

1   2   3   4   5   x5_____

Mathematical Significance

Strategies and models/representations are increasingly used over time to identify big mathematical ideas.

1   2   3   4   5   x5_____

Total _____

Comments:

4: "Polya Was a Mathematician" Songsheet

Sung to the tune of *Joy to the World*, by Three Dog Night

Part 1:
Polya was a mathematician.
Was a good friend of mine.
Never understood a single student he had,
So he asked himself why,
Yes he asked himself why.

> [Chorus, Part 1]
> Understand the problem.
> Plan what you will do.
> Carry out your plan,
> And then look back.
> Yes, don't forget to look back.

Ask yourself these questions:
This is what I'd do.
Have I ever seen a problem like this before?
How's the unknown related to the known?
Define your variables.
(Understanding will help you plan.)

> [Chorus]

Can you draw a picture?
Can you make a chart?
Can you write down an equation,
Or find a pattern?
These will help you start.
(Yes, a plan will help you start.)

> [Chorus]

Can you add things to your picture?
Write a formula for your chart?
Solve your equation,
Or extend your pattern further?
Aesthetically this is art.
(Yes, aesthetically this is art.)

> [Chorus]

Does your answer make sense?
What have you learned?
Can you find an easier way now that you
     are done?
What's the meaning of your answer?
(Yes, interpret the meaning of your answer.)
* * *What's the new question??* * *

> [Chorus]

Part 2:
But Polya was a Mathematician!
Not a pedagogue.
All he ever did was train for problem types,
And he always used whole class instruction.
(Yes, he always used direct instruction.)

> [Chorus, Part 2]
> Now we work in groups.
> Share our strategies.
> Redefine the problem when we get inspired.
> Light the mathematics fire.
> (Yes, light the mathematics fire.)

Polya, what can I do?
I don't have a question to pose!
All I ever did was follow your rules,
Never math'matized outside of school.
(I just learned the problem types.)

> [Chorus 2]

What if I didn't get an answer?
What if I changed the problem?
What if I think non-linearly,
Learnin' on the way to the prize?
(Learning new things all of the time.)

> [Chorus 2]

After I got an answer.
I changed the question I posed.
My solution raised a question 'bout another
     type of math,
An investigation I want to pursue.
(An exploration I devised myself.)

> [Chorus 2]

But you know I love to lecture!
Love to have my fun!
I'm a high night flyer
And a Rainbow rider
And a straight shootin' son of a gun.
(Yes, a straight shootin' son of a gun.)

<So sing it now!>

Understand the problem.
Plan what you will do.
Carry out your plan,
And then look back.
Yes, don't forget to look back.

<repeat> <whisper:>Yo' look back now, ya heah!

5: Taking Action

Taking Action Assignment

Action: the most vigorous, productive, or exciting *activity* in a particular field, area, or group

Now that you have over six weeks of intense experience working with and teaching children, it's time to take action on what you have learned. We have seen how this teaching experience has stirred up many thoughts, questions, and ideas that you have around teaching and learning. This final piece provides you the opportunity to spend the last few weeks of our class reflecting on what you have learned and what you are thinking about now, and then designing a way to take action on an aspect of your teaching. Here's another way of looking at this final assignment: What's the next step in your teaching? As a teacher, what do you feel you need or want to do with what you have learned about teaching in this class?

Criteria

Your final action must:

- document all of the work that leads up to taking this action (notes, drafts of questions, emails, reflections, etc.)
- give you the opportunity to push your thoughts about teaching even further
- illustrate a pedagogical idea that connects to the philosophy and practice of this course
- either influence you or influence other people or both around a pedagogical issue or idea coming from this course
- include an artifact that documents you taking action on what you have learned

- include a page-long reflection after your action that addresses the following questions:
  - o Why this action?
  - o What would one of the writers from this course say about your action? Why do you think they would say that?
  - o What did you learn from taking this action?
  - o What pedagogical questions does your action raise for you? Why these questions?

Here are some possible forms your action could take:

- reconnecting with your students in a way that builds on the work you have already done
- drafting an article for an education journal based on your teaching experience
- writing a letter to a former teacher that has influenced you and is somehow connected to the work that you have done this semester
- a photographic essay that illustrates a significant finding from your teaching
- a dialogic piece between teaching partners
- writing a preface to the book of riddles that your students have written, describing the work that you have done with the group, giving the philosophical and pedagogical significance behind riddles
- meeting with a teacher/principal/parent and having a conversation about an idea in your teaching from this semester
- meeting with one of your students to explore an aspect of your teaching that is on your mind now
- your own idea that meets the criteria described above

Please bring all the work leading up to your action (including your artifact and reflection) to our final event. At that time, you will have a chance to present your action to a small group.

Suggested starts:

- Look over the five artifacts you selected to put in the front of your portfolio. How could they be the start to an interesting action?
- Write a reflection with the objective of coming to an idea for an action by the end of it.
- Go back through the reflections that you have written during your teaching. What are the patterns? What are the issues that seem to be pressing in them? How can you use your action to work through or explore one of these issues?
- Got an idea? Make a plan of action that you can follow.

Taking Action Workshop

1. How is your action project helping you to push your thoughts about teaching even further? List three different ways, with at least two examples for each way:

2. Who is being influenced as part of your action? How is your action designed to make sure this influence happen?

3. Find two quotes from our readings that you could use to describe the "why" behind your action.

Your final action must:

- document all of the work that leads up to taking this action (notes, drafts of questions, emails, reflections, etc.)
- give you the opportunity to push your thoughts about teaching even further
- illustrate a pedagogical idea that connects to the philosophy and practice of this course
- either influence you or influence other people or both around a pedagogical issue or idea coming from this course
- include an artifact that documents you taking action on what you have learned
- include a page-long reflection after your action that addresses the following questions:
  - o Why this action?
  - o What would one of the writers from this course say about your action? Why do you think they would say that?
  - o What did you learn from taking this action?
  - o What pedagogical questions does your action raise for you? Why these questions?

4. Find two quotes from our readings that you could use to interpret the ideas that are developed in your action.

5. What pedagogical questions does your action raise for you? Why these questions?

6. How is this action specifically about teaching and learning mathematics?

7. What would be the best use of class time next week, as a celebration of this action work, and as an opportunity to interact with guests?

6: MathWorlds Hints and Comments

Bernadette Bacino

MathWorlds 1: Reverse Answer to Questions

1. Averages
    a. average = 16
    b. some examples (although the possibilities are endless)
- 24, 15, 8, 17, 11, 21
- 17, −5, 28, 62, −14, 9, 34, −3
- 27, ¾, −5, 51, −½, $^{45}\!/_2$

    c.
- groups must be an EVEN number of numbers unless "16" is a member of the group, making it an ODD number of numbers
- the groups must add up to be a multiple of 16
- there are endless ways to group together numbers that are multiples of 16

    d. *Mathematician Questions.* What if the average is an odd number? How would this change the outcome of part (c.)? How could we describe an average as a "center" of a particular group of numbers? Could we come up with a way of describing groups of numbers that seem to have more than one "center," that cluster in subgroups that are not around the calculated "average"?

    Teacher Questions. How do these discoveries about averages affect different data in statistics? Do you think that scientists or researchers ever manipulate data in order to reach certain averages, knowing that there are many different ways to reach the same average? How accurate are data and statistical analyses?

2. Shapes
    a. Any shape that falls under the category of a parallelogram, for example, a rectangle, rhombus, or square.
    b. *Square.* Equilateral and equiangular parallelogram.
    Parallelogram. Quadrilateral with two pairs of parallel sides.
    Trapezoid. Quadrilateral with exactly one pair of parallel sides.

Rhombus. Equilateral parallelogram.

Kite. Quadrilateral with two different pairs of consecutive congruent sides.

Isosceles trapezoid. A trapezoid whose non parallel sides are congruent.

c. Basically all that you know is that there are two sets of parallel sides! Otherwise, the polygon could have any number of sides and any additional sets of parallel sides.

Mathematician Questions: The types of shapes that we were examining in part (b.) fall under the category of being a "regular polygon." Could we give "clues" to describe particular **irregular polygons**? Why/why not?

d. *Teacher Questions*. Take a look around objects in your classroom, things around the house, or objects outside. What shapes tend to be more prevalent than others? Why do you think this is the case?

3. Sewing Shapes Together

a. Two common shapes (some examples):
- two squares can make a rectangle
- two triangles can make a square
- two triangles can make a rectangle
- two trapezoids can make a concave quadrilateral
- two rhombi can make a rectangle
- two kites can make a concave quadrilateral.

b. Three common shapes (some examples):
- three squares can make a rectangle
- three rectangles can make a square
- three triangles can make a trapezoid
- three trapezoids can make a bigger trapezoid
- three rhombi can make a parallelogram
- three rhombi can make a rectangle.

c. Four common shapes (some examples):
- four squares can make a bigger square
- four rectangles can make a bigger rectangle
- four rectangles can make a square
- four triangles can make parallelogram
- four triangles can make a rectangle
- four trapezoids can make a bigger trapezoid
- four rhombi can make a parallelogram
- four rhombi can make a bigger rhombi.

d. Are there more results for shapes made from an even number of shapes or an odd number? Why do you think this is the case?

4. Game Dice

a. A game that comes to mind where you would *definitely* need two six-sided cube dice is RISK (in fact you need five total). You could not play this game with just one twelve-sided die. When players are "attacking," both the attacker and defender must roll (attacker having the ability to roll three dice and defender rolls two dice) and the outcome of the "win" is dependant upon having more than one six sided die. A game where it would not matter as much if you used one twelve-sided dice instead of two six-sided dice would be MONOPOLY, since each player moves the total of the sum of the two six-sided dice. However, without the option of using the two six-sided dice, players would not be able to get the bonus of "snake eyes" when rolling two "1s", which gives the player either $100 or $500.

b. i. *Spinners.* Risk could work the same way if you had spinners that went from one to six, but then you would need five spinners which might make the game less "fun" in a way since you can't spin three spinners at the same time (or at least this would be very hard to do). Monopoly would be fine with spinners, but would also eliminate the chance of rolling "snake eyes" and getting that extra bonus.

ii. *Coin Flips.* Coin flips would not work very well for either Risk or Monopoly, although it would be possible to do, but this would definitely slow down the pace of each game. When you are "attacking" in Risk you would have to agree that, for instance, "heads beats tails," but the chance of getting a tie would be more likely than using six-sided dice. In Monopoly, you could just add up the total of the coin flips, for example, saying that heads = 1 and tails = 2, but then the most spaces you could move at a time would be four instead of twelve, slowing down the pace of the game.

iii. *Something Else.* One thing that comes immediately to mind is having the numbers 1 through 6 listed twice on little tiles in a cup and the players can pick out each tile from the cup, which will dictate how many spaces they go, who wins in an "attack," and so on. However, I just don't think this would be as fun for the players as rolling dice, and would, again, slow down the pace of each game.

c. *Mathematician Questions.* What if the six-sided dice that are used in these games are not "regular," that is, the opposite sides of the dice *do not* add up to 7, as is the case with all manufactured game dice today? How would this affect the game? Or would it at all? Would the probability change or would a roll of the dice still "be a normal roll"?

Teacher Questions. What are some games that successfully use something other than dice to dictate many of the strategies and rules of the game? Why do you think this game is successful in not using dice?

(An example here that comes to mind is the game Twister, which uses the spinner to dictate where you hands and feet should be placed.)

5. Story Graphs

a. This will vary depending on each individual's distance from school and how long it takes to get to school. You must take into account traffic, speed limits, and road closings as well! This will *not* be a constant graph, unless you are traveling at the same velocity the entire time! (Remember that velocity is distance traveled per unit time.)

For example, it takes me anywhere from 20 to 30 minutes to get from home to school, and a few times, it even took me less than 20 minutes (traveling at night and above the speed limit). But the overall distance to school is constant, always 14.11 miles. However, the time will be the variant.

So let's say, for example, that it takes me 5 minutes to travel the first 3 miles of my trip to school. Then it takes 7 minutes to travel another 5 miles, making my total trip take 12 minutes so far. Then the last 6.11 miles takes 13 minutes, making my total trip take 25 minutes! Here is an example of a graph that would display this:

distance	time	velocity
0	0	0
3	5	0.6
8	12	0.714
14.11	25	0.47

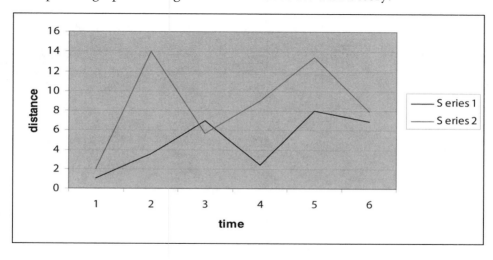

b. Graphing volume versus weight. The graph of this scenario would be constant, since as the volume of an object increases, the weight increases. Given this, the graph would look similar to the graph above of distance versus time, since it is a linear graphical representation. Perhaps we're filling a container with materials that have different densities/weights, so that, while we are pouring the materials in, they suddenly change and the weight increases at increased rates.

c. How would distance versus time be different if you were traveling in an airplane, where once you reached the required speed and altitude for the trip, everything remains constant? What if there is a storm and the plane receives some turbulence?

How would volume versus weight be different if you were in space where everything is weightless? Could you even graph such a scenario?

d. Example of a graph that might be hard to associate with a story:

Actually, at first I thought this would be hard to associate with a story, but now, re-looking at this graph, it is very plausible that the story could be of two cars driving to the same place, but during different traffic times. Once one uses some imagination, it seems very hard to make a graph that could not be connected with a creative story! Can you think of one?

e. Consider volume versus time: Think of a huge jug, perhaps something shaped like a soda bottle, but a lot bigger. For most of the bottle, the graph of volume versus time would remain constant, since as you pour water into the jug, it is falling at a normal rate since the volume of the jug is constant at the bottom of the jug and would take the same time to fill it. But think about once the jug starts to fill up near the top. The shape of the jug is narrower at the top, so the water will fill up more quickly since the volume is not as great at the top of the jug! What would a graph of this scenario look like?

6. Subtraction Heaven

a. *Mersin.* This student subtracted the "normal" way, using borrowing to get the answer.

Legza. Here Legza put the 7 above the 40 and then rounded the 29 up to a 30. Then to get the answer, you have to subtract 30 from 40 to get 10, add, 7 from the "1s" column to the extra 1 that it takes to get from 29 to 30, to get 8, and lastly, 10 + 8 = 18! Basically here, Legza sees the numbers as being on a number line; she adds the numbers that are either above or below the "multiples of ten" and then adds that to the difference between the multiples of ten!

Pabko. This student made the 7 a 17 in order to subtract the "1s column" and then instead of "borrowing" from the 4 to make it a 3, he added 1 to the 2, which still gives the correct answer!

b.

c. Obviously Mersin's strategy using "borrowing" would work for three-digit subtraction. Pabko's strategy would work as well, but would be a bit more complicated:

example :

$$\begin{array}{r} 425 \\ -248 \\ \hline \end{array} \rightarrow \begin{array}{r} 4\ 2\ (15) \\ -\ 2\ (5)\ 8 \\ \hline \end{array} \rightarrow \begin{array}{r} 4\ (12)\ (15) \\ -\ (3)\ (5)\ 8 \\ \hline 1\ 7\ 7 \end{array} = \boxed{177}$$

Bernie's strategy would work for three-digit subtraction, but would also take extra maneuvering:

example:

$$\begin{array}{r} 425 \\ -248 \\ \hline \end{array} \rightarrow \begin{array}{r} 450\,(25) \\ -\ 250\ (2) \\ \hline \end{array} \rightarrow \begin{array}{r} 450 \quad\ (25) \\ -\ 250 \quad -(2) \\ \hline 200 \quad\ \ 23 \end{array} \rightarrow 200-23 = \boxed{177}$$

but would probably have to do this again!

$$\begin{array}{r} 200 \\ -\ 23 \\ \hline \end{array} \rightarrow \begin{array}{r} 200\ (0) \\ -\ 20\ (3) \\ \hline \end{array} \rightarrow \begin{array}{r} 200 \quad (0) \\ -\ 20 \quad -(3) \\ \hline 180 \quad\ 3 \end{array} \rightarrow 180-3 = \boxed{177}$$

again, have to ignore negatives

d. *Mathematician Questions.* From a mathematician's standpoint, do you think that it is good to know more than one way to solve a problem or equation? Why/why not? Think of different ways, similar to the alternative methods to do subtraction, which could be applied to addition, multiplication, and division. Is there any basic knowledge that you need to know about these particular operations that are necessary to perform any of these methods?

Teacher Questions. What kind of strategies similar to these could do you for multiplication? What about division? Why are these strategies really helpful for students or do you think that the "traditional" methods of teaching subtraction, multiplication, and division are better in the long run?

MathWorlds 2: Multiple Answers

1. Grass Seed

a. Since the grass seed covers two square feet, is area is two feet^2. So if the plot of lawn is three feet by five feet, your total area would be fifteen feet^2.

Choices of what to buy:
- eight bags with enough seeds to cover one square foot left over
- seven bags and then another smaller bag that would be enough for one square foot of grass
- seven bags and just spread your seeds a little thinner
- seven bags and ask your neighbor to share a little extra ☺
- fifteen bags and then save the ones unopened and the bag that is half-way filled until next year!

b. Ways to work this problem:
- draw a picture
- make a chart
- count (for example, 1 bag = 2, 2 bags = 4, 3 bags = 6 . . .).

c. Examples: The grass seed covers ten square feet and your backyard is 125 square feet. Use a realistic shape of a yard that is irregular.

d.

some examples:

examples	grass seed coverage	area of coverage	dimensions of plot	area of plot	numerical calculations	bags purchased	bags left over
1.	2ft x 2ft	4ft²	3ft x 5ft	15ft²	$\frac{15ft^2}{4ft^2} = 3\frac{3}{4}$	4	¼
2.	10ft x 10ft	100ft²	10ft x 12.5ft	125ft²	$\frac{125ft^2}{100ft^2} = 1\frac{1}{4}$	2	¾
3.	2ft x 2ft	4ft²	circle w/ radius 5ft	$(A=\pi r^2)$ $=\pi(5)^2$ $=25\pi$ $=78.54ft^2$	$\frac{78.54ft^2}{4ft^2} =$ 19.64	20	0.36 or 9/25
4.	4ft x 4ft	16ft²	triangle w/ base=10ft height=6.4ft	$(A=\frac{1}{2}bh)$ $=\frac{1}{2}(10)(6.4)$ $=32ft^2$	$\frac{32ft^2}{16ft^2} = 2$	2	0
5.	5ft x 5ft	25ft²	"L" shape: □ = 4ft x 4ft ▭ = 6ft x 8ft	A = sum of shapes A(□)=16ft² A(▭)=48ft² = 64ft²	$\frac{64ft^2}{25ft^2} =$ 2.56	3	0.44 or 11/25

The variables that make the most difference are the area that the grass seed covers and the total area of your lawn! It might be more profitable to buy a certain amount of the bigger bags of seed, see how much you would have left over if you bought another bag but didn't use it fully, and try to find a bag that is smaller that would come close to the "leftover area."

e. *A comparison problem.* A football field is a rectangle that is 360 feet by 160 feet. A bag of grass seed costs $9.95 per bag and each bag covers five square feet. A square foot of Astro Turf costs $11.75. If you chose to use real grass, the football field would have to be re-seeded every year and the cost of maintenance (lawn mowing, weeding, fertilizing, etc.) is $1500 a month. The Astro Turf only has to be replaced every ten years, and its maintenance cost is $3300 every year.

- Using this information, which way would be a more profitable way to cover the football field initially?
- How about after five years?
- Ten years (before the Astro Turf had to be replaced)?
- Make a chart that organizes your calculations to solve this problem.
- Make a chart that shows how much it would cost from years one to ten to maintain the field using grass versus Astro Turf.

  f. How much would it cost to re-carpet your entire house? What steps would you have to take to figure this out? What if you decided to cover your floors with wood-flooring instead of carpet? Would this be more or less expensive? Research the price ranges for the different types of carpets that you like, as well as the type of wood flooring you could use, in order to figure out this problem. What if you need to pour concrete to fill a space in front of your building? What if the shape is not a nice rectangle; how would you determine the area?

2. Division & Decimals

  a. By hand:

$$\frac{7}{8} = \quad 8\overline{)7.000} \quad \Rightarrow \quad \frac{7}{8} = 0.875$$

  b. You know you've gone far enough because you have no more remainders!
  c. By hand:

$$\frac{2}{7} = \quad 7\overline{)2.000000000} \quad \text{repeating pattern} \quad \Rightarrow \quad \frac{2}{7} = 0.\overline{285714}$$

  You can hopefully see the pattern in the numbers for this fraction, so you know that it is a repeating decimal. Once you see this pattern, you know you've gone far enough!

  d. *Halfs.* Any number divided by 2 will either go in evenly if it is an even number or will have 0.5 after the whole number if it is an odd number.
  Thirds. Any EVEN number divided by 3 (but not a multiple of 3) will have 0.33333333333 repeating after the whole number; any ODD number divided by 3 (that is not a multiple of 3) will have 0.6666666666666 repeating after the whole number, and any multiple of 3 will, of course, divide evenly by 3.
  Quarters. Any number divided by 4 that is NOT a multiple of 4 will have the following pattern: one number after a multiple of 4 will have 0.25 after the whole

number, two numbers after a multiple of 4 will have 0.5 after the whole number, and three numbers after a multiple of 4 will have 0.75 after the whole number.

Fifths. Any number divided by 5 that is *not* a multiple of 5 will have the following pattern: one number after a multiple of 5 will have 0.2 after the whole number, two numbers after a multiple of 5 will have 0.4 after the whole number, three numbers after a multiple of 5 will have 0.6 after the whole number, and four numbers after a multiple of 5 will have 0.8 after the whole number.

Sixths. Any number divided by 6 that is *not* a multiple of 6 will have the following pattern: one number after a multiple of 6 will have 0.166666 repeating after the whole number, two numbers after a multiple of 6 will have 0.333333 repeating after the whole number, three numbers after a multiple of 6 will have 0.5 after the whole number, four numbers after a multiple of 6 will have 0.666666 repeating after the whole number, and five numbers after a multiple of 6 will have 0.833333 repeating after the whole number.

We already did sevenths and eighths above, they are special cases!

Ninths. Any number divided by 9 that is *not* a multiple of 9 will be that number repeating as the decimal after the whole number. For example, $1/9$ is 0.11111111111 repeating; $3/9$ is 0.33333333 repeating, and so on. For numbers after 9, 10 = 1, 11 = 2, 12 = 3, so that 12 = 1.33333333 repeating. This is kind of like a "Base 9" conversion system, such that any number past 9 loops around to be represented as numbers from 1 to 9, which is why 10 = 1, 11 = 2, 12 = 3, and so on.

e. This implies that any *rational* decimal can be written in the form a/b. In fact, that is the definition of a rational number! *irrational* numbers cannot be written in the form a/b. Irrational numbers do not terminate or have patterns like we saw in $2/7$.

f. As we just determined, decimals that don't stop and don't repeat are called *irrational* numbers and cannot be expressed in fraction form!

g. You can consider time as an example of the relationship between fractions and decimals. For example, someone might say "it's a quarter past two", but you might say, "it's 2:15." Since there are 60 minutes in an hour, one quarter, or one fourth of 60 is 15 minutes, which is why 2:15 is the same as a quarter past 2 o'clock. In the case of time, the system that is used is a "Base 60" conversion system, similar to the Base 9 conversion system we discussed for the situation of "ninths" above. Once a number goes past 60, it loops around back to 1 and starts the pattern all over again!

h. What other conversion systems do we use in our everyday life besides time? How do these different conversion systems relate to decimals and fractions in the ways that we have been discussing?

3. **Iterative Functions**
   a. for the function $f(x) = 3x + 7$, start by plugging in $x = 2$. You come up with $f(x) = 13$. Then plug in 13 for x and you get f(x) = 46. Next plug in 46 for x and you get $f(x) = 145$ . . . this scenario continues indefinitely so the function $f(x) = 3x + 7$ when looked at as an iterative function will go on forever, with the "answers" approaching infinity.

b. For example: g(x), h(x), and K(x) all approach zero when you start with $x = 2$ and evaluate it as an iterative function. However, J(x) produces negative numbers going farther and farther away from zero, and M(x) approaches infinity when you start with $x = 2$ and evaluate it as an iterative function.

c. One function that would not be categorized into either approaching zero or going farther away from zero as a negative value is the function:

$$f(x) = \frac{x-1}{x^2 - 2x + 1}$$

When you start with $x = 2$ and evaluate it as an iterative function, it does not seem to follow any pattern, but does approach negative numbers after you evaluate it at x = 2, which equals 1, and then at $x = 1$, which equals 0.

Another function which does not follow the patterns mentioned above is the function:. When evaluated iteratively starting at $x = 2$, $f(x) = 0$, and then will always equal zero after that.

d. If you start evaluating these functions iteratively starting at a different number than 2, the outcome for most of these will be different at first but then eventually approach the same value in the end. For example, if you started evaluating the original $f(x) = 3x + 7$ at $x = 100$ instead of at $x = 2$, the answers will be much higher than starting at $x = 2$, but the overall result will be that this function approaches infinity, just the same as when we started with $x = 2$.

e. How would these functions differ if you did not evaluate them iteratively, but instead, as a limit, as it is done in Calculus? What if you evaluated them as a summation from 0 to a specific number? What if you evaluated them iteratively as a function within themselves? For example, take the original function, $f(x) = 3x + 7$. Then put $f(x)$ wherever "x" is, so $f(f(x)) = 3 * f(x) + 7$ which equals $3 * (3x + 7) + 7$ which equals $9x + 28$. If you do this a bunch of times, will you see a pattern? For example, using this same function $f(x) = 3x + 7$, $f(f(f(f(f(f(x)))))) = 729x + 2548$. Could you predict this without doing the algebra? Pick a function and try this a few times to see if there is a pattern that will allow you to predict what function would come next without having to do the "tedious algebra"!

4. Fermi Problems
a. A few examples of "Fermi Problems":
- How many water balloons would it take to fill up your school gymnasium?
- How many piano tuners are there in the city of Chicago?
- What is the mass in kilograms of the whole student body of your school?

Fermi Questions need:
- Communication among those solving the problems.
- Students to utilize estimation.
- Students to emphasize the thought processes instead of getting an "answer."

A good website to find out more about Fermi Problems and see many examples of questions: www.jlab.org/~cecire/garden/fermiprob.html

b. Some "Fermi Problems" of my own (just some examples):
   - How many people rent a shore house in Sea Isle City, NJ the week of July 4?
   - How many *highway miles* would you drive taking the most direct routes from New York City, NY, to Shenandoah National Park, VA, to Omaha, NE, to Arches National Park, UT, and ending in Seattle, WA?
   - What is the volume of the Grand Canyon?

5. Chairs and Tables

If three points determine a plane, why do chairs and tables usually have four legs? A few possible answers:
   - A table technically *could* stand with just three legs, similar to a tri-pod or stool, but it could easily fall over. If one leg of the three became unstable, then the table could not stand with only two legs due to the placement of the legs in the first place. So in this instance, four legs create more *stability*.
   - Four legs also create more equilibrium with the weight of the chair when a person is sitting in it, or with the weight put on top of a table; in essence, they provide *two* planes of stability, in case you lean away from one of the legs, creating a center of gravity that defines a plane using two of the legs and a hypothetical third leg opposite the one that actually exists.
   - The tables and chairs developed by the early Greeks and Romans almost always had four legs and were used for nobility, so perhaps the reason many tables and chairs still do today is a matter of precedence and tradition set by our ancient forefathers.

6. Angle Bisectors

a. Angle bisectors in triangles bisect the three angles of the triangle, cutting each angle into two congruent angles. The three angle bisectors intersect at a point called the *Incenter*. This point called the *Incenter* is equidistant from the three *sides* of the triangle. The Incenter is also the center of the *Incircle*, which is a circle that can be drawn *inside* the triangle that is *tangent* to each of the 3 *sides* of the triangle. Also, no matter what type of triangle one might examine, the Incenter is always *inside* the triangle. (A good way to remember this is since it is called the "*in*center," it must be *in*side the triangle; with other special points of a triangle, this is not always the case.)

b. No matter what type of triangle one examines, the incenter is *always* equidistant from the three sides of the triangle. This statement is the basis for the *Angle Bisector Theorem*, which states, "any point on an angle bisector is equidistant from the sides of the angle, *and* this distance is always perpendicular."

c. What about Perpendicular Bisectors in triangles? What is the "special name" of the point where these three bisectors intersect? Can you find anything similar to the facts stated in part (a.) between Angle Bisectors and Perpendicular Bisectors? (Some answers to these questions: The Perpendicular Bisector in a triangle is perpendicular to each side though its midpoint. These three bisectors intersect at a point called the *Circumcenter*. Similar to the angle bisectors, perpendicular bisectors are equidistant from the 3 *vertices* of the triangle. The Circumcenter is also the center of the *Circumcircle*, which (yup you guessed it) is a circle that can be drawn *outside* the triangle that is *tangent* to the three *vertices* of the triangle. Finally, the Circumcenter *can be outside* the triangle, depending on what type of triangle one is examining (in fact, in a *right triangle*, the *circumcenter* is the *midpoint* of the *hypotenuse* of the triangle!). Pretty COOL!!!!!!)

MathWorlds 3: Reading and Writing Mathematics

Answers for questions 1 and 2 will vary depending on the student or teacher reading and analyzing these questions.

MathWorlds 4: Pitching Questions at Various Levels

Answers for these questions will vary depending on the location of the school and the state standards associated with these topics, as well as each individual teachers' ability to deviate from the curriculum and determine what he/she thinks is important to cover regarding these topics for each grade level. Common assumptions about the trajectory of mathematics education across the life-span posit "simpler" problems early on, with increased complexity and abstraction in higher grades. Readers of this book are encouraged to challenge their assumptions in this exercise; do not presume the traditional "building block" approach to mathematics education!

MathWorlds 5: Turning "Puzzles" into "Problems" or "Exercises"

1. Park Paths
  a. It is possible to draw three lines that go to the prospective gates without intersecting. The large house could build a path that weaves around the house on the left, then up and around the house on the right and then down to the middle gate. The house on the left can create a path that weaves above and around the house on the right and then to the right-hand gate, being parallel to the large house's path. Finally, the house on the right can just build a straight path to the gate on the left!
  b. Some questions to ask:
   • What type of path set-up would be the most cost efficient?
   • What type of path set-up would use up the least amount of grass?
   • What type of path set-up would be the most aesthetically pleasing?
   • Make some variations to this scenario: change the number of houses, change the number of gates, allow a certain number of intersections, and/or change the locations of the houses.
  c. The question that appears the most interesting to me is, which path set-up would be the most cost efficient? You would have to take into account which path set-up would be the shortest overall and the cost of how you want to "pave the path."
  d. I believe that knowing which path set-up would be the most cost efficient would help to answer the other questions.

2. Hanky Knot

a. How??!?!?! Try it yourself to see if this is possible!!!
Solution. First put one arm above the hanky and one arm below. While still keeping the hanky in this position, cross your arms. Then grab the respective ends of the hanky that are right below your left and right hands. Make the knot!!!!!

b. Some questions:
- Can you tie a knot in a hanky with just one hand?
- What is the fastest way to tie a knot in a hanky?
- Do you think that the length and/or thickness of the hanky play a part in these answers?
- What about a towel? Would this be easier or harder to tie a knot with? Why?
- Could you tie a different kind of knot? How about a square knot?
- Is there another way to solve this problem? What if you had to cross your arms first and then try to tie the knot? By changing the order of this algorithm around, would it still be able to work in many different ways?

c. The most interesting question in (b), to me, is the last question. By changing the order of the algorithm you come up with an entirely different problem!

d. Again, I feel that answering the last question would help to answer the other questions, as well as a variety of others!

3. Whythoff

a. For the first version of this game, you can take four buttons from both saucers first, and then four buttons from the second saucer to win the game. My next quesion is, what if there are fewer than four buttons?

b. For the trickier version, you can take two buttons from all three saucers, then one button from the first and second saucer, then 3 buttons from the first saucer to win the game. But will this *always* work?

c. The strategy that I used to win this game is to try to "empty" one saucer at a time, starting with the saucer with the least amount of buttons and finishing with the saucer with the greatest amount.

d. You could do any number of things to change this game:
- Add more saucers.
- Players must claim an equal number of buttons from *all* saucers, unless the saucer is empty.
- The saucers shown contain buttons that are multiples of each other; how would it be different if the amount of buttons in each saucer were NOT multiples of each other?

e. The question that seems the most interesting to me is the last question. The game would still work, but it is interesting to me that the two examples given contained buttons that were multiples of each other.

4. Tossing Dice

a. This is a very interesting trick, if you can't figure it out by simply reading it on paper, try it yourself with three dice and you will see how the "magician" does his trick right away!

Here is how the trick works: (don't read this if you want to try and figure it out yourself first!). The dice are rolled three times, and then you are asked to take the sum of the numbers showing on the top of the dice. Then you are asked to pick *one* die and add the number on the bottom. Here is the trick: since the opposite sides of a six-sided die *always* add up to 7, the "magician" knows that no matter what, the sum of the one die's top and bottom number will always add to 7. Then you roll that same die again and finally add the top number to the sum. So all the "magician" has to do is add the total of the top numbers showing at the end and then add 7 more, and he/she will *always* have the correct grand total!

b. Some questions to ponder:
- How would this trick work if we rolled four dice in the beginning, chose two dice and added the numbers on the bottom of those, and then rolled those dice again?
- How would this trick work if you used, say, three eight-sided dice instead? What about three twelve-sided dice?
- Would this trick still work if you still did the first and second "rules" of the trick, but then instead of rolling the same die that you added the bottom to the sum, chose a different die to roll?

c. The question that seems the most interesting to me is the second one. This trick would still definitely work, as long as the eight-sided dice and twelve-sided dice were set up the same way as the six-sided dice. For example, the sum of the opposite sides of the eight-sided dice would have to add up to 9, and the sum of the opposite sides of the twelve-sided dice would have to add up to 13.

d. I believe that all of the "questions to ponder" would lead to development of other related questions after knowing how to answer each one of them.

5. Cutting Cakes
a. To divide the blank square c into *five* equal parts, you can do this any number of ways. One obvious way is to make five equal horizontal rectangles. Another way is to make five equal vertical rectangles.

b. Some questions:
- What if the question was exactly the same, but specifically said that the shapes could not be rectangles? Would you still be able to do the problem?
- What if the question asked to divide the square into four equal parts? Do think there would be more answers to the question if this was the case?
- How about dividing the square into five shapes that are not equal or similar? (i.e., if you divide one part into a triangle, none of the other four parts may be a triangle as well.) Would this work?

c. All of these questions seem interesting to me in different ways because it forces the reader to think about the square separate from its normal attributes.

d. I think that the first and third questions, once answered, would help to develop further questions and thoughts.

6. Chain of Fools
a. The minimum numbers of cuts to do this job would be to cut four links.

b. Here is one question; can you think of others? What is the minimum amount of cuts you would need to make in order to link all of the chains, but if they did not have to be connected into a circle?

c. These are my favorites; can you imagine why? What if you changed the number of links that were already together? Would this change the outcome? What if you had to create a certain number of links that were then all linked to some sort of big loop? Think of an earring or some other type of "dangling" jewelry.

d. Responses here will vary depending on the questions invented. When I discussed this with other teachers, we made connections to number relationships and geometrical patterns that were not directly related to anything with the initial jewelry investigation. The connections across mathematics curriculum topics felt more important than answers to the basic questions.

7. The Big Race

a. If both run the exact same speed as before and Musky starts ten yards behind, then since he finished ten yards ahead before, now they would finish at the same time.

b. Some questions:
- What if Musky starts ten yards behind, but this time he runs ¼ of a yard faster than he did the first time? Now who would win and by how much?
- Suppose the second scenario is still the case, but this time Quashanda runs ½ of a yard faster and Musky runs ¾ of a yard faster? Who would win this time? And by how much?
- Is it really feasible to say that these runners would run a constant speed for the whole race? Wouldn't you think that runners, and anyone racing for that matter, be it swimmers or cyclists, usually increase their speed once they are approaching the finish line?

c. The question that seems the most interesting is the second one. You could set up a nice system of equations to solve this problem, and once you know how to do this type of problem, you could develop many different types of scenarios similar to this one.

d. I think that once the second question is answered, this would lead to the development of many other types of scenarios. Also, considering the third question could also lead to the development of other types of questions that would involve the races being run at different speeds throughout.

MathWorlds 6: Same Math, Different Metaphors

1. Group Work

a.

$$\frac{3x^2 - 19x - 14}{6x - 11x - 10} =$$

$$\frac{(3x + 2)(x - 7)}{(3x + 2)(2x - 5)} =$$

$$\frac{x - 7}{2x - 5}$$

$-5\left|3x-5\right| = 10$

$\left|3x-5\right| = 2$

$3x - 5 = 2$	OR	$3x - 5 = -2$
$3x = 7$		$3x = 3$
$x = \frac{7}{3}$		$x = 1$

b. Point: $(0, 3)$
   Line: $8x + 7y = 40$

To solve this problem, first solve for y in the above equation:

$7y = -8x + 40$
$y = -\frac{8}{7}x + \frac{40}{7}$

From this we can conclude that the *slope* of this line is $-\frac{8}{7}$.

i. To find the equation of a line *parallel* to the original line going through the point $(0, 3)$:
   ○ The slope of this line would also be $-\frac{8}{7}$ since the lines are parallel.
   ○ Use point slope form:

   $y-y_1 = m(x-x)$

   $y-3 = -\frac{8}{7}(x-0)$

   $y-3 = -\frac{8}{7}x$

   $y = -\frac{8}{7}x + 3$

ii. To find the equation of a line perpendicular to the original line going through the point $(0, 3)$:
   ○ The slope of this line would be $\frac{7}{8}$ since a line perpendicular to another line has a slope that is the negative reciprocal.
   ○ Use point slope form:

   $y-y_1 = m(x-x)$

   $y-3 = -\frac{7}{8}(x-0)$

   $y-3 = \frac{7}{8}x$

   $y = \frac{7}{8}x + 3$

c. Three aspects of circles that are not true about hyperbolas:
   • Any point on a circle is equidistant to the center of the circle. This is not true for hyperbolas

- In a circle, you can find the angles of the central angle and an inscribed angle given an arc measurement. You cannot do this with hyperbolas.
- A circle is a "regular" shape; a hyperbola is not.
- You can inscribe or circumscribe a circle in or around any regular polygon; you cannot do this with a hyperbola
- A funny one: you can make a wheel the shape of a circle, but a "hyperbolic wheel" might not work so well!

d. Yes you could do this, but it might not be that sturdy. Think of a beach bungalow or hut, or a camping tent. Most of the angles in such structures are obtuse angles, which is good for rain drain-off in camping tents and huts. However, if you built a house or structure like this, the ceilings might not be that high, or you would need a large plot of land since the building would be very wide!

e. In more layman's terms, all of these "sets" have infinite amounts of numbers or points within each set, so looking at this scenario this way, all of these sets, the cardinality is infinite. However, if you look at this situation in a more "technical-way", the following is true:

The cardinality of the *natural numbers* is denoted *aleph-null* ($\aleph_0$), while the cardinality of the *real numbers* is denoted c. It can be shown that $c = 2^{\aleph_0} > \aleph_0$ by Georg Cantor's Diagonal Argument. The *continuum hypothesis*, which was also developed by Cantor, states that there is no *cardinal number* between the cardinality of the reals and the cardinality of the natural numbers, that is, $c = \aleph_1$.

Looking at these mathematics problems using the following metaphors of a "workspace":

a. *Internet Social Space.* If you were to do such problems in a "workspace" that was, for example, like Facebook or MySpace, you could post these questions or send emails to your classmates and teachers and have their answers or suggestions on how to solve these problems posted in a thread that others can read and build upon.

b. *School Playground.* In a school playground, there is more freedom for the students. Adults supervise, but the children have more of a choice of what to do. To make an example of this "workplace," the teacher could set up stations where each of these problems are being addressed, but students may go and come from each station as they please. Also similar to a school playground, if a student has not completed a certain task, like an original assignment or homework, he/she is not allowed to browse the "stations" until he/she has done his/her original work.

c. *Fashion Section of an On-Line Newspaper.* In this "workplace" students can go around and examine each other's ways/fashion of doing a problem, since students are able to take different approaches to the same problem and still come up with the same answer. Students could observe each other's social patterns within the math classroom and in other aspects of school. For example, if you know that one of your classmates is good at math, you could observe his/her social patterns in, say, study hall, to see what he/she does there. Does he/she choose to work on math problems, other problems, or simply put his/her head down to rest?

d. *Allergist's Office.* In this "workplace," you could use a questioning method similar to what doctors do to find out what sort of symptoms their patients are showing. In relation to the mathematics classroom, students could ask each other questions and "probe" one another to find answers and possible solutions to the problem.

2. Social Issues Mathematics

a. Global and Local Distribution of Wealth

Some good websites:

sociology.ucsc.edu/whorulesamerica/power/wealth.html

www.ined.fr/fichier/t_publication/141/publi_pdf2_pop_and_soc_english_368.pdf

www.gp.org/platform/2004/economics.html

Some good books:

Community Health Analysis: A Global Awareness at the Local Level by G. E. Alan Dever

Ecology of the New Economy by Nigel J. Roome, Jacob Park

Public Finance in Theory and Practice by Richard Abel Musgrave, Peggy B. Musgrave

A good "expert" for this topic would be anyone who is proficient with Fathom since there is much data to do with these topics in Fathom. Also, anyone who is good with statistical analysis would be a good "expert."

b. *Infectious diseases*

Some good websites:

cartercenter.org/health/index.html

www.cdc.gov/ncidod/

www.nlm.nih.gov/medlineplus/infectiousdiseases.html

Some good books:

Infectious Diseases by Sherwood L. Gorbach, John G. Bartlett, and Neil R. Blacklow

Infectious Diseases in 30 Days by Frederick S. Southwick

Colour Atlas of Infectious Diseases by Ronald Emond, Philip Welsby, and H.A. Rowland

A good "expert" on infectious diseases would be a doctor or someone who does such research

c. *Consumption of oil*

Some good websites:

www.nationmaster.com/graph/ene_oil_con-energy-oil-consumption

www.thirdworldtraveler.com/Oil_watch/Oil_ReservesProducConsump.html

maps.unomaha.edu/Peterson/funda/Sidebar/OilConsumption.html

www.scaruffi.com/politics/oil.html

Some good books:

Oil and Gas in the Environment, published by The Stationery Office Books (Agencies)

A Thousand Barrels a Second: The Coming Oil Break Point and the Challenges Facing an Energy Dependent World by Peter Tertzakian

A good "expert" on this topic could again be someone who is good with statistical analysis and with maneuvering Fathom.

d. *Number of deaths attributed to genocide*

Some good websites:

users.erols.com/mwhite28/warstats.htm

www7.nationalgeographic.com/ngm/0601/feature2/index.html

baltimorechronicle.com/2006/101406Roberts.html

Some good books:

Facing Evil: Confronting the Dreadful Power Behind Genocide, Terrorism, and Cruelty by Paul B. Woodruff and Harry A. Wilmer

Power, Violence and Mass Death in Pre-Modern and Modern Times by Joseph Canning, Hartmut Lehmann, and Jay M. Winter

With Intent to Destroy: Reflections on Genocide by Colin Martin Tatz and Colin Tatz

A good "expert" on this topic would be any sociologist or statistician who studies and can provide information regarding this topic.

e. Volunteers who contribute to their community, globally and locally

Some good websites:

www.globalvolunteers.org/?gclid=CO3R46Sy5owCFSWQGgodxUXX7A

www.partners.net/partners/Default_EN.asp

www.thevolunteerfamily.org/Volunteers/Default.aspx

Some good books:

Volunteer Tourism: Experiences that Make a Difference by Stephen Wearing

Make a Difference: America's Guide to Volunteering and Community Service by Arthur I. Blaustein

Giving from Your Heart: A Guide to Volunteering by Dr. Bob Rosenberg and Guy Lampard

A good "expert" in regards to this topic would be a social worker or someone involved with the school's community service organization.

3. **Social Issues in Curricular Content**

a. All of the resources above could be used and manipulated into a Statistics, Data Analysis or Algebra class.

i. If the resources were used at the beginning of a lesson/unit, the students could see the real-life applications of statistics and algebra.

ii. If the resources were used in the middle of a lesson/unit, they could be used to reinforce some knowledge and information learned in the classroom by application to real-life situations.

iii. If the resources were used at the end of a lesson/unit, they could be used as a final project to show applications and connections between what the students learned in class and the real world.

b. The students could use this information on Fathom, which utilizes statistical data to form charts, graphs, and comparison of the student's choice. This could be introduced as a part of the lesson/unit, or it could be previously known by the students and just have to be applied to the work in the class. The students may also apply their past knowledge of graphing and charts to these lessons/units.

c. NCTM Standards Addressed:

   i. *Data Analysis and Probability Standard*

   Instructional programs from pre-kindergarten through grade 12 should enable all students to—

   - Formulate questions that can be addressed with data and collect, organize, and display relevant data to answer them.

   ii. *Representation*

   Instructional programs from pre-kindergarten through grade 12 should enable all students to—

   - create and use representations to organize, record, and communicate mathematical ideas
   - select, apply, and translate among mathematical representations to solve problems
   - use representations to model and interpret physical, social, and mathematical phenomena

   iii. *Connections*

   Instructional programs from pre-kindergarten through grade 12 should enable all students to—

   - recognize and use connections among mathematical ideas
   - understand how mathematical ideas interconnect and build on one another to produce a coherent whole
   - recognize and apply mathematics in contexts outside of mathematics.

   iv. *Problem Solving*

   Instructional programs from pre-kindergarten through grade 12 should enable all students to—

   - build new mathematical knowledge through problem solving
   - solve problems that arise in mathematics and in other contexts
   - apply and adapt a variety of appropriate strategies to solve problems
   - monitor and reflect on the process of mathematical problem solving.

d. The following topics might cause concerns among parents, administration, and even the teacher due to the nature of the topics, depending on the grade level of the students:

Infectious Diseases;

Number of Deaths Attributed to Genocide

   Before applying math to either of these researchable topics, the teacher could send home a letter letting the parents know that these topics will be taught and mathematically analyzed in the classroom and there could be an "consent" form attached for the parent to sign saying that it is "ok" for their child to be exposed to this type of information. Also, before the teacher begins planning lessons related to these topics, he/she could get it approved from the administrative standpoint to avoid potential conflict in that regard.

Bernadette Bacino

I am a graduate student in Mathematics Education at Arcadia University in Glenside, PA. I graduated from Loyola College in Maryland in 2002 with a BS in Mathematical Sciences. After college I worked in a service program through the Jesuit Volunteer Corps in Seattle, WA, working with adults with disabilities. This life-changing experience led me to pursue working with students with disabilities; teaching at Wordsworth Academy in Fort Washington, PA, an alternative school for children with behavioral and emotional disabilities, was my most important educational experience. I plan to continue my education by becoming certified in Special Education as well. I enjoy hiking and camping, and I am an Assistant Coach at Upper Dublin Aquatic Club, in Fort Washington, PA.

Bibliography

Albert, Linda. 1989. *A Teacher's Guide to Cooperative Discipline: How to manage your classroom and promote self-esteem*. Lebanon, IN: Ags Publishing

—— 2003. *Cooperative Discipline: Teacher's handbook*. Lebanon, IN: Ags Publishing.

Appelbaum, Peter. 1995. *Popular Culture, Educational Discourse, and Mathematics*. Albany, NY: SUNY Press.

—— 1998. Target: Number. In *The Post-Formal Reader*, Joe Kincheloe and Shirley Steinberg (eds.), 423–48. New York: Garland.

—— 1999. The Stench of Perception and the Cacophony of Mediation. *For the Learning of Mathematics*, 19 (2): 11–18.

—— 2004. Where is the Mathematics? Where are the Mathematicians? In *Multiple Intelligences Reconsidered*, Joe Kincheloe (ed.), 70–83. New York: Peter Lang.

Appelbaum, Peter, and Rochelle Kaplan. 1998. An Other Mathematics: Object relations and the clinical interview. *Journal of Curriculum Theorizing*, 14 (2): 35–42.

Bachman, Kellie. 1996. Education for All Students: A case study. *Educational Transitions*, 1 (1): 16–19.

Baker, Dave, Cheryl Stemple, and Tony Stead. 1990. *How Big is the Moon? Whole maths in action*. Portsmouth, NH: Heinemann.

Banyai, Istvan. 1998. *Zoom*. New York: Puffin Books.

Bauersfeld, Heinrich. 1995. "Language Games" in the Mathematics Classroom: Their Function and Their Effects. In *The Emergence of Mathematical Meaning: Interaction in classroom cultures*, Paul Cobb and Heinrich Bauresfeld (eds.), 72–92. Hillsdale, NJ: Lawrence Erlbaum.

Beck, Henry. Undated webpage. *Frequently asked questions about psychoanalytic therapy*. http://users.erols.com/henrywb/Psyan.html#pa9

Belenky, Mary Field, Blythe McVicker Clinchy, Nancy Rule Goldberger, and Jill Mattuck Tarule. 1986. *Women's Ways of Knowing: The development of self, voice, and mind*. New York: Basic Books.

Berlinski, David. 2000. *The Advent of the Algorithm: The 300-year journey from an idea to the computer*. New York: Harvest Books.

—— 2008. *Infinite Ascent: A short history of mathematics*. New York: Modern Library Chronicles.

Bettelheim, Bruno. 1979. *Surviving and Other Essays*. New York: Vantage Books.

bhabha, homi. 1994. *The Location of Culture*. New York: Routledge.

Block, Alan. 1997. *I'm Only Bleeding: Education as the practice of violence against children*. New York: Peter Lang.

—— 1999. Curriculum from the back of the bookstore. *Encounter: Education for Meaning and Social Justice*, 12 (4): 17–27.

—— 2004. *Talmud, Curriculum, and the Practical: Joseph Schwab and the rabbis*. New York: Peter Lang.

Britzman, Deborah. 1989. Who has the floor? Curriculum, teaching, and the English student teacher's struggle for voice. *Curriculum Inquiry*, 19 (2): 143–62.

—— 1998. *Lost Objects, Contested Subjects: Toward a psychoanalytic inquiry of learning*. Albany, NY: SUNY Press.

—— 1999. On becoming a "little sex researcher": Some comments on a polymorphously perverse curriculum. In *Contemporary Curriculum Discourses: Twenty years of JCT*, W. Pinar (ed.), 379–97. New York: Peter Lang.

Brown, Stephen I. 1993. Mathematics and humanistic themes: Sum considerations. In *Problem Posing: Reflections and applications*, Stephen Brown and Marion Walter (eds.), 249–78. Hillsdale, NJ: Erlbaum.

—— 1996. Posing mathematically: A novelette. In *Mathematics, Pedagogy, and Secondary Teacher Education*, Thomas Cooney, Stephen Brown, John Dossey, Georg Schrage, and Erich Wittman (eds.), 281–370. Portsmouth, NH: Heinemann.

—— 2001. *Reconstructing School Mathematics: Problems with problems and the real world*. New York: Peter Lang.

Brown, Stephen, and Marion Walter. 1983. *The Art of Problem Posing*. Philadelphia, PA: Franklin Institute Press. New edition 2005. Hillsdale, NJ: Erlbaum.

—— 1993. *Problem Posing: Reflections and applications*. Hillsdale, NJ: Erlbaum.

Buber, Martin. 1965. *Between Man and Man*. New York: Macmillan.

Carraher, Terezinha Nunes. 1989. Material embodiments of mathematical models in everyday life. In *Mathematics, Education, Society*, Christine Keitel *et al.* (eds.), 8–9. Paris: UNESCO Document Series No. 35.

CIEAEM (International Commission for the Study and Improvement of Mathematics Teaching). 2006. *Manifesto 2000 for the Year of Mathematics*. Retrieved February 19, 2006, from CIEAEM website: www.cieaem.net

Clifford, James. 1988. *The Predicament of Culture: Twentieth-century ethnography, literature, and art*. Cambridge, MA: Harvard University Press.

Clinchy, Blyth McVicker. 1996. Connected and separate knowing: Toward a marriage of two minds. In *Knowledge, Difference, and Powers: Essays inspired by women's ways of knowing*, Nancy Goldberger, Jill Tarule, Blyth Clinchy, and Mary Belenky (eds.), 205–47. New York: Basic Books.

Cline-Cohen, Patricia. 1982. *A Calculating People: The spread of numeracy in early America*. Berkeley, CA: University of California Press.

Consortium of National Arts Education Associations. 1994. *National Standards for Art Education: What every young American should know and be able to do in the arts*. Reston, VA: Music Educators National Conference.

Costa, A.L. (ed.). 1985. *Developing Minds: A resource book for teaching thinking*. Alexandria, VA: ASCD.

Daniels, Harvey. 2001. *Literature Circles: Voice and choice in book clubs and reading groups*. Portland, ME: Stenhouse.

Davis, Brent. 1996. *Teaching Mathematics: Towards a sound alternative*. New York: Garland.

—— 1997. Listening for differences: An evolving conception of mathematics teaching. *Journal for Research in Mathematics Education*, 28 (3): 355–82.

—— 1998. Studying the shapes of knowledge. *Journal of Curriculum Theorizing*, 14 (3): 36–38.

Davis, Philip J. 1995. Mathematics and common sense—cooperation or conflict? In *Mathematics (Education) and Common Sense: The challenge of social change and technological development* (proceedings of the 47th CIEAEM meeting), Christine Keitel, Uwe Gellert, Eva Jablonka, and Mirjam Müller (eds.), 29–38. Berlin, Germany: Freie Universität Berlin.

Davis, Philip, and Reuben Hersh. 1981 *The Mathematical Experience: The world according to Descartes*. Boston, MA: Houghton Mifflin.

de Beaugrande, R. 1988 In search of feminist discourse: The "difficult" case of Luce Irigaray. *College English*, 50: 253–72.

de Lauretis, Theresa. 1987. *Technologies of Gender: Essays on theory, film, and fiction*. Bloomington, IN: Indiana University Press.

Deleuze, Gilles. 2002. *Dialogues*. New York: Columbia University Press.

Deleuze, Gilles and Felix Guattari. 1991. *What is Philosophy?* New York: Columbia University Press.

Dembo, Myron H. 2004. Don't lose sight of the students. *Principal Leadership*, 4 (8): 37–42.

Efran, Jay, Michael Lukens, and Robert Lukens. 1990. *Language, Structure and Change: Frameworks of meaning in psychotherapy*. New York: W.W. Norton.

Egan, Kieran. 1988. *Teaching as Storytelling: An alternative approach to teaching and curriculum in the elementary school.* Chicago, IL: University of Chicago Press.

—— 1990. *Romantic Understanding: The development of rationality and imagination, ages 8–15.* New York: Routledge.

—— 1992. *Imagination in Teaching and Learning: The middle school years.* Chicago, IL: University of Chicago Press.

Eglash, Ron. 1997. When math worlds collide: Intention and invention in ethnomathematics. *Science, Technology and Human Values*, 22 (1): 79–97.

—— 2001. How to tell the difference between multicultural mathematics and ethnomathematics. Annual meeting of the National Council of Teachers of Mathematics, Orlando, FL, April 4–7. www.rpi.edu/~eglash/isgem.dir/texts.dir/eth_mlt.htm

Elbow, Peter. 1986. *Embracing Contraries: Explorations in learning and teaching.* New York: Oxford University Press.

Ennis, R.H. 1985. Goals for critical thinking. In *Developing Minds: A resource book for teaching thinking*, A.L. Costa (ed.), 54–57. Alexandria, VA: ASCD.

Fasheh, Munir. 1989. Mathematics in a social context: Math within education as praxis versus within education as hegemony. In *Mathematics, Education, Society*, Christine Keitel (ed.), 84–86. Paris: UNESCO Document Series No. 35.

—— 1990. Community education: To reclaim and transform what has been made invisible. *Harvard Educational Review*, 60 (1): 19–35.

Fawcett, Harold. 1938. *The Nature of Proof*, NCTM Yearbook. New York: Columbia University Teachers College Bureau of Publications.

Felman, Shoshana. 1987. *Jaques Lacan and the Adventure of Insight: Psychoanalysis in contemporary culture.* Cambridge, MA: Harvard University Press.

Fiske, John. 1989. *Understanding Popular Culture.* Boston, MA: Unwin Hyman.

Gardner, Howard. 1999. *Intelligence Reframed: Multiple intelligences for the 21st century.* New York: Basic Books.

Gellert, Uwe and Eva Jablonka (eds.). 2007. *Mathematization and Demathematization: Social, philosophical and educational ramifications.* Rotterdam: Sense Publishers.

Gerofsky, Susan. 2001. Genre analysis as a way of understanding pedagogy in mathematics education. In *(Post) Modern Science (Education): Propositions and alternative paths*, John Weaver, Peter Appelbaum, and Marla Morris (eds.), 147–76. New York: Peter Lang.

Ginsburg, Herbert, Rochelle Kaplan, and Arthur Baroody. 1992. *Children's Mathematical Thinking: Videotape workshops for educators.* Evanston, IL: Everyday Learning Corporation.

Ginsburg, Herbert, Susan Jacobs, and Luz Stella Lopez. 1998. *The Teacher's Guide to Flexible Interviewing in the Classroom: Learning what children know about math.* Boston, MA: Allyn & Bacon.

Glazer, Evan. 2001. *Using Internet Primary Sources to Teach Critical Thinking Skills in Mathematics.* Westport, CT: Greenwood.

Gough, Noel. 1998. "If This Were Played Upon a Stage": School laboratory work as a theatre of representation. In *Practical Work in School Science: Which way now?* Jerry Wellington (ed.), 69–90. London: Routledge.

Greene, Maxine. 1973. *Teacher as Stranger: Educational philosophy for the modern age.* Belmont, CA: Wadsworth.

—— 1986. In search of a critical pedagogy. *Harvard Educational Review*, 56 (4): 427–41.

Grumet, Madeleine. 1988. *Bitter Milk: Women and teaching.* Amherst, MA: University of Massachusetts Press.

Gustavson, Leif. 2007. *Youth Learning on Their Own Terms: Creative practices and classroom teaching.* New York: Routledge.

Gutstein, Eric. 2006. *Reading and Writing the World with Mathematics: Toward a pedagogy for social justice.* New York: Routledge.

Gutstein, Eric and Bob Peterson. 2005. *Rethinking Mathematics: Teaching social justice by the numbers.* Milwaukee, WI: Rethinking Schools.

Haraway, Donna. 1989. *Primate Visions.* New York: Routledge.

—— 1992. Situated knowledges: The science question in feminism and the privilege of partial perspective. In *Simians, Cyborgs, and Women: The reinvention of nature*, D. Haraway, 183–202. New York: Routledge.

Hawkins, David. 1980. *The Informed Vision.* New York: Pantheon.

Helfenbein, Robert. 2004. Economies of Identity: High-school Students and a Curriculum of Making Place. Paper presented at the annual meeting of the American Association for the Advancement of Curriculum Studies, San Diego, CA, April 9–12.

—— 2006. Economies of identity: Cultural studies and a curriculum of making place. *Journal of Curriculum Theorizing*, 22 (2): 87–100.

Hersh, Reuben. 1997. *What is Mathematics, Really?* New York: Oxford University Press.

Howell, Russell, and James Bradley (eds.). 2001. *Mathematics in a Postmodern Age: A christian perspective*. Grand Rapids, MI: Eerdmans.

Jardine, David. 1988. There are children all around us. *The Journal of Educational Thought*, 22 (2A): 178–86.

—— 2006. Unable to return to the gods that made them. In *Curriculum in Abundance*, David Jardine, Sharon Friesen, and Patricia Clifford, (eds.), 267–78. Mahwah, NJ: Erlbaum.

Kaplan, Rochelle G. 1987. The development of mathematical thinking as a function of the interaction between affective and cognitive factors. *Genetic Epistemologist*, 15 (3/4): 7–12.

—— 1991. Teaching philosophy as a factor in curriculum change. *Focus on Education Journal*, 1991 Edition: 13–19.

Keitel, Christine. 1989. Mathematics education and technology. *For the Learning of Mathematics*, 9 (1): 103–20.

Keitel, Christine, Ernst Klotzmann, and Ole Skovsmose. 1993. Beyond the tunnel vision: Analyzing the relationship between mathematics, society and technology. In *Learning from Computers: Mathematics education and technology*, Christine Keitel and Kenneth Ruthven (eds.), 243–79. New York: Springer-Verlag.

Kincheloe, Joe. 1993. *Toward a Critical Politics of Teacher Thinking: Mapping the post-modern*. Westport, CT: Bergin & Garvey.

Kincheloe, Joe, Shirley Steinberg, and Aaron Gresson (eds). 1996. *Measured Lies: The* Bell Curve *examined*. New York: St. Martin's Press.

Kirshner, David. 2000. Exercises, probes, puzzles: A crossdisciplinary typology of school mathematics problems. *Journal of Curriculum Theorizing*, 16 (2): 9–36.

—— 2002. Untangling teachers' diverse aspirations for students learning: A crossdisciplinary strategy for relating psychological theory to pedagogical practice. *Journal for Research in Mathematics Education*, 33 (1): 46–58.

Kohl, Herb. 1994. *I Won't Learn from You and Other Thoughts on Creative Maladjustment*. New York: The New Press.

Kohn, Alfie. 1993. *Punished by Rewards: The trouble with gold stars, incentive plans, A's, praise, and other bribes*. Boston, MA: Houghton Mifflin.

Ladson-Billings, Gloria. 1995. Making mathematics meaningful in multicultural contexts. In *New Directions for Equity in Mathematics Education*, Walter G. Secada, Elizabeth Fennema, and Lisa Bird Adajian (eds.), 126–45. Reston, VA: NCTM/Cambridge University Press.

Lakatos, Imré. 1976. *Proofs and Refutations: The logic of mathematical discovery*. New York: Cambridge University Press.

Lappan, Glenda, James Fey, William Fitzgerald, Susan Friel, and Elizabeth Phillips. 1996. *Getting to Know Connected Mathematics: A guide to the connected mathematics curriculum*. White Plains, NY: Dale Seymour.

Lave, Jean. 1988. *Cognition in Practice: Mind, mathematics and culture in everyday life*. New York: Cambridge University Press.

—— 1997. The culture of acquisition and the practice of understanding. In *Situated Cognition: Social, semiotic and psychological perspectives*, David Kirshner and James Whitson (eds.), 17–35. Mahwah, NJ: Lawrence Erlbaum.

Lave, Jean, and Etienne Wenger (eds.). 1991. *Situated Cognition: Legitimate peripheral participation*. New York: Cambridge University Press.

Lefebvre, Henri. 1991. *The Production of Space*. Cambridge, MA: Blackwell.

Lochhead, Jack. 1987. Thinking about learning: An anarchistic approach to teaching problem solving. In *Thinking Skills Instruction: Concepts and techniques*, Marcia Heiman and Joshua Slomianko (eds.), 174–82. Washington, DC: NEA.

Lock, Graham. Undated webpage. *Space is the Place: Interview with Sun Ra*. www.cs.uchicago.edu/AACM/ITUTUSITE/SCIFI.html

Loring, Ruth M. 2001. Music and skillful thinking. In *Developing Minds: A resource book for teaching thinking*, Arthur Costa (ed.), 332–36. Alexandria, VA: Association for Supervision and Curriculum Development.

MacDonald, James. 1995. The quality of everyday life in schools. In *Theory as a Prayerful Act*, James MacDonald, edited by Bradley MacDonald, 111–26. New York: Peter Lang.

McLaren, Peter. 1995. *Critical Pedagogy and Predatory Culture: Oppositional politics in a postmodern era*. New York: Routledge.

Martusewicz, Rebecca. 1992. Mapping the terrain of the post-modern subject: Post-structuralism and the educated woman. In *Understanding Curriculum as Phenomenological and Deconstructed Text*, William Pinar and William Reynolds (eds.), 131–58. New York: Teachers College Press.

Mason, John, Leone Burton, and Kaye Stacey. 1985. *Thinking Mathematically*. Reading, MA: Addison-Wesley.

Mellin-Olsen, Stieg. 1987. *The Politics of Mathematics Education*. Dordrecht, Holland: D. Reidel.

Miller, Janet. 1980. Women: The evolving educational consciousness. *Journal of Curriculum Theorizing*, 2 (1): 238–47.

—— 1990. *Creating Spaces and Finding Voices: Teachers collaborating for empowerment*. Albany, NY: State University of New York Press.

Mirochnick, Elijah. 2003. I wanna be a Kennedy: A visually and musically mediated conversation on owning one's self image in mass culture. JCT/Bergamo Conference on Curriculum Theory and Classroom Culture. Dayton, OH, October 2–4.

Mizer, Robert. 1990. *Mathematics: Promising and exemplary programs and materials in elementary and secondary schools*, Mathematics Education Information Report, ERIC Document No. 335230. Washington, DC: Office of Educational Research and Improvement.

Moon, Jean, and Linda Shulman. 1995. *Finding the Connections: Linking assessment, instruction, and curriculum in elementary mathematics*. Portsmouth, NH: Heinemann.

Moses, Robert. 2001. *Radical Equations: Math, literacy and civil rights*. Boston, MA: Beacon Press.

National Council of Teachers of Mathematics (NCTM). 1989. *Curriculum and Evaluation Standards for School Mathematics*. Reston, VA: NCTM.

—— 1991. *Professional Standards for Teaching Mathematics*. Reston, VA: NCTM.

—— 1995. *Assessment Standards for School Mathematics*. Reston, VA: NCTM.

—— 2000. *Principles and Standards for School Mathematics*. Reston, VA: NCTM.

—— 2004. *Principals and Standards for School Mathematics*. Retrieved February 19, 2006, from NCTM website: http://standards.nctm.org/document/chapter7/data.htm

—— 2006. *Curriculum Focal Points for Prekindergarten Through Grade 8 Mathematics: A quest for coherence*. Reston, VA: NCTM.

—— 2007. *Questions and Answers*. www.nctm.org/standards/focalpoints.aspx?id=274

Nespor, Jan. 1997. *Tangled up in School: Politics, space, bodies, and signs in the educational process*. Mahwah, NJ: Erlbaum.

Noddings, Nel. 1984. *Caring*. Berkeley, CA: University of California Press.

O'Daffer, Phares G., and Bruce Thornquist. 1993. Critical thinking, mathematical reasoning, and proof. In *Research Ideas for the Classroom: High School Mathematics*, Patricia S. Wilson (ed.), 39–56. New York: Macmillan/NCTM.

Ohanian, Susan. 1992. *Garbage Pizza, Patchwork Quilts, and Math Magic: Stories about teachers who love to teach and children who love to learn*. New York: Freeman.

Olson, Margaret. 1989. Room for learning. *Phenomenology + Pedagogy*, 7: 173–86.

Paul, Richard, A.J.A. Binker, and Daniel Weil. 1990. *Critical Thinking Handbook K-3rd Grade: A guide for remodeling lesson plans*. Rohnert Park, CA: Center for Critical Thinking and Moral Critique, Sonoma State University.

Pennsylvania Department of Education. 2001. *Academic Standards for Mathematics*. Retrieved February 20, 2006, from Pennsylvania Department of Education website: www.pde.state.pa.us

Pérez, David Callejo, Stephen Fain, and Judith Slater. 2004. *Pedagogy of Place: Seeing space as cultural education*. New York: Peter Lang.

Piaget, Jean. 1952. *The Child's Conception of Number*. London: Routledge & Kegan Paul.

Pinar, William. 1995. *Understanding Curriculum*. New York: Peter Lang.

Pinxten, Rik, Ingrid van Dooren, and Frank Harvey. 1983. *Anthropology of Space*. Philadelphia, PA: University of Pennsylvania Press.

Pinxten, Rik, Ingrid van Dooren, and E. Soberon. 1987. *Towards a Navajo Geometry*. Ghent, Belgium: K.K.I.

Pitt, Alice. 2003. *The Play of the Personal: Psychoanalytic narratives of feminist education*. New York: Peter Lang.

Polya, G. 1945. *How to Solve it: A new aspect of mathematical method*. Princeton, NJ: Princeton University Press.

Popkewitz, Thomas. 2004. The alchemy of the mathematics curriculum: Inscriptions and the fabrication of the child. *American Educational Research Journal*, 41 (1): 3–34.

Povey, H. 1998. "That spark from heaven" or "of the earth": Girls and boys and knowing mathematics. In *Gender in the Secondary Curriculum: Balancing the books*, A. Clark and E. Lillard (eds.), 131–44. New York: Routledge.

Powell, Arthur, and Marilyn Frankenstein. 1997. *Ethnomathematics: Challenging eurocentrism in mathematics education*. Albany, NY: State University of New York Press.

Reys, Robert, Mary Lindquist, Diana Lambdin, Nancy Smith, and Marilyn Suydam. 2004. *Helping Children Learn Mathematics: Active learning edition with integrated field activities*. Hoboken, NJ: John Wiley & Sons.

Robertson, Judith. 1997. Fantasy's confines: Popular culture and the education of the female primary-school teacher. In *Learning Desire: Perspectives on pedagogy, culture, and the unsaid*, Sharon Todd (ed.), 75–96. New York: Routledge.

Romagnano, Lew. 1994. *Wrestling with Change: The dilemmas of teaching real mathematics*. Portsmouth, NH: Heinemann.

Rorty, Richard. 1981. *Philosophy and the Mirror of Nature*. Princeton, NJ: Princeton University Press.

Rosenthal, Bill. 2004. Why 0.9999 . . . is and is *not* equal to 1. Lecture given at Arcadia University, Glenside, PA, March 2.

Rotman, Brian. 1993. *Ad Infinitum: The ghost in Turing's machine . . . taking the god out of mathematics and putting the body back in*. Stanford, CA: Stanford University Press.

—— 2000. *Mathematics as Sign: Writing, imagining, counting*. Stanford, CA: Stanford University Press.

Sangalli, Arturo. 1992. Forum: Mathematics for everyone—Arturo Sangalli on the challenge of mathematical education. *New Scientist*, 14 November: 49.

Schoenfeld, Alan (1989) Mathematical thinking and problem solving. In *Toward the Thinking Curriculum: Current cognitive research*, Lauren Resnick and Leopold Klopfer (eds.), 83–103. ASCD Yearbook. Alexandria, VA: ASCD.

Secada, Walter G., Elizabeth Fennema, and Lisa Bird Adajian (eds.). 1995. *New Directions for Equity in Mathematics Education*. Reston, VA: NCTM/Cambridge University Press.

Shapiro, Svi. 1993. Curriculum Alternatives in a Survivalist Culture: Basic skills and the "minimal self." In *Critical Social Issues in American Education: Toward the 21st century*, H. Svi Shapiro and David Purpel (eds.), 288–304. New York: Longman.

Shor, Ira.1980. *Critical Teaching and Everyday Life*. Chicago, IL: University of Chicago Press.

—— 1996. *When Students have Power: Negotiating authority in a critical pedagogy*. Chicago, IL: University of Chicago Press.

Skovsmose, Ole. 1994. *Towards a Philosophy of Critical Mathematics Education*. Dordrecht, Netherlands: D. Reidel.

—— 2000. Aporism and critical mathematical education. *For the Learning of Mathematics*, 20 (1): 2–8.

—— 2005. *Travelling through Education: Uncertainty, mathematics, responsibility*. Rotterdam: Sense Publishers.

Smith, S. 1990. The riskiness of the playground. *The Journal of Educational Thought*, 24 (2): 71–87.

Smyth, J. 1989. A critical pedagogy of classroom practice. *Journal of Curriculum Studies*, 21 (6): 483–502.

Swetz, Frank J. 1987. *Capitalism and Arithmetic: The new math of the 15th century*. La Salle, IL: Open Court.

Tanner, Laurel. 1985. The path not taken: Dewey's model of inquiry. *Curriculum Inquiry*, 18 (4): 471–79.

Todd, Sharon. 1997. Looking at pedagogy in 3-d. In *Learning Desire: Perspectives on pedagogy, culture, and the unsaid*, Sharon Todd (ed.), 237–60. New York: Routledge.

U.S. Census Bureau. 2006. *Income Distribution to $250,000 or More for Households: 2005*. Retrieved February 13, 2006 from website: http://pubdb3.census.gov/macro/032006/hhinc/new06_000.htm

Values Institute. 2006. Glossary. *Ethics Update*. San Diego, CA: University of San Diego. http://ethics.sandiego.edu/LMH/E2/Glossary.asp

Van Dyke, Frances, and Alexander White. 2004. Examining students' reluctance to use graphs. *Mathematics Teacher*, 98 (2): 110–17.

van Manen, Max. 1991. *The Tact of Teaching: The meaning of pedagogical thoughtfulness*. Albany, NY: State University of New York Press.

van Manen, Max, and Bas Levering. 1996. *Childhood's Secrets: Intimacy, privacy, and the self reconsidered*. New York: Teachers College Press.

Vygotsky, Lev.1986/34. *Thought and Language* Cambridge, MA: MIT Press.

Wagner, Barbara H. Undated webpage. *Educator's Pledge*. www.inspiringteachers.com/inspirations/pledge (last visited, 6.17.05).

Walkerdine, Valerie. 1987. *The Mastery of Reason: Cognitive development and the production of meaning*. New York: Routledge.

—— 1989. *Counting Girls Out*. London: Virago.

—— 1990. *Schoolgirl Fictions*. London: Verso.

—— 1996. Redefining the subject in situated cognition theory. In *Situated Cognition: Social, semiotic and psychological perspectives*, David Kirshner and James Whitson (eds.), 57–70. Mahwah, NJ: Lawrence Erlbaum.

Walkerdine, Valerie, and Helen Lucy. 1989. *Democracy in the Kitchen: Regulating mothers and socializing daughters*. London: Virago.

Walshaw, Margaret. 2004. *Mathematics Education within the Postmodern*. Charlotte, NC: Information Age Publishing.

Walter, Marion. 1996. Curriculum topics through problem posing. In *Talking Mathematics: Supporting children's voices*, Rebecca B. Corwin *et al.* (eds.) 141–47. Portsmouth, NH: Heinemann.

Wang, Hongyu. 2002. The call from the stranger: Dwayne Huebner's vision of curriculum as a spiritual journey. In *Curriculum Visions*, William Doll and Noel Gough (eds.), 287–99. New York: Peter Lang.

Watson, Dorothy, Carolyn Burke, and Jerome Harste. 1989. *Whole Language: Inquiring voices*. New York: Scholastic.

Weems, Lisa and Patti Lather. 2001. A psychoanalysis we can bear to learn from. Review of Britzman, Deborah. 1998. *Educational Researcher*, 29 (6): 41–42.

Weinstein, Matthew. 1996. Towards a cultural and critical science education. Paper presented at the annual meeting of the American Educational Research Association. New York, April 8–12.

Weisglass, Julian. 1990. Constructivist listening for empowerment and change. *Educational Forum*, 54: 351–70.

—— 1994. Changing mathematics means changing ourselves: Implications for professional development. In *Professional Development for Teachers of Mathematics, 1994 Yearbook*, Douglas B. Aichele and Arthur F. Coxford (eds.), 67–78. Reston, VA: NCTM.

Wells, David. 1995. *You are a Mathematician: A wise and witty introduction to the joy of numbers*. New York: John Wiley & Sons.

Wenger, Etienne, Richard McDermott, and William Snyder. 2002. *Cultivating Communities of Practice*. Cambridge, MA: Harvard Business School Press.

West, Cornell. 1994. *Race Matters*. New York: Vintage Books.

Whitin, David, and Robin Cox. 2003. *A Mathematical Passage: Strategies for promoting inquiry in grades 4–6*. Portsmouth, NH: Heinemann.

Wikimedia Foundation, Inc. 2006. *Household Income in the United States*. Retrieved February 13, 2006, from Wikipedia website: http://en.wikipedia.org/wiki/Household_income_in_the_United_States

Willinsky, John. 1992. Of literacy and the curriculum in Canada. *Journal of Curriculum Studies*, 24 (3): 273–80.

Winnicott, D.W. 1986. Sum, I am. In *Home is Where We Start From: Essays by a psychoanalyst*, D.W. Winnicott, 55–64. New York: W.W. Norton.

Yackel, Erna. 1995. Children's talk in inquiry mathematics classrooms. In *The Emergence of Mathematical Meaning: Interaction in classroom cultures*, Paul Cobb and Heinrich Bauresfeld (eds.), 131–62. Hillsdale, NJ: Lawrence Erlbaum.

Name Index

Note: page numbers in **bold** refer to illustrations where the name is mentioned.

Subject Index

Note: page numbers in **bold** refer to illustrations.

democracy 33, 84, 133, 138, 142–3, **168**, 176–7, 240–2; bourgeois 233; new 264

desire 17, 21, 31, 45, 175–80, 198, 209, 225, 240, 252, 256, 277, 281–7, 295; and consumer culture 160; and psychoanalysis 51–2, 54–8, 79–85, 145

developmental theories 52, 69, 81, 93, 166–7, 171–2, 202, 230–9, 249

dialogic participation 174, 177, 307

difference 175–80, 203–5, 225, 254, 281

direct instruction 22–3, 226–30, 305

disparity 175–7, 225, 281, 289

diversity 144–5, 176

division 89, 182–3, 317–18

Donald Duck in Mathemagic Land **191**

empowering pedagogies 167, 203, 248–51, 255, 267

enculturation 146, 172, 290–3

entering 236–7, 245

equity 128, 143–4, **144**, 148–58

ethics 32, 53, 210, 240; ethical stance 286–95; moral choices 233–5

ethnomathematics 145, 163–6, 234, 293

events, designing 199–202

excorporation 168

exercises 36, 47, 119–21, 175–8, 184–6, 195–6, **196**, 215, 244, 254–5, 260–9, 280–5, 291, 321–4

exponentiation 12, 109–14, 149–51, 157, 194

factoring 212

factors 99, 110–14

family involvement 8, 32, 67, 107–10, 137, 148, 155, 170, 183–4, 221–4, 234, 244, 254, 279, 307, 328–9

fantasy of control 54, 145

Fermi problems 90, 319–20

Five Chips 195–6, **196**

five-part unit 21–5, 37, 120–1, 144, 183, 298–9

Flow Structures 230–43

Focal Points 290–1

fractions 40–5, 74, 89, 120, 127, 137, 158, 182–3, 200–1, 246, 277, 280, 317–18

Free-math 261

functional relationships 2–12, 123, 135–6, 149, 158, 217, 258–81; cultural functions 160–3; iterative functions 89–90, 318–19; recursive 170

gatekeepers 155, 221–2

gaze, the 171

gendered analysis 140, 151, 160, 166–7, 203–7, 287–8

generalizing 2, 50, 70, 106, 109, 130, 134, 138–41, 234, 253, 260, 298, 300–4

geometry: coordinate systems 50, 150–4, 158, 216–21, 287; and critical thinking 131, 141; in elementary curriculum 181–3; international comparisons of middle school 191; in secondary curriculum 179; as strand of integrated unit 148–51, 162, 216, 234; transformational 39, 94–5, 123–4, 168; two and three-dimensional Euclidean 7–8, 24, 49, 90, 93–100, 123–4, 133–5, 236; van Hiele levels 92–4, **94**; visual and modeling 9–12, 95–100, 123–5, 152, 184–5, 265, 285

graph paper 122

graphs 28–30, 50, **116**, 136–8, 149–58, 163, 182–3, 193–4, 206, 216–22, 257, 280, 312–13, 328–9

group work 23–34, **53**, **54**, 57, **133**, **144**, 172–7, 213–15, 226–43, 247–8, 259–66, **268**, **276**, 285, 296, 299–307, 324; examples of 2–10, 38–9, 88, 95, 134–56, 182–3, 216–23; and interviewing 67, 79–88; metaphors for 188–208; places for 209–10; and popular culture 160–6

habituation 291–3

habitus 173–5

holding 62–6, 79–81, 173, 256–7; and response 57–9

holding back 54–5, 265; withholding 256

hundreds chart 5, 16

Illuminations 197, 217–18

infinity 158, 253

infrastructure 13, 33–5, 51, 121, 225, 298–9

inquiry, conditions for 32–5

instructive interaction, myth of 273–9

interculture 163–8

interviewing 25, 137–8, 156; *see also* assessment

inverse functions 4

investigative lessons 228–9, 242–6

irony 19–20

it see *I-Thou*

I-Thou relationship 58, 62–8, 236

journals 29–30, 66, 87, 127–9, 134, 151–6, 173, 192, 198–9, **263**; student peer review 13

kitchen front and back 107–13, 126

knowing, separate and connected 269–73, 285

KWL chart 4

257–8, 267–71, 291, 300; absurd 72; standard 26–7, 71–80, 119
proportion 8–11, 137, 258
Pulp Fiction Test 19–21
Punnett square 36
"Purloined Letter, The" 53
puzzle questions 16–25, 30, 119–20, 167, 184–6, 201, 229, 237, 260–1, 302, 321–4

ratio 8–11, 137, 258
reading 26–9, 66, 107–14, 122, 129, 142, 155, 170–3, 197, 229, 236–9, 321; of popular culture 162–8; as professional reflection 37, 86, 127–8, **144**, 172, 215–16, **228**, 241–5, 265, 273–4, 302, 308–9; *see also* literature circles
reason 48, **68**, 74, 90, 141–2, 155, 166–7, 180–1, 194, 201, 210, 233–4, 249, 255, 269, 275–85; reasonableness 28–9, 53, 75–8, 91–6
Reason's Dream 130–3, 167, 291
recreational mathematics 83, 161–2
reification 169, 291
representation 12
resistance 55, 75, 156, 255–69, 279–84, 294
rewards 16, 33, 55, 77, 85, 122–4, 132, 187, 235, 251, 259

sabermetrics 15
secrets 8
self-esteem 164, 171
sense-making 18, 27
Sketchpad® 24
skip counting 2–6
social justice 130, 148–58, 170–6, 216–22, 236
space *see* learning
space-off 123, **205**, 207–9
specializing 2, 106, 109, 134, 141, 260, 298, 300–4
"spiritual journey" 287, 293
standardized tests 14–15, 36, 88, 119, 129, 138, 146–8, 178–9
standards 13, 69, 130–3, 141–9, 155–6, 178, 183, 188–9, 197, 211–25, 232–50, 278–91, 321, 329

strategy 28–30, 50, **64–7**, 116, 129, 134–42, 194, 239, **272**, **276**, 299, 315; games 115, 136, 185, 236, 322; pedagogical 33, 67, 145, 208, 270
student: "good" 61, 83–5, 226, 259, 267, 279; as teacher 14–15
studenting 225–6, 238, 246
subtraction 4, 50, 74, 111, 158, 182–3, 193–4, 233, 314–15
symmetry 38–9, 48, 135, 168

taking action 24–5, 306–9
talking back 145
Talking Stick 194–6, **196**, 214
teacher: "good enough" 81–2; as learner 14–15; as modernist 20–1
teaching: as confrontation with limits 32; as inquiry 32–5; as storytelling 17–18, 232–6, 247–50, 254, 285
technology 214, 236, 284; and commodity culture 162–4; of power 166–71; social impact 152
tessellations 39, 123–4
textbooks 7, 16, 23, 36, 83, 116–19, 125–6, **133**, 160, 170, 178–9, 221, 235–9, 246–7, 264, 284
thinking classroom 63–6, **67**, 133, 140–5
time: as curriculum strand 50, 162–3, 183, 197, 313–14, 318; as pedagogical challenge 128, 262
TIMSS (Third International Mathematics and Science Study) 190–1, **191**
topology 7–8, 12
transference **61**, **62**, 172–3

University of Cape Town 264

variables 18, 50, 73, 89, 102–12, 136, 169, 189, 193–4, 285, 300, 305, 316; *see also* Malcolm X
vision 170–1; technology of 171–5
voice 133, **35**, 39–41, 148, 156–7, 175–7, 203–4, 239–43, 248, 272, 281, 287

what if not? 116–17, 300–4
Write a Book 262

3684 065